SAMPLE SELECTION, AGING, AND REACTIVITY OF COAL

SAMPLE SELECTION, AGING, AND REACTIVITY OF COAL

Edited By

RALPH KLEIN
National Bureau of Standards
Gaithersburg, MD.

ROBERT WELLEK
National Science Foundation
Washington, D.C.

WILEY

A WILEY-INTERSCIENCE PUBLICATION

JOHN WILEY & SONS

New York · Chichester · Brisbane · Toronto · Singapore

Library of Congress Cataloging-in-Publication Data:

Sample selection, aging, and reactivity of coal / edited by Ralph
  Klein, Robert Wellek.
      p.   cm.
    "A Wiley-Interscience publication."
    Bibliography: p.
    Includes index
    ISBN 0-471-87555-4
    1. Coal—Sampling.   2. Coal—Analysis.   I. Klein, Ralph, 1918–
    II. Wellek, Robert.
  TP325.S26 1989
  662.6'22—dc 19   88–23574
  Printed in the United States of America   CIP

  10 9 8 7 6 5 4 3 2 1

CONTRIBUTORS

L. G. AUSTIN, Mineral Processing Section, The Pennsylvania State University, University Park, Pennsylvania.

N. BERKOWITZ, Department of Mineral Engineering, University of Alberta, Edmonton, Alberta, Canada.

R. J. GRAY, 303 Drexel Drive, Monroeville, Pennsylvania.

P. M. GY, Consulting Engineer. (Sampling), Cannes, France.

D. S. HOOVER, Air Products and Chemicals, Inc., Allentown, Pennsylvania.

R. S. HSU-CHOU, Department of Civil and Environmental Engineering, University of Southern California, Los Angeles, California.

D. E. LOWENHAUPT, Research Division, Conoco Coal Development Co., Library, Pennsylvania.

O. P. MAHAJAN, Corporate Research Department, Amoco Research Center, Naperville, Illinois.

R. N. MILLER, Air Products and Chemicals, Inc., Allentown, Pennsylvania.

H. C. G. DO NASCIMENTO, Department of Civil and Environmental Engineering, University of Southern California, Los Angeles, California.

C. E. ROBERTS, JR., Henry Krumb School of Mines, Columbia University in the City of New York, New York, New York.

R. S. C. ROGERS, Standard Oil Company of Ohio, Cleveland, Ohio.

F. K. SCHWEIGHARDT, Air Products and Chemicals, Inc., Allentown, Pennsylvania.

P. SOMASUNDARAN, Henry Krumb School of Mines, Columbia University in the City of New York, New York, New York.

W. SPACKMAN, College of Earth and Mineral Sciences, The Pennsylvania State University, University Park, Pennsylvania.

T. F. YEN, Department of Civil and Environmental Engineering, University of Southern California, Los Angeles, California.

CONTENTS

PREFACE

Without power sources, modern civilization could not exist. Population growth and technological development place ever increasing demands on world resources, and, except for the promise of energy from fusion, these resources are finite and limited. Apart from a small proportion of hydroelectric, fission, geothermal, and solar energy in the total energy budget, the fossil fuels—coal, oil shale, oil, and natural gas—are the mainstays of power sources. Coal constitutes the greatest potential resource and will still be in abundant supply when oil reserves have been severely depleted.

Because of the importance of coal, the discipline properly referred to as coal science has matured. Even though coal is a complex and rather ill-defined material, increasingly sophisticated and quantitative experimental techniques are being applied to its study. Development of engineering processes such as the use of coal slurries, fluidized-bed combustion, liquefaction, and gasification are greatly extending present and future applications. The coal science literature, as might be expected in view of this intense interest, is indeed vast. The inspiration for this book had as its source the requirement of the coal science community for a premium coal sample bank. Standard quality coal samples are needed by researchers to minimize sample variations in comparing experimental results among laboratories. A proposal to initiate such a sample bank was made by Heinz Sternberg at a panel meeting sponsored by the Gas Research Institute. Subsequently, the Department of Energy with the Gas Research Institute held a workshop in Atlanta, Georgia (March 1980) to discuss the concept. The need for such a sample bank was recognized, and a premium sample bank was instituted at the Argonne National Laboratories in Chicago, Illinois.

The primary task in initiating a sample bank is the appropriate choice of coal, that is, sample selection. This is followed by preparation and storage. Preparation of coal for storage and the storage of coal, whether for a sample bank on a small scale or for industrial purposes on a large scale, require consideration of the preservation of chemical and physical properties. Oxidation, with its deleterious effects on coking, beneficiation, and other processes, is a prime example. In addition to chemical changes, coal aging is usually accompanied by alterations in the pore structure. Less emphasized, but not to be ignored, is microbial action. Preparation of coal for storage most often involves grinding. The prepared and stored coal requires monitoring; and such surveillance involves not only appropriate sampling techniques but also sensitive analytical procedures.

The sensitivity of the rate and product ratios in coal hydrogenation (or other coal processes) has not been critically examined with respect to variables such as moisture and oxygen. It is an arguable point on how much effort should be expended in providing superpreserved coal and exquisitely characterized coal samples if in fact *basic* science experiments on the one hand and engineering processes on the other are insensitive to *small* variations in coal properties. Unfortunately, in the generation and preservation of coal samples, considerations such as sample selection, preparation, aging, storage, surveillance, and reactivity have not always been buttressed by experimental evidence, either because of the absence of evaluated and convincing experimental results or the lack of sufficient data.

We believed that it would be useful to have a book with chapters devoted to subjects pertinent to coal selection, sampling, aging, and reactivity, including those that previously have not received much emphasis. We hope that insights and an appreciation of the processes will be afforded to the reader. We recognize with gratitude not only the contributions of the chapter authors, but also the advice and counsel of many in the coal science community. We particularly acknowledge the encouragement, support, and patience of the editorial staff of John Wiley and Sons.

RALPH KLEIN
ROBERT WELLEK

Gaithersburg, Maryland
Washington, D.C.
January 1989

SAMPLE SELECTION, AGING, AND REACTIVITY OF COAL

1

SAMPLE SELECTION

W. Spackman
College of Earth and Mineral Sciences
The Pennsylvania State University

1.1   STATEMENT OF THE PROBLEM

The intended use of a coal sample should govern the selection of the sample. This statement of an obvious truth may seem unnecessary to the reader, yet the selection of inappropriate samples is by no means an infrequent phenomenon. Although the analogy is far from perfect, the selection of coal samples has been somewhat reminiscent of the fabled blind men, each of whom attempted to describe an elephant but each from a different sampling point. The blind man feeling the trunk arrived at a quite different conclusion than the man examining a massive leg, and the latter's description deviated appreciably from the descriptions arrived at by the man examining the massive expanse of the animal's side and the strange configuration of its tail. Each man understood he was studying the sample of an elephant, hence each could reasonably assume that the four descriptions would be essentially similar. Such an assumption is not justified, of course. How does this relate to coal sampling? Suffice it to say that the newcomer to coal science and technology often has "reasonably assumed" that if he obtained a sample of "Pittsburgh Coal," he would be analyzing or researching the same material that others already had described, and if he used similar methods, his results would be "essentially similar." The elephant displays various characteristics depending on the sampling point, and the same can be said of the Pittsburgh coal seam. A sample from eastern Ohio contrasts markedly from one obtained in south-central Pennsylvania, and these differ, in turn, from samples obtained in the vicinity of the city of Pittsburgh. Lateral changes in coal seam composition represent just one of the several matters to consider when one addresses the question: "From which coal seam and from what geographic location should my coal sample be derived?"

A person concerned with supplying coal samples frequently receives such requests as, "Please send me a sample of high-volatile A bituminous coal," or "I would like to have a sample of lignite." Such requests are perplexing to the supplier, and he wonders if the requestor's needs will be met by a splint coal or a bright coal, one with a large amount of exinite or a small amount, an Illinois coal with a high concentration of sulfur, or West Virginia coal that displays excellent coking properties. Should the lignite be a uraniferous sample from South Dakota or an Alabama coal formed in a very saline peat-forming environment? In these cases, the requestors think they are asking for a sample of *a particular kind of coal* when, in fact, they are requesting a sample of *a sizable class of coals.*

Thus, the diversity of the materials involved, coupled with a sometimes deceiving and unfamiliar terminology, results in a situation in which the selection of coal samples becomes a challenging enterprise to say the least.

1.2   THE MATERIAL TO BE SAMPLED

Full knowledge of the population being sampled (hence, knowledge of what is available) would be extremely helpful in insuring that a given sample is suited for its intended use. Acquiring such knowledge would involve the effort of a lifetime and even then might not be attained. Acquisition of an awareness of the overall content of the population can be acquired, however, and this can be of considerable use in framing questions that will assist in identifying the most appropriate material for study or analysis.

It is interesting to note that coal seams occur on all of the Earth's continents, including Antarctica. They are by no means equally distributed (Figure 1.1), but

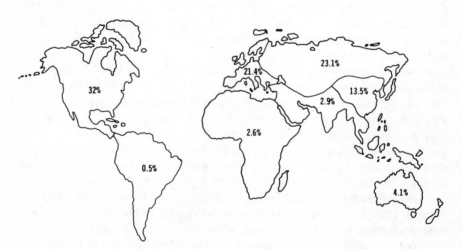

Figure 1.1   Distribution of recoverable coal reserves.

their widespread occurrence suggests that a variety of conditions may have attended their origins and subsequent developmental histories. In North America, commercially important coals occur in nine so-called provinces as shown in Figure 1.2. These are the:

1. Alaskan Province,
2. Northern Canadian Province,
3. Eastern Province,
4. Interior Province,
5. Northern Great Plains Province,
6. Rocky Mountain Province,

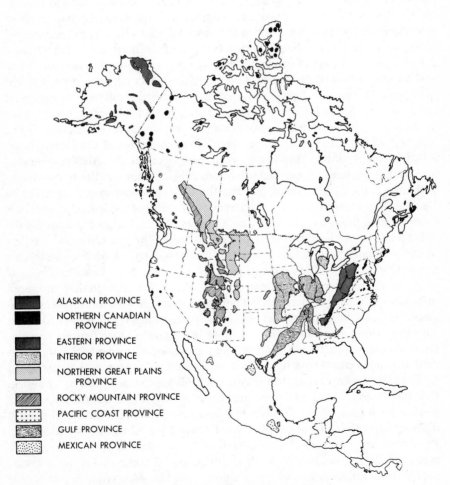

ALASKAN PROVINCE
NORTHERN CANADIAN PROVINCE
EASTERN PROVINCE
INTERIOR PROVINCE
NORTHERN GREAT PLAINS PROVINCE
ROCKY MOUNTAIN PROVINCE
PACIFIC COAST PROVINCE
GULF PROVINCE
MEXICAN PROVINCE

Figure 1.2 Coal provinces of North America.

7. Pacific Coast Province,
8. Gulf Province, and
9. Mexican Province.

Each province consists of two or more *Coal Regions* and often many *Coal Fields*. This spatial distribution of coal seams is of significance in selecting coal samples for several reasons. First, the time of origin of the coals can differ greatly from province to province, region to region, and even field to field. This carries with it the possibility of changes in the precursor materials involved in coal formation as organic substances evolved in the plant kingdom. Generally speaking, there were three main geologic periods during which coal formed on the North American continent: the Pennsylvanian Period (about 250 million years ago), the Cretaceous Period (about 80 million years ago) and the Tertiary Period (about 30–60 million years ago). Lesser occurrences of coal were formed at other times such as the Triassic coals of Virginia and the Mississippian coals of northwestern Alaska. On a worldwide basis, coal seams have been encountered in the rocks of every geologic period from the Devonian to the Quaternary, hence in terms of time of origin, samples ranging in age from 1 million to 350 million years could be assembled if one so desired. Second, coals from different geographic areas may have been formed in decidedly different depositional environments in which there are dissimilar arrays of plant species, microorganisms, precursor substances, and chemical and physical conditions. Thus, although of the same general age, the freshwater lignites of the Great Plains Province may contrast markedly when compared to the Alabama lignites formed under saline depositional conditions subject to the conditions associated with an advancing sea. Third, because the U.S. coal seams occur in different *Geologic Provinces* as well as in different *Coal Provinces*, it is likely that coals from different areas have experienced a variety of different geologic conditions subsequent to the burial of their precursor peats. The pressures and temperatures associated with differing depths of burial will have a profound effect on the characteristics of the resultant coal, as will the rate of heating. The coal seams in northwestern Colorado and northeastern New Mexico that have been influenced by the rapid increases in temperature, and the sometimes intense heat associated with the intrusion of molten rock material, display characteristics that vary from those associated with the more usual temperature regimes encountered by coals in other areas. Similarly, in a geologic province in which the rocks (and associated coals) have been subjected to the shearing forces that attend the folding of the layers of the Earth's crust, the coals are characterized by a physical structure rarely found in other areas. Accordingly, the anthracites of the Anthracite Region of the Eastern Province can be expected to be physically different from the anthracites of Crested Butte Field of the Rocky Mountain Province. Lastly, the rocks forming the Earth's surface in the various geographic areas can differ considerably in their mineralogical composition. As a result, they exert an influence on the composition of the inorganic components of coal seams. Their erosional substances that are washed into the coal-forming swamps and marshes affect the geochemical conditions in the environment and at the same time may influence the trace element content of the resultant coal seam. It is evident that the mineralogy and certain aspects of a coal's chemistry will be different in a province dominates by volcanic rocks from that of seams associated only with carbonate-rich sediments.

1.2.1 Class of Rocks Called *Coal*

At the outset, it is important to recognize that *coal* is not a single substance but a term applied to *a class of rocks*. Another point to be made is that rocks, including coals, are composed of particles or grains of various compositions that are bonded together to form a solid mass. In the case of coals, some of the grains are composed of one or more mineral materials (which are crystalline in structure and usually noncarbonaceous), but most of the particles consist of one or more maceral substances (which are noncrystalline in structure and carbonaceous in composition). The size of the individual particles also is of interest because of its effects on sampling procedures. In coals, particle sizes vary from submicroscopic to hundreds of cubic centimeters. Rocks usually are formed of only a small number of different substances that are present in other than trace amounts, whereas coals typically are formed of many different materials. In this, and other ways, coals make up the most complex class of rocks encountered in the Earth's crust.

How shall we define this material? Most definitions fail to include one or more of the points made above. A standard dictionary states that coal is "a black or brownish-black solid, combustible mineral substance...."[1] The American Society for Testing Materials (ASTM) defines coal as "a brown to black sedimentary rock (in the geologic sense) composed principally of consolidated and chemically altered plant remains."[2] One of the more adequate definitions was published by J. M. Schopf, who said, "Coal is a readily combustible rock containing more than 50 percent by weight and more than 70 percent by volume of carbonaceous material, formed from compaction or induration of variously altered plant remains similar to those of peaty deposits."[3] For present purposes, it will be useful to regard coal as a class of complex combustible rocks containing both maceral and mineral substances, with the maceral materials comprising more than 50 percent of the rock mass on a weight basis.

1.2.2 Inorganic Constituents of Coal

A large number of different minerals have been encountered in coals (in the vicinity of 60 or more), but only a fraction of these occur in significant concentrations. (What is significant can be determined, however, by the intended use of the sample.) The commonly encountered minerals are listed in Table 1.1. They occur as discrete grains of pure mineral substance embedded in a maceral matrix or as concentrations of mineral grains occurring as lenses (partings) or

TABLE 1.1 Minerals Commonly Encountered in Coal

Silicates

Clay minerals
  Sericite $KAl_2(AlSi_3O_{10})(OH)$
  Kaolinite $Al_2Si_2O_5(OH)_4$
  Smectite $Al_2Si_2O_5(OH)_4$
Quartz SiO_2

Sulfides

Pyrite FeS_2 (cubic)
Marcasite FeS_2 (orthorhombic)
Sphalerite ZnS
Galena PbS
Chalcopyrite $CuFeS_2$

Sulfates

Gypsum $CaSO_4 \cdot 2H_2O$
Barite $BaSO_4$
Jarosite $K_2O \cdot 3FeO_3 \cdot 4SO_3 \cdot 7H_2O$

Carbonates

Siderite $FeCO_3$
Ankerite $CaFeMg(CO_3)_3$
Calcite $CaCO_3$
Dolomite $CaMg(CO_3)_2$

Oxides

Haematite Fe_3O_4
Goethite $FeOOH$

Phosphates

Apatite $Ca_5(PO_4)_3(F, Cl, OH)$
Phosphorite   No definite composition

Accessory minerals

Zircon $ZrSiO_4$
Rutile TiO_2
Orthoclase $KAlSi_3O_8$
Plagioclase $(NaCa)Al(AlSi)Si_2O_8$

fissure fillings in the coal seam. Their size and mode of occurrence influences the manner in which the coal seam is sampled and the manner in which the sample is handled to produce representative subsamples.

Usually the clay minerals occur in greatest concentration. It is common for them to occur in submicron particle sizes completely encased in maceral material, rendering them almost impossible to remove should one wish a *mineral-free* sample. At the other extreme, the clay minerals may form thick (measured in millimeters or centimeters) layers in the seam that can have large lateral extents. Quartz is common in coals but usually is present in small concentrations (1–3 percent). The water velocities in peat-forming environments are seldom great enough to carry sand grains into the system. There are exceptions to this, however, particularly in those cases in which the wetlands occur in a sandy terrain. Opaline quartz is frequently introduced into an accumulating peat deposit as plant-formed *phytoliths* and as sponge *spicules*. The long-term fate of the opaline quartz materials is not clear, and they have not been reported as constituents of coal samples. Pyrite is another mineral that occurs in a variety of sizes and modes of occurrence. Like the clay minerals, it often occurs in small particle sizes of 1–10 microns. Unlike clay minerals, it commonly is found as spherical clusters of crystals called *framboids* and in planar deposits in fissures formed after consolidation of the peat. Its origin is related to brackish and saline peat-forming environments; hence, it is preferentially concentrated in coal types derived from such settings. Normally it is absent in the case of freshwater coal types. Therefore, samples of anthracite from Pennsylvania and subbituminous coal from Wyoming will have little or no pyrite present.

In addition to the true minerals discussed above, lignites and subbituminous coals can have cations such as Na, Ca, and K attached to the carboxyl units in certain of the maceral substances. In the combustion process, these elements report to the ash as oxides or sulfates, and they are sometimes regarded as part of the total *mineral matter* in coals. In selecting samples of low-rank coals, their presence may be a matter for consideration.

Minerals play significant roles in carbonization, combustion, liquefaction, and gasification. In selecting samples for tests or studies related to these processes, their presence and concentrations need to be considered. In studies focused solely on the organic materials in coal, their removal may be a critical factor in sample selection. Accordingly, knowledge of identity and mode of occurrence of these *inorganic* materials is useful in the selection process.

1.2.3   Organic Constituents of Coals

The principal fraction of coals is, of course, the carbonaceous portion. This is made up of a variety of maceral substances occurring in a wide range of particle sizes. Micrinite, for example, occurs in particles on the order of 1–3 μm in diameter, the greatest dimension of sporinitic particles ranges from about 5 to more than 200 μm, and vitrinitic *grains* can involve masses involving hundreds of cubic centimeters.

Three classes of maceral materials are encountered in coals, and all three normally are represented in any given coal sample. The liptinite (sometimes the term *exinite* is used) suite consists of a group of relatively hydrogen-rich macerals derived from lipoid, waxy, and resinous plant precursors, including a spectrum of materials from the contents of plant cells and the materials impregnating their walls. The proportions of particular liptinitic macerals are affected by the environments in which the peat forms and by the geologic time of formation. Thus, resinite and suberinite are present in greater concentrations in Cretaceous and Tertiary coals, and sporinite is often found in greater concentrations in Pennsylvanian coals. The relationship with environment is reflected in the concentration of alginite in boghead coals and sporinite in canneloid coal types.

Contrasting with the macerals of the liptinite suite are those of the inertinite group. These are *carbon-rich* materials containing small amounts of hydrogen. The term inertinite stems from the fact that these substances often behave as inert materials in the carbonization process. Inertinite particles occur in a variety of sizes and textures, ranging from the 1 μm micrinite particles on the small end to sizable (e.g., 2 mm $\times$ 2 cm $\times$ 4 cm) chips of fusinite on the large end. Fusinite is a very friable, porous material, the porosity being the direct result of the cellular construction of the original plant tissue, whereas macrinite is a relatively dense material, occurring in the form of relatively small irregularly shaped particles. As normally classified, semifusinite, sclerotinite, and macrinite are more variable in composition than the other inertinite macerals.

Macerals of the vitrinitic suite form the major portion of most Northern Hemisphere coals, commonly composing 60–80 percent of the total volume. This contrasts with some Australian and African coals in which the inertinite materials may dominate. Vitrinitic macerals contain less carbon than the inertinitic substances and less hydrogen than the liptinitic macerals. Because of their preponderant occurrence, they tend to determine the coal's overall properties. For this reason, the chemical and physical properties described for particular coal samples more often than not are describing the properties of the vitrinite as opposed to the properties of the other macerals in the sample. The vitrinitic suite of macerals consists of two principal groups of macerals, the huminites, which predominate in lignites and some low-rank subbituminous coals, and the vitrinites, which predominate in higher-rank subbituminous and bituminous coals. Because precursor plant substances are more numerous and varied than the substances found in coal, it is not surprising to find that those coals containing the least altered plant substances (the brown coals and lignites) contain the greatest variety of vitrinitic macerals. Thus, six different huminite macerals are usually recognized in describing lignites, whereas only two or three vitrinites may be recognized in higher-rank coals. It is generally agreed that this classification of vitrinitic materials is inadequate to describe the actual array of these materials that are present in coals. As the result of this, and because of the need to utilize the information on a coal's maceral composition in commercial coal utilization practices, a provisional and arbitrary method of stating the concentrations of vitrinitic macerals in coals has been adopted. This differentiates between various vitrinitic materials, using their light-reflecting capacities as determined microscopically.[4] Accordingly, 20 or more vitrinitic maceral materials are recognized using subscripts to indicate their level of reflectivity, for example, V_6, V_7, V_8, and so on.

Vitrinitic particles give bituminous coals their striated or *banded* appearance and their shiny, black, *vitreous* luster. Their absence, or their presence only in the form of very small particles, results in a *nonbanded* coal type displaying a matte luster. In the case of *banded* coal types in which the particles of vitrinite are in excess of 1 mm in thickness, grains of *pure* vitrinite will be generated when the sample is crushed. Various size fractions of the crushed coal will contain various quantities of the vitrinites, depending on their friabilities. Therefore, one should not assume that different size fractions of a given coal sample will always be of

TABLE 1.2   Macerals Commonly Encountered in Coal

Huminite–Vitrinite Suite

Huminite	Vitrinite
Textinite	Telinite
Ulminite	Telocollinite
Attrinite	Desmocollinite
Densinite	Vitrodetrinite
Gelinite	
Corpohuminite	

Inertinite Suite

Fusinite
Semifusinite
Micrinite
Macrinite
Sclerotinite

Liptinite Suite

Sporinite
Cutinite
Resinite
Alginite
Suberinite
Fluorinite
Exudatinite
Bituminite

the same composition. This is particularly true when the original sample is a blend of coals or coal types.

In summary, as in the case of minerals, coals contain a large assortment of dissimilar macerals. Table 1.2 serves to identify the more important of the latter. It is now evident that coals are heterogeneous mixtures of organic and inorganic materials, made more complex by the variety of both the macerals and the minerals that can participate in numerous permutations to form the rock mass. The variety of mixtures available may, on the one hand, make the sample selection process more difficult, but on the other hand, they provide a valuable population of differing materials for study.

1.2.4 Chemical Constituents of Coal

Conventional coal analyses produce a number of different kinds of quasi-chemical data that have proven useful in coal utilization and helpful in sample selection. The most common set of data is generated in the course of a *proximate analysis*.[5] This produces three analytically determined values: moisture content, volatile matter yield, and ash yield upon combustion. *Fixed carbon* is calculated by difference, and the numbers may then be transformed to moisture-free values, ash-free values, and moisture- and ash-free values if desired. Table 1.3 displays a typical example of proximate analysis values. If the amount of mineral matter and/or the organically bound cations are pertinent in sample selection, the ash yield will provide a useful guide in this connection. Volatile matter yield is employed in classifiying coals by *rank*.[6] This classification was devised for practical purposes in evaluating coals for combustion and carbonization but is valuable in sample selection because it provides a general indication of the level of coalification achieved by at least some of the maceral materials, principally the vitrinitic materials. Table 1.4 presents this classification scheme and Figure 1.3 displays the areal distribution of coals of various ranks in the United States. As may be seen from the table, if one wishes to select a "low volatile bituminous coal" for study or testing, coals with volatile matter yields between 14 and 23 percent (fixed carbon contents between 78 and 86) are included in the

TABLE 1.3 Typical Types of Data Produced by a Proximate Analysis

	As Received (%)	Dry (%)	DAF[a] (%)	DMMF[b] (%)
Moisture	4.19			
Ash	15.20	15.86		
Volatile matter	33.68	35.15	41.78	40.02
Fixed carbon	46.93	48.99	58.22	59.98

[a]Dry, ash free
[b]Dry, mineral matter free

TABLE 1.4 Classification of Coals by Rank According to the American Society for Testing Materials

Rank Group	Fixed Carbon (% DMMF[a]) Equal or Greater Then	Less Than	Calorific Value (Btu/lb: MMMF[b]) Equal or Greater Than	Less Than	Agglomerating Character
Meta-anthracite	98	—			nonagglomerating
Anthracite	92	98			
Semianthracite	86	92			
Low volatile bituminous coal	78	87			
Medium volatile bituminous coal	69	78			
High volatile A bituminous coal		69	14,000	—	agglomerating
High volatile B bituminous coal			12,000	14,000	
High volatile C bituminous coal			11,500	13,000	
			10,500	11,500	
Subbituminous A coal			10,500	11,500	nonagglomerating
Subbituminous B coal			9,500	10,500	
Subbituminous C coal			8,300	9,500	
Lignite A			6,300	8,300	
Lignite B			—	6,300	

[a]Dry, mineral matter free
[b]Moist, mineral matter free
Source: Data from ASTM, Standard Method D388-84: Standard classification of coals by rank. *Annu. Book ASTM Stand.* Vol. 0505, (1986).

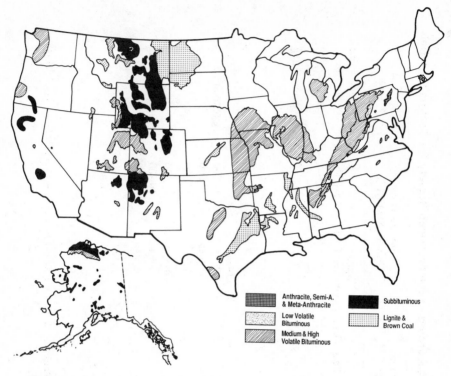

Figure 1.3 Areal distribution of coals of various ranks in the United States.

population. The other primary value used in this classification system is the calorific value or British thermal unit (Btu) content per pound of coal. This is used in distinguishing between different kinds of bituminous, subbituminous, and lignitic coals (see Table 1.4). It now can be seen that the selection of merely "a bituminous coal" for research or experimentation fails to establish, except in a very imprecise way, the kind of sample intended for study. In other words, the sample population is so large and internally so diverse that selecting a *representative* sample is impossible.

Another set of chemical values is produced through the *ultimate analysis* of a coal sample.[7] This determines the ash yield and the carbon, hydrogen, nitrogen, sulfur, and oxygen contents of the sample, the last value usually determined by difference. A typical analysis with the values calculated to various bases is presented in Table 1.5. The carbon content probably is more useful as an index of the level of coalification than the previously mentioned volatile matter yield, and ratios such as the H/C ratio are also helpful. Organic chemists should note that the *sulfur content* includes the *total sulfur* present, which may be largely derived from mineral sulfides and sulfates as opposed to being a constituent of the maceral materials. In this connection, another set of analytical procedures results in a sulfur forms analysis.[8] This produces three analytically determined

TABLE 1.5   Typical Types of Data Produced by an Ultimate Analysis

	As Received (%)	Dry (%)	DAF[a] (%)	DMMF[b] (%)
Ash	15.20	15.86		
Carbon	64.65	67.48	80.20	83.64
Hydrogen	4.62[c]	4.82	5.73	5.97
Nitrogen	1.17	1.22	1.45	1.51
Sulfur	3.81	3.98	4.73	
Oxygen	6.36[c]	6.64	7.90	8.89

[a]Dry, ash free
[b]Dry, mineral matter free
[c]Excludes moisture

TABLE 1.6   Typical Types of Data Produced by a Sulfur Forms Analysis

	Dry (%)	Dry, Ash Free (%)
Pyritic Sulfur	2.16	2.57
Sulfatic Sulfur	0.27	0.32
Organic Sulfur	1.55	1.84
Total Sulfur	3.98	4.73

values: total sulfur content, pyritic sulfur, and sulfide sulfur contents. With these three values, the organic sulfur content is determined by difference. Table 1.6 presents a typical set of sulfur values. Pyritic sulfur values may be as low as zero and as high as 6 percent. Organic sulfur normally lies between 0.01 and 3.00 percent and in rare cases may be as high as 10 percent. Sulfate sulfur typically is small in its concentration, seldom exceeding 1 percent.

Knowledge of the elemental composition of the ash in a coal sample has practical application in predicting a coal's fouling and slagging tendencies and in evaluating its impact in steel-making, hence, another set of data is often generated and referred to as *ash chemistry* or *major and minor elements* in coal ash.[9] Table 1.7 presents such a set of data. The values normally are reported as a percentage of the oxide of the element in the ash yielded by the sample. This does not display the mode of occurrence in the coal for, as previously mentioned, cations such as calcium, potassium, and sodium may be present in the maceral materials as a part of the carboxyl structure. In certain coal provinces, the sodium may be present as NaCl and, if this is pertinent to sample selection, an analysis of the chlorine content of the coal will be useful.

TABLE 1.7 Typical Types of Data Produced by an Analysis of Ash Chemistry

	Percent of Oxide in HTA	Percent of Element in Total Dry Coal
SiO_2	43.0	2.75
Al_2O_3	20.10	1.45
TiO_2	0.93	0.98
Fe_2O_3	23.70	2.27
MgO	0.73	0.06
CaO	3.68	0.36
Na_2O	0.39	0.04
K_2O	1.80	0.20
P_2O_5	0.10	0.01
SO_3	3.70	0.20

In recent years, the environmental significance of various trace elements contained in coals has become progressively more evident. This coupled with machines that mass produce information on trace element concentrations has resulted in an abundance of information on the subject.[10-12] Certain trace elements (e.g., B, Be, Ge, U) appear to be most commonly contained within organic structure, whereas elements such as Mg, Zn, Cd, As are more often found in the crystalline minerals. The concentrations of some trace elements reflect specific depositional environments, hence, specific geochemistries; whereas the elevated concentrations of trace elements in general may reflect the formation of coals near the margins of the peat-forming basins. In some instances, the distribution of individual trace elements in seams may be unrelated to basin margins and depositional environments. Uranium, for example, often is concentrated in the upper portions of the lignite seams of North Dakota and South Dakota, and vanadium occurs in increased concentrations at the base of some West Virginia coals.

From the above, it is evident that chemical data on coals are of major importance in sample selection. It is also evident that the data are not always what they seem to be. For research purposes, a thorough and critical evaluation of such data can materially improve suitability and quality in the sample selection process. The more specific one can be, based on understanding the limits of the possibilities, the greater the chances of receiving the desired sample from the sample supplier.

1.2.5 Physical Properties of Coals

Specific gravity, hardness, grindability, porosity, surface area, bulk density, and size consist are among the physical properties of coals and coal products that are commonly measured. These are of interest in sample selection because of their effects in sample processing and because of their potential effects in studies involving reactions with gaseous and liquid reagents.

Coal cleaning methods frequently involve crushing, sizing, and flotation. The crushed coal is sieved to produce several size fractions, which may then be subjected to flotation in liquids of various densities to reject some of the relatively heavy mineral material or to otherwise modify the composition of the prepared product. Hardness and specific gravity would appear to be key properties in these operations; however, it is *friability* plus *grindability* and particle settling velocity that play the significant roles.

The hardness of the mineral constituents obviously vary from species to species. This hardness is expressed using the mineralogist's Moh's scale, which is a means of ranking minerals on a scale of 1 (talc) to 10 (diamond). For example, the hardness of pyrite is 6 and that of quartz is 7. One way of determining the hardness of macerals involves dropping a diamond point onto the maceral surface and measuring (microscopically) the size of the resulting indentation.[13] As in the case of minerals, different macerals may display quite different hardnesses, ranging from elastic sporinite to the relatively hard but friable fusinitic material.

A coal's toughness or resistance to crushing is only partly related to the inherent hardness of individual macerals and minerals. The rock mass (i.e., the coal type) possesses a friability, or lack thereof, that is not readily predicted from a knowledge of the hardnesses of the component minerals and macerals. The ease with which a given coal type is crushed is strongly influenced by the extent to which planes or zones of weakness have developed in the mass. Figure 1.4 diagrammatically illustrates the principal planes of weakness in a well-consolidated but banded coal type. As the result of overburden pressures, such coal types, like many other rocks, are characteristically cut by two intersecting

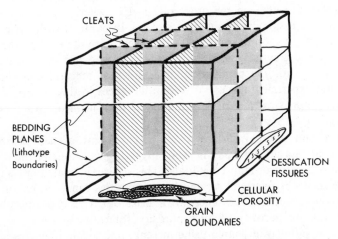

Figure 1.4 Principal planes of weakness in a well-consolidated banded coal.

sets of vertical fractures, which miners have termed *cleat*. One set is usually better developed than the other, and this often determines the direction of mining. As indicated in the figure, other potential breakage planes exist at lithotype interfaces and grain boundaries. Microfractures in vitrinite lenses and poorly consolidated fusain layers represent zones of weakness through which breakage may occur. The various coal types differ widely in their resistance to crushing; hence, when several are present in a coal seam, this will affect the composition of samples derived from various size fractions.

A common method of measuring the *grindability* of a coal is referred to as the *Hardgrove grindability test*.[14] This entails the use of a ball-mill type grinding device into which is placed a 16 × 30 mesh size fraction of the coal to be tested. After sixty revolutions of the machine, the ground coal is sieved to determine the amount of degradation that has occurred. Unfortunately, the 16 × 30 mesh test sample will be biased in favor of the harder lithotypes in the coal, and the results may not be definitive enough for some purposes. In spite of this, the Hardgrove grindability index may be useful in selecting coal samples when physical strength is a pertinent factor (see Chapter 3 for more on the subject of coal grinding).

Bulk density is expressed in pounds per cubic foot of crushed coal. The weight is, of course, a function of the inherent weight of the coal and the particle size of the coal product being measured.[15] Bulk densities are of interest in connection with transporting, storing, and carbonizing coals. Selection of samples for use in investigations relating to carbonization and spontaneous combustion can benefit from an understanding of the impact of bulk densities on these processes.

As in the case of hardness, the specific gravity of both minerals and macerals varies considerably. Values for minerals are well established and as might be expected are high relative to the specific gravities of macerals. To the extent that the latter are known, the information can be found in the *International Handbook for Coal Petrology*.[16] The liptinitic macerals tend to have the lowest specific gravities (often on the order of 1.1–1.2), whereas inertinitic macerals such as fusinite display values in excess of 1.4. In contrast with these gravities, quartz possesses a specific gravity of 2.2 and that of pyrite is 5.02. Although the range of specific gravities displayed by the macerals seems small, it has proved sufficient to permit the production of samples of maceral concentrates for analysis and experimentation.[17,18] Recent advances in this area hold promise for greatly improving our understanding of the chemical structure of coal constituents.

Porosity and internal surface area are clearly properties of concern in selecting coal samples for investigations involving reactions of the coal when treated with gaseous or liquid materials. (This subject is treated in Chapter 4 to which the reader is referred.) Again, these properties can, and do, vary widely from one coal type to another. There are at least four types of porosity displayed by coals: fracture porosity, cellular porosity, intergrain porosity, and micro-porosity. The latter is measured using mercury and alcohol penetration and is not visible to the unaided eye or with the use of a normal light microscope. The most conspicuous type of fracture porosity is formed by the previously mentioned process of cleat formation. On a smaller scale, the intragrain dessication fractures characteristic of certain types of vitrinitic material also contribute to a coal's gross porosity. Cellular porosity is, by definition, a feature of such macerals as fusinite, semifusinite, textinite, and, to a certain extent, ulminite. The size and shape of such pores is a reflection of the size and form of the cell lumina in the precursor plant tissue coupled with any modifications introduced by the coalification processes. The size of individual pores will be measured in microns (tens to hundreds or more) and their length may greatly exceed their cross-sectional dimensions. In contrast to this, a coal's micropores are measured in angstroms. The micropores themselves usually are defined as pores less than 7 Å in diameter, and these account for the great bulk of a coal's pore volume. Transitional pores are on the order of 7–14 Å in size and the pores larger than 14 Å are sometimes spoken of as *macropores*. The macropores and transitional pores help to provide access to the micropores in the microporous system, hence, are important in the diffusion of reagents into the coal structure.

Intergrain porosity results from the imperfect bonding of the irregular surfaces of two juxtaposed grains. It, together with the cellular porosity of fusinite, accounts for the high permeability of fusain strata in coal seams. These, together with fracture porosity (mainly cleat), account for the fact that coal seams in semiarid regions often are regarded as aquifers.

In summary, knowledge of the physical properties of minerals, macerals, and coal types not only assist in the sample selection process but at the same time alerts one to the need for proper sample processing and permits the evaluation of sample quality when the processing steps are known.

1.2.6   Other Properties Pertinent to Sample Selection

In addition to the properties already discussed, there are others that may influence the sample selection process. One of these is the propensity of many coals to undergo compositional changes as the result of a phenomenon referred to as *weathering*. This subject is dealt with, in other chapters, hence, it is treated only briefly here. Weathering occurs when a coal is exposed to the oxygen in the air or oxygen in percolating ground waters. Samples taken from a natural outcrop or from a mine face that has been exposed for some time commonly contain *weathered* coal. Lignites and low-rank bituminous coals are particularly susceptible to weathering, whereas anthracites normally show little effect. In the case of lignite seams that are thinly overlain by poorly consolidated sediments (as many are), the coal may be weathered even though it is far removed from any outcrop and has never been exposed to air. Because of this, obtaining unweathered samples of certain seams may prove to be a difficult task.

One of the effects of exposing lignites to air is to induce a rapid and irreversible dehydration of the coal, resulting in what is termed *slacking*. The latter is a physical disruption or decrepitation of the coal that can transform a solid block into a pile of fine particles in a relatively short period of time. Some compositional changes must also attend the transformation and a thin slice of translucent huminite can be converted to an opaque mass in a matter of seconds.

Low-rank coals (lignites through high volatile C bituminous) may also exhibit tendencies for spontaneous combustion. This can occur within the seam prior to mining, and it can occur in coal samples that are not properly stored. This usually does not become involved in the sample selection process unless the attendant studies concern the phenomenon itself.

When vitrinitic macerals achieve a certain level of coalification (that associated with bituminous coals), they acquire the ability to soften and become plastic when heated in an inert atmosphere. This property is exceedingly important in coal carbonization and liquefaction and may have a negative impact in certain gasification processes. The nature of the plasticity and the characteristics of the carbonized residues vary greatly and are determined by the kind of vitrinite involved. A coal's *plastic properties* often are measured employing a device known as a *plastometer*.[19] This serves to determine: (1) the temperature at which the coal mass begins to soften, (2) the maximum level of fluidity achieved, (3) the temperature at which the maximum fluidity is reached, (4) the solidification temperature, and (5) the temperature range through which the coal has remained plastic. The *coke button test* is another means of evaluating a coal's plastic behavior and observing the extent to which a swollen carbonized product is produced as the result of being heated in the absence of air.[20] Oxidation and/or weathering depresses and may even destroy the ability of the vitrinitic macerals to develop a fluid phase. The plastometer and the coke button test have both been used to detect even incipient weathering, the plastometer being particularly useful in this connection. In the sample selection process, knowledge of a coal's plastic properties will provide a clue as to the type of vitrinitic materials that are present and may also be helpful in screening out samples that have been weathered. It should be noted, however, that although a coal's plastic behavior is determined mainly by the vitrinitic material present, this behavior is modified by the amounts and types of other macerals and minerals in the coal mix. In general, high volatile A bituminous coals will tend to display the greatest degree of fluidity, and low volatile bituminous coals will produce the most dense and anisotropic carbonized residue.

In summary, coals exist in the Earth's crust in both an unweathered and a weathered state. Weathering in nature and oxidation during processing and storing in the laboratory can have a profound effect on coal composition. Except for special purposes, unweathered samples should be sought and care taken to prevent any subsequent alteration. The huminitic and vitrinitic macerals are the ones most susceptible to oxidation, and their response varies depending on the particular maceral species involved. Because these macerals form the bulk of most coal samples, it is important that they be given prime consideration in sample selection and that they be adequately protected against oxidation.

1.2.7   Coal Seam Composition and Its Significance in Sample Selection

The different maceral and mineral grains composing a coal seam often are not uniformly distributed through the thickness of the seam or throughout its lateral extent. Instead, distinctive mixtures occur, each containing its own particular concentrations of various macerals and minerals. These naturally occurring mixtures reflect an origin in a particular ecological and geochemical setting and are here termed *lithotypes* or *coal types*. The coal seam usually is a *stratified rock body*, that is, it is made up of a number of superposed layers, each layer being composed of a mixture of maceral and mineral grains that is different from the mixtures forming the juxtaposed layers (Figure 1.5). Each layer is typically lenticular in form and its areal extent commonly covers many acres to many square miles. These layers are referred to as *lithobodies*; some are only a few inches thick, whereas others may be many feet thick. Laterally, they may vary in thickness, and this is particularly true as their margin is approached. At their very margin, they gradually thin and finally wedge-out, changing the composition of the seam by their loss.

Coal seams, then, are collections of laminar (lenticular) lithobodies whose number and thicknesses may change from place to place. Each lithobody is formed of a lithotype that is different from the lithotypes forming overlying and underlying lithobodies, but the same lithotype may form several of the seam's

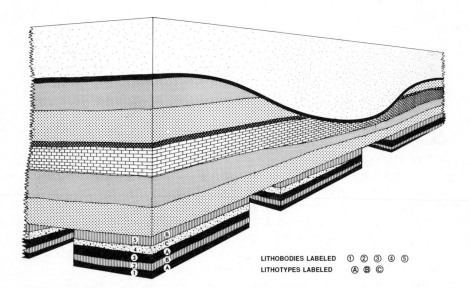

LITHOBODIES LABELED   ① ② ③ ④ ⑤
LITHOTYPES LABELED   Ⓐ Ⓑ Ⓒ

Figure 1.5   Diagrammatic representation of the stratified construction of coal seams.

lithobodies. In other words, coal seams are stratified mixtures of mixtures, and the mix is not likely to remain constant over the entire areal extent of the coal bed.

The importance of identifying coal samples with specific sites becomes evident when the above is recognized. It follows that the mere designation of a sample as "Pittsburgh coal" or "Illinois #6" does not guarantee a particular set of compositional characteristics. Accordingly, sample selection often requires identifying both the name of the seam and the geographic location from which the sample derives.

Another fact to be noted is that the lithotypes represented in a coal seam at a given site can display very different physical properties, notably their hardnesses. This will have a significant effect if coal samples are to be selected from crushed coal products. The harder lithotypes will tend to be concentrated in the coarser-size fractions, whereas very friable lithotypes will be overrepresented in the fine-size fractions. This can sometimes be used to select coal samples of particular compositions from a whole seam product.

1.3   THE ORIGIN OF COAL SEAM STRATIFICATION AND ITS SIGNIFICANCE IN SAMPLE SELECTION

An extended discussion of coal origins is not required in this book, but several points can usefull be made in the interests of demonstrating that the compositional variations in coals are neither infinite nor random. Instead, they are limited in number and rigidly controlled by reasonably well-known biological, ecological, geochemical, and geological processes. This control begins in the peat-forming swamps and marches.

1.3.1   Pertinent Characteristics of Coal-forming Swamps and Marshes

Any coal seam of significant lateral extent (more than a few acres) will have had its origin in a complex of swamps and/or marshes as opposed to being derived from a single ecological site. Each ecologically distinct area within the complex possesses its own plant community, hence, its own array of coal precursor substances. Each possesses its own geochemical environment with its particular collection of microorganisms. Each is subject to a certain hydrological regime. As the result of the interaction of all of these factors, each ecologically distinct area in the swamp/marsh complex produces its own particular peat type, which will subsequently develop into a specific coal lithotype. Thus, the precursors of a number of different lithotypes are being formed concurrently in a swamp/marsh complex of any size, but the number is by no means limitless and they bear an ecologically controlled relationship to one another. The braskish marsh environments shown in Figure 1.6A never occur on the seaward side of the saline mangrove forest that lines the coast, and the water lily–cattail marsh in the

Figure 1.6 Photographs of contrasting peat-forming environment in a single swamp/marsh wetland complex. A. Two different brackish marsh environments bordered (along the shore) by a saline swamp forest displaying the well-defined boundaries of the coal-forming areas. B. Three freshwater marsh environments in front of a freshwater swamp forest, each producing a particular peat type.

foreground of Figure 1.6B always indicates a wetter environment than the saw grass community behind it, and the forest community in the background always identifies a still drier environment. A predictable relationship exists between these swamp and marsh environments, hence, a predictable relationship must exist between their resultant lithotypes.

1.3.2 Dynamics of Peat Formation

For plant materials to accumulate and form peat, the agencies of decay and decomposition must be seriously inhibited. In swamp and marsh environments, this is accomplished by maintaining the water table at or above the surface for much if not all of the year. Even so, the thick peat deposits required to form minable coal seams will not form unless a proper balance exists between the rate of production and deposition of organic debris and the rate of subsidence of the surface on which the debris is accumulating. An improper balance will result in the transformation of the site into a place no longer suitable for the plant community occupying the site, and the community will be replaced by one more adapted to the wetter or drier environment produced by the imbalance. In extreme cases of imbalance, peat formation ceases completely.

Superimposed on this sensitive balancing of subsidence, water level mainten-ance, and organic matter production is a phenomenon known as *plant succession.* As the result of its occupancy of a site, a plant community in a wetland environment will tend to modify its site in a fashion that results in its replacement by another plant community. For example, an area may be dominated by aquatic plants growing in 3 ft of water at the outset, and, through the accumulation of organic debris from the aquatics, the site may be rendered suitable for a marsh vegetation, which in turn may convert the area into a site that is dry enough to support a swamp forest. In passing from an open water area congested with water lilies to a hardwood swamp forest, several different vegetational assemblages will have occupied the site, each forming its own particular peat type in its own microbiological and geochemical setting. The result is a *layer cake* of peat types that is capable of being transformed into a stratified coal seam after burial.

Notice that plant succession can proceed without subsidence in wetland environments. When differential rates of subsidence are added to the system, along with variations in the hydrological regime and variable rates of organic matter production (which may be dependent on the plant community involved), we begin to comprehend the dynamics of the initial phase of coal seam formation. Learning that coal seams typically contain several coal types now comes as no surprise. In fact, it would appear to be inevitable.

1.3.3 Phenomena Producing and Controlling Coal Seam Stratification

In the preceding section, the development of stratification in a peat sequence as the result of plant succession was described. This process can be interrupted or reversed by changes in the rates of subsidence or alteration of the hydrological regime. In addition, a number of other mechanisms operate in swamp/marsh complexes to produce laminated coal seams. The most important of these are the processes known to the geologist as transgression and regression.

Laterally extensive coal seams originate on low-lying coastal plains and deltas where sea level strongly influences the level of the water table on land. An increase in the rate of subsidence of the land will cause the sea to transgress onto the land surface at the coastline. This transgression of saline water destroys the swamp forest at the coast and causes migration of the forest in a landward direction. This, in turn, forces inland migration of the brackish-water plant communities onto areas formerly occupied by the more salt-tolerant fresh-water communities. The process involves long periods of time and as each community moves inland, it is depositing a layer of distinctive peat on top of the layer formed by the community that previously occupied the site. The end product is a stratified blanket of peat in which the individual layers form sheets of peat with areal extents that are far in excess of the areas occupied by the source plant community at any one point in time. Figure 1.7 diagrammatically illustrates the effect of transgression on a swamp/marsh complex. A standstill of sea level will result in the vertical build-up of each peat type, and a gradual lowering of sea level reverses the process causing a regression and seaward migration of the various plant communities.

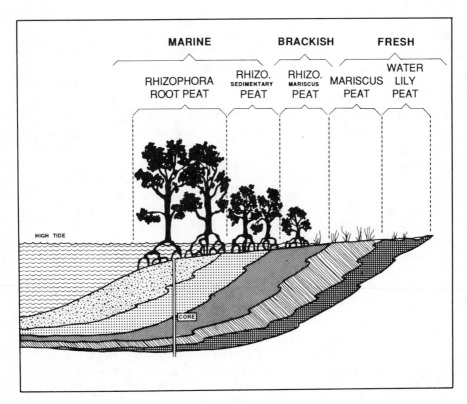

Figure 1.7   The stratified peat deposit formed in Southwestern Florida as the result of the transgression of the Gulf of Mexico.

Strata dominated by mineral material (partings) can be introduced into the sequence in several ways. In regions characterized by volcanic eruptions, ash falls can cover large areas and form fine-grained intraseam strata. Along a coast, hurricanes can carry sediment from the marine waters onto the surface of the accumulating peat, temporarily interrupting peat formation. On deltaic plains, streams commonly are contained within their banks by a ridgelike levee that borders the channel. During periods of excessive flooding, the level may be breached, allowing sediment-laden water to flow onto the adjacent peat surface, depositing its load as an inorganic mud. Such *splay* deposits normally are of limited areal extent, but they can result in the presence of numerous lenticular partings in coal seams and these become part of the as-mined coal.

1.3.4 Diagenetic Processes' Impact on a Coal's Maceral and Mineral Composition

Processes operating during the peat stage of coal formation are referred to as being *diagenetic*. The most important of these is the collection of processes that result in the differential destruction of certain plant substances and the concentration of others. Through the operation of these processes, a peat sediment dominated by lignin-derived macerals tends to be created. As the less resistant plant tissues and substances are destroyed, materials such as resins, waxes, cutins, and sporopollenin become a more conspicuous part of the peat. Swamp and marsh fires not only affect the standing vegetation since their products accumulate on the peat surface; the fires themselves commonly extend into the upper layers of the peat. Some plant communities are fire-prone, the saw grass marsh of the Florida Everglades being one example. Lightning-generated fires in the saw grass result in charcoal being produced, and fusinite is a characteristic component of saw grass peat.

Bacteria and fungi operating in the uppermost layer of the peat have a profound effect on both the composition and the texture of the peat produced. The assemblage of organisms involved and the duration and effectiveness of their operation is influenced by the geochemical conditions associated with the host plant community. Thus, the characteristics of each peat type are partly determined by the plant substances generated in the environment, partly by the impact of bacteria and fungi on these substances, and partly by substances generated by the microorganisms themselves. The latter can include mineral materials as well as maceral material. Pyrite, for example, is formed well below the peat surface by anaerobic bacteria that serve to combine the iron and sulfur in the geochemical system into iron disulfide. The bacteria frequently occupy the cavities of uncollapsed plant cells (Figure 1.8B), producing micron-size pyrite grains and spherical clusters of crystals that are only a few microns in diameter (see Figure 1.8B). Upon further compaction of the maceral materials, the cell cavities are destroyed; and the fine-grained pyrite becomes encased in an organic matrix. Removal of this pyrite to produce mineral-free coal samples can prove to be a very difficult task. In addition to requiring the proper bacterial species,

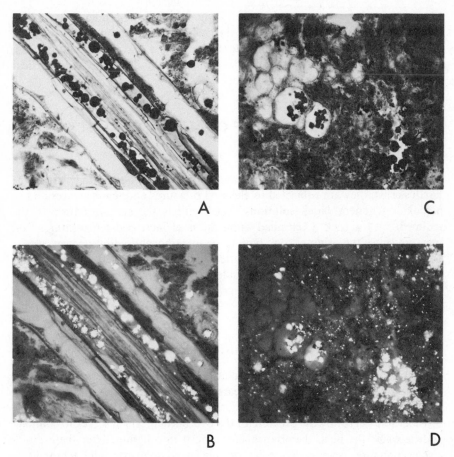

Figure 1.8 Modes of occurrence of pyrite in peat show in transmitted (A,C) and reflected light (B,D). A,B: Fromboidal pyrite, $5-10\,\mu$ in diameter, often will end up encased in vitrinite. C,D: Micron size pyrite grains that will become encased in a maceral matrix upon compaction of the peat.

pyrite formation also requires that iron and sulfur be present in the system in the correct form. Because the source of sulfur usually is sea water, peat types formed in saline and brackish water environments possess pyrite as a component of their mineral fraction, whereas fresh-water peat types characteristically do not. Diagenetic processes, therefore, are of major consequence in determining the composition and texture of the strata forming the components of a peat sequence.

In summary, coal seam stratification is a consequence of the multicommunity composition of peat-forming systems and the reaction of these plant communities to geological and hydrological changes in the systems. The stratification is not random, hence, is predictable once the biological and geological principles are understood. Underscored, however, is the fact that the prudent selection of coal samples requires: (1) awareness of the lack of vertical homogeneity in a coal seam, (2) awareness of the probable lateral change in a coal seam's composition, and (3) awareness of the marked dissimilarities in chemical and physical properties displayed by superposed coal types.

1.4   THE PHENOMENON OF COAL AND COAL SEAM METAMORPHISM

When sediments such as sands, silts, and clays become buried and part of the Earth's crust, they are subjected to elevated pressures and temperatures and to percolating mineral-laden solutions. The result is the transformation of the sediment into a rock. Continued application of increased temperatures and pressure cause the original rock type to be metamorphosed into a new type.

Similarly, when a peat seam is buried under a thick overburden of other sediment types, it is subjected to a considerable confining pressure; and as depth of burial increases, the temperature gradually rises. It is at this point that *coalification* begins.

1.4.1   Factors Affecting Coalification

In contrast to the importance of percolating, mineral-rich waters in cementing other sediment types, the process appears to have little or no role to play in the consolidation of peat into coal. Instead, the consolidation is brought about to a considerable degree by the pressure of the overburden, which serves to compact and de-water the peat, thereby transforming it into lignite. After these initial effects, pressure assumes a secondary role in coalification, and temperature becomes the driving force for metamorphism. Although the effects are profound, the temperatures required to reach even the highest levels of coalification (i.e., anthracites) are comparatively low. Most coal seams have not experienced temperatures in excess of 150°C, and usually the maximum temperature involved will be much less. The factor that operates to increase the impact of temperature is geologic time. The same compositional transformations can be brought about by either the application of a low temperature for a long period of time or the application of a higher temperature for a shorter period of time.[21]

1.4.2   Impact on Macerals and Minerals

The organic substances composing the coal macerals are very sensitive to increases in temperature, which is, of course, the reason why only low temperatures are required for their continued metamorphosis. All macerals do not follow the same *coalification path*, however. Schopf[22] described these variable responses of different organic materials as "variable coalification"

(Figure 1.9). By this, he meant that inertinitic macerals such as fusinite have undergone a large and rapid change in their initial formation and are only slightly affected by the normal temperatures of coalification. (This is probably true because they have already seen higher temperatures in their formation). In contrast, sporinitic macerals undergo only modest changes with the application of small temperature increases but begin a rapid transformation when a certain temperature threshold is reached. Intermediate in response are the huminitic and vitrinitic macerals, which are quite sensitive to temperature increases throughout coalification, following a reasonably steady pathway in their metamorphosis. Dormans et al.[23] and Teichmüller and Teichmüller[21] have presented a similar picture of maceral metamorphosis in their H/C versus O/C diagram (Figure 1.10). The diagrams of Dormans et al., the Teichmüllers' synthesis, and Schopf's diagram had best be regarded as oversimplifications of the facts. It is probably closer to the truth to regard each maceral as comprising, in fact, a series of genetically related macerals with end members that can be radically different in composition. This is well illustrated by the huminite–vitrinite series in which the end member found in anthracite bears no similarity to the end members found in low-rank lignites. As mentioned earlier, a provisional means of distinguishing between the members of this series has been

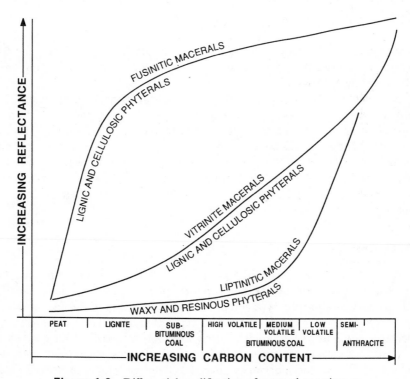

Figure 1.9　Differential coalification of maceral constituents.

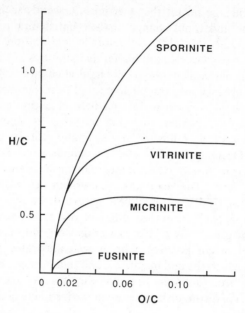

Figure 1.10   Coalification pathway of four macerals based on their H/C versus O/C ratios.

devised using the reflectance properties of the macerals as a device to call attention to differences in composition.[24] This procedure can and is being used with other maceral series.

The inorganic substances comprising the coal minerals are much less sensitive to small increases in temperature. Many of these require an elevation in temperature to several hundred degrees before metamorphosis occurs, hence, their composition remains constant in most coal seams. Exceptions to this are associated with certain of the clay minerals. Kaolinite can be transformed into pyrrophyllite and in some cases into chlorite, but those conversions are encountered only in high-rank anthracites so they become matters for consideration in sample selection in relatively few cases.

1.4.3   Impact on Coal Lithotypes

As previously mentioned, coal lithotypes exhibit properties that are not readily predicted from a knowledge of the properties of the constituent macerals and minerals. The durainic lithotype composed principally of sporinite and micrinite is an exceedingly hard and tough coal type, characteristics not necessarily expected of a mixture of granular micrinite and elastic sporinite. Also, we have seen that the several swamp and marsh environments each yield different mixtures of organic and inorganic materials. Upon consolidation each of these becomes a coal lithotype of low rank.

The impact of temperature and pressure on the various lithotypes is not well known and is poorly documented. This is most unfortunate as it is the characteristics of these mixtures that often control the behavior of coals in beneficiation and utilization. It is clear, however, that the properties of lithotypes of lignitic rank are quite unlike those found in low volatile bituminous coal. Therefore, it may be useful to regard each lithotype as being part of a metamorphic series, just as individual macerals are parts of metamorphic series.

1.4.4   Chemical and Physical Effects

The chemical and physical effects of progressive coalification have generally been described for *whole coal* samples, meaning that coal from all of the seam's lithobodies (at one site) have been mixed together to form the sample. Because most seams consist principally of macerals of the huminite–vitrinite series, the characteristics described tend to relate to these materials to the exclusion of other maceral groups. The most notable exception to this is the work of M. and R. Teichmüller[21,25] in which vitrites representing various metamorphic levels were studied.

The previously described classification of coals by rank (Table 1.4) recognizes four of the important coal properties that change during progressive coalification. The calorific value increases from an arbitrary 6300 Btu/lb to more than 14,000 Btu/lb; the volatile matter yield gradually decreases to less than 2 percent on a dry, mineral matter-free (DMMF) basis with a correlated increase in fixed carbon. Coals are transformed from a nonagglomerating material through an agglomerating phase and back to a nonagglomerating material as coalification proceeds. Moisture and oxygen contents are particularly high in low-rank coals but both diminish as the high volatile bituminous stage is reached. The moisture content of German soft brown coals ranges between 35 and 37 percent, and this is reduced to less than 10 percent in high volatile B bituminous coal. After reaching the bituminous coal level of coalification, these properties (along with others) continue to change but in a nonparallel fashion as shown in Figure 1.11.

One of the best summaries of the chemical and physical effects of coal metamorphism was presented by Teichmüller and Teichmüller,[21] and a portion of this is presented in modified form in Figure 1.12. Although it is not all inclusive, it serves to emphasize the fact that the phenomenon of coal seam metamorphism produces a diverse array of coal types and maceral substances.

1.5   SUMMARY OF THE IMPACT OF THE COMPOSITION OF COAL AND COAL SEAMS ON SAMPLE SELECTION

The large number of species in the coal rock class makes proper sample selection a difficult task. Presumably, one should understand the population being sampled as a prerequisite to intelligent sample selection, but in many cases this is impractical without the help of a specialist in coal characteristics. Even after the specifications for the sample have been defined, there can be a problem in translating those specifications into a seam name and geographic location.

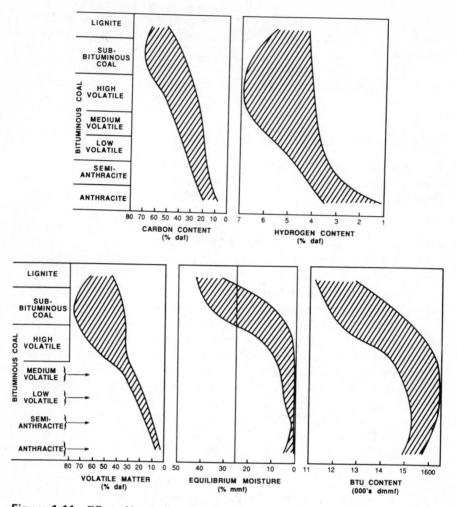

Figure 1.11   Effect of level of coalification on several chemical parameters (modified from Teichmüller and Teichmüller[21]).

(There are 119 coal seams in West Virginia and 45 coalfields in Wyoming—how does one go about choosing one seam and one area?)

On the other hand, the large number of coal types available affords the researcher with a wealth of material and an array of compositions to satisfy his needs. In the form of the coal lithotypes, nature has created a spectrum of mixtures from which sample suites can be selected to permit examination of the effects of varying the concentrations of component x or y. The effects of progressive coalification on the macerals and lithotypes add another dimension

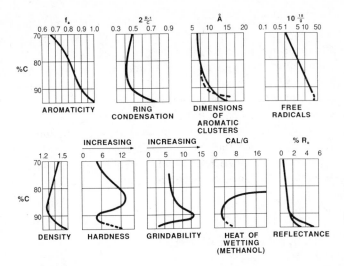

Figure 1.12   Physical and chemical effects of progressive coalification (modified from Teichmüller and Teichmüller[21]).

to the variety of compositions available, extending still further the opportunities to match sample availability with sample specifications.

Coal seams are not uniformly heterogeneous, hence, randomly chosen samples are not likely to be representative of the seam and may vary considerably from sample to sample. Thus, sample selection involves specifying a kind of coal required as well as seam name and location. Although not uniformly heterogeneous, coal seams are composed of layers of various coal types, and the compositional characteristics of the layers and the stratified construction of the seams are rigidly controlled by the interaction of biological, geochemical, and geological factors. Accordingly, lateral changes in coal seam composition and construction may and do occur but always in an orderly and predictable manner. Again, one is confronted with both a curse and a blessing. In sample selection, one must be alert to possible lateral changes and specify a geographic location within the seam area if compositional control is critical. This may require knowledge of the lateral changes associated with the seam in question that, in turn, can be outside the expertise of the sample requestor. On the blessing side is the fact that lateral changes afford a number of opportunities to study the compositional impacts of changing environmental sequences on total seam composition and, in some instances, to study the effects of progressive coalification on a single seam.

One aspect of the impact of coal composition on sample selection not usually mentioned is the fact that the need for fresh, unweathered samples automatically reduces the size of the available sample population. Although drill core samples are sometimes available, most samples come from operating mines. For good reason, mining companies normally try to avoid mining seams with high concentrations of sulfur or ash-forming minerals. Also, thicker seams are preferred over thin ones and coals with undesirable combustion or coking characteristics tend to be programmed for later mining. The impact of this may be small, but it may be of consequence in particular cases.

Thus, the selection of coal samples for basic and applied research is made challenging by the variety of compositions available and the numerous sites and seams from which samples may be obtained. Even in those restricted cases in which the seam and area are determined by direct industrial use, compositional variation must be taken into account in determining sample type and sampling strategies.

1.6 TYPES OF COAL SAMPLES

At some point a decision must be made concerning the type of sample being sought in addition to specifying its compositional or physical characteristics. In this connection, a simple description of each of the various types may be useful for the new entrant into coal science and technology.

1.6.1 Seam Samples

Seam samples are taken directly from the coal seam and are of several types.

1. *Drill cores* are usually rod-shaped samples measuring 1–3 in. in diameter by the thickness of the seam. If the coring has not fractured the coal and if none of the seam has been lost in the operation, drill cores provide excellent samples of the seam with all seam components in their original position and orientation.

2. *Channel samples* are bulk, crushed samples of the entire thickness of the seam taken so as to have each of the seam's strata represented in proper proportion to its thickness (see Figure 1.13).

3. *Column or Columnar samples* are solid pillars of the seam, approximately 1 ft by 1 ft by the seam thickness (Figure 1.13), that are marked as to orientation and that retain all seam components in their original position relative to each other. A column sample may be taken as a single pillar but more often is taken from the seam as a series of blocks, each marked as to position and orientation.

4. *Grab samples* consist of a lump or lumps of coal randomly collected from the seam (or other source). Usually no information is available on seam position, orientation, or the extent to which the sample is representative of the full seam thickness.

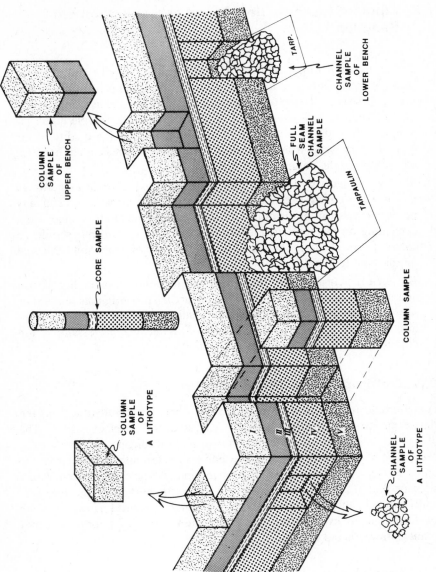

Figure 1.13 Source and character of several types of coal samples.

33

1.6.2 Tipple, Stockpile, Barge, Rail car, and Shiphold Samples

1. *Tipple and preparation plant samples* are bulk samples of crushed coal, taken in increments over a period of time at a coal tipple or preparation plant.[26] The sample may represent coal from a single site in a single seam, multiple sites in a single seam, or multiple seams. The sample usually is obtained from the product as it leaves the plant, but the quality of the material can vary from as-mined to a highly beneficiated product, depending on the nature of the facility.

2. *Stockpile samples* are bulk samples of crushed coal taken from a storage pile that may consist of coal from a single seam or multiple seams and may be either cleaned or as-mined coal. Sampling methods should vary, depending on the construction of the pile (e.g., a layered pile should be cored at several sites). The age of the pile may be of concern if fresh coal samples are required.

3. *Barge, rail car, and shiphold samples* are taken for industrial quality control and may not be suitable for scientific research. Their origin as to seam or seams may be poorly defined and their qualities will have been adjusted to meet purchase specifications. Methods of taking such gross samples are described in ASTM Standard D-2234-82.[26]

1.6.3 Outcrop Versus Mine Samples

An *outcrop sample* is taken from the coal seam where it naturally intersects the surface of the ground, usually in a hillside, cliff, stream channel, or roadcut. The coal typically has been exposed to the atmosphere for a long period of time, even decades. The effects of weathering may extend several feet into the seam from the exposed surface. A *mine sample*, on the other hand, will have been taken from a seam face that was produced during the mining operation. If the face was recently exposed, a fresh, unweathered sample can be obtained. However, some mine faces remain exposed for years so a mine sample in itself does not guarantee unweathered material.

1.6.4 Fresh Versus Weathered Samples

The term *fresh sample* would appear to refer to a coal sample that has recently been obtained, but usually it refers to a sample taken from a newly exposed coal face or some other source (e.g., a core) where it is known that the coal has been exposed only very recently to the atmosphere. A *weathered sample* is one in which the properties of the coal have been altered as the result of exposure to the atmosphere or to percolating ground water that contains sufficient oxygen to produce changes in the coal. The detection of the effects of weathering is discussed in Chapter 6 to which the reader is referred. It should be noted that, whereas many studies require unweathered samples, there are several types of analyses and/or research in which the effects of weathering are of no consequence or may even be beneficial. In palynological research, for example, the coal sample is oxidized as a part of sample processing, so any natural oxidation may serve to reduce processing time.

1.6.5   Raw, As-mined, and Cleaned Samples

A *raw sample* is one that has not been subjected to any cleaning operation that would have removed a portion of the coal or other rock from the sample. The term *as-mined sample* describes a sample of the coal product as it comes from the mine. It can differ from a raw coal sample in that it inevitably contains rock from the roof and/or floor of the seam, and the former may contain only the rock partings and in some cases the larger of these may have been removed. A *cleaned coal sample* (or clean coal sample) is one that has been processed to remove unwanted material. Samples of clean coal usually are derived from coal preparation plants that are being operated to reject material containing large concentrations of mineral material and/or sulfur-bearing materials.

1.6.6   Crushed Versus Lump Samples

These terms are self-explanatory. Most conventional analyses and the majority of research studies utilize crushed coal, but a few require samples in lumps or blocks. Caution must be exerted in restricting studies only to lump coal as this can bias results toward the tougher lithotypes at the expense of the more friable material in the seam.

1.6.7   Fractionated Samples

Several analytical procedures utilize only a particular fraction of a coal sample in the analyses. The Hardgrove grindability test, for example, utilizes a 16×30 mesh size fraction for the test.[14] Although an effort is made to retain as much of the sample as possible, the inevitable result is a bias toward the tougher lithotypes. Porosity and surface area studies are other examples of instances in which specific size fractions are employed (see Chapter 4). In those studies in which only a specific portion of the coal sample is employed, it is important to be aware of the possibility of differential fractionation and a bias toward specific coal lithotypes.

1.6.8   Premium Versus Conventional Samples

A *premium coal sample* has been collected, processed, and stored in a manner designed to prevent or minimize any chemical alteration of the coal as the result of its extraction from the seam. Premium samples are always taken directly from a seam, although the material may subsequently be processed to produce lithotype or maceral concentrates. All handling and processing of premium samples (except for the initial extraction from the seam) should be done in an inert atmosphere and the final samples sealed in such an atmosphere in a glass container. Whereas certain types of chemical research must have premium coal samples, most research and analytical work require only *conventional coal samples*. In the production of conventional samples, it is standard practice to minimize the possibility of chemical alteration in the extraction, handling, and processing operations, and the resultant sample usually is stored either in argon or nitrogen as an inert atmosphere.

As is evident from the above, coal samples are available in a variety of types and qualities. In research work in particular, the sample selection process may involve a judgment as to which kind and quality is required in order to satisfy the demands of the research and its objectives.

1.7 SELECTION, HANDLING, AND CHARACTERIZATION

It is obvious that the selection of improper samples defeats all subsequent effort to produce a particular result. Sample selection, then, is a critical step in everything from routine testing to sophisticated research. Four matters require thorough consideration:

- Sample qualities,
- Sample source,
- Sample type,
- Sample condition.

1.7.1 Sample Qualities

The first subject to be addressed should be that which concerns the sample qualities required by the research objectives or experimental testing. The reason for this is that the qualities of the sample will automatically reduce the number of source areas to be considered. If the research requires samples of anthracite, for example, the choices are narrowed to Pennsylvania, Colorado, and Virginia. If lignites are required, the Northern Great Plains Province and the Gulf Province become prime target areas.

As just implied, the first quality to consider may be coal *rank*. As discussed previously, the level of coalification achieved has a profound effect on a coal's chemical and physical properties and, as a result, on its behavior in utilization. Research on coal carbonization will normally employ coals in the high volatile bituminous B range through low volatile bituminous coals, whereas research in the area of coal liquefaction may exclude only low volatile bituminous coals and anthracites. It should be noted that within any given ASTM rank class, the level of coalification can differ significantly, hence, if more precise control on the level of coalification is required, the vitrinite reflectance value should be used as this is a relatively precise measure of coalification. If this is not available, the carbon content from the ultimate analysis converted to a DMMF basis can be used.

After establishing the rank level or levels suited to the research objectives or the program of experimental testing, the other aspects of coal quality must be examined for possible effects on the proposed studies. These qualities include:

1. Chemical characteristics of the *whole coal* as determined by proximate, ultimate, and sulfur forms analysis,
2. Chemical characteristics of the *coal ash*, including both major and minor elements,
3. Trace element concentrations,
4. Types and concentrations of macerals present,
5. Types and concentrations of minerals present,
6. Physical properties such as hardness, grindability, porosity, surface area, and permeability,
7. Behavioristic properties such as fluid behavior as measured by plastometers, slagging and fouling behavior as indicated by ash fusion tests, and charring or coking behavior as measured by various tests and usually followed by microscopy.

The relative importance of each of these qualities will vary depending on the purpose associated with the use proposed for the sample. Mineralogical studies with geologic objectives may find the *fixed carbon content* and *Gieseler plasticity* totally irrelevant, and studies of a coal's *molecular structure* will find little pertinence in the amount of superficial moisture originally present.

Because of the numerous qualities that must be considered and because coals with qualities that are optimal in all respects may not exist, it may be necessary to assign priorities to the relevant properties. An obvious method is to prepare a check list of the important qualities arranged in order of importance as a means of rapidly evaluating potential samples.

1.7.2   Sample Source

Consideration of sample source requires attention to the geologic as well as the geographic aspect of sample location. In the case of multiple samples from the same site, the question of the spatial relations of the samples must be addressed. The principal geologic matter to be settled is the identity of the seam or seams to be sampled. This may require consideration of such questions as: "Is geologic age of concern?" "Is the character and/or thickness of the overlying strata of interest?" "To what extent is cleat developed in the seam?" "Is the seam subject to splitting and the inclusion of thick partings?" "Where is the margin of the coal-forming basin?"

Selecting the geographic source for the coal sample will often be aided by the qualities required, as discussed above. In most cases, this will only serve to identify the coal provinces or regions where the sought after sample may be found. Because of the size of these areas and the lateral variation in the characteristics of coal seams, it is necessary to be more specific about the sample site or sites. Seam quality information is often available on a county basis (see *Keystone Coal Buyers Guide* published annually by McGraw-Hill), and this may be helpful in being more precise in identifying the sample location.

In addition to the restraints that coal quality requirements place on source areas, another constraint is normally involved in that most samples are collected in active mines. The productive approach, therefore, would appear to involve the evaluation of the quality data for the mines in the prospective areas. Two problems exist in this connection. First, mine specific data tends to be very limited in availability and scope and often is regarded as proprietary. Second, information on the existence of small operating mines is not readily obtained, although state geological surveys may be of assistance in this connection.

In spite of the problems involved, identification of the geographic source for samples should be as specific as possible. Whenever practical, a particular mine should be designated; or, in lieu of this, a restricted area such as a township or county should be identified. If core drilling is proposed as an alternate to mine or outcrop sampling, the spatial distribution of samples must be specified as another geographic parameter. The distribution required will vary with the objectives involved and may be constrained by funds available.

1.7.3   Sample Type

The merits of the various types of samples should be considered as part of the sample selection process. Of these, the whole seam channel sample is the one most frequently used, as it is representative of the seam's composition at the site of sampling. Channel samples also can be restricted to individual lithobodies or benches within the seam; and such samples can be useful in studying lithotype composition and in establishing the differential concentration of various materials in the seam, for example, sulfur. A crushed, drill core sample is a form of channel sample if complete. Its total volume is much smaller than the ordinary channel sample and it encompasses a smaller area (usually 1–3 in. in diameter). Except for these characteristics, cores are very desirable samples as they are likely to be composed of fresh, unaltered coal. Channels taken at coal outcrops should be avoided, if possible, because of the likelihood of leaching and oxidation.

Columnar samples can also be taken of the whole seam, of individual lithobodies, and of seam benches (see Figure 1.13). These are special purpose samples used when the components of the seam must be retained in their original relative positions and their orientations known. This consideration is most common in geological work and in testing programs involving evaluation of compressive strengths and permeabilities. An intact drill core sample is a form of column sample and, as such, is very useful in coal seam correlation studies.

Samples of bulk, coal mine products taken from tipples, preparation plants, stockpiles, barges, rail cars, and shipholds are essential in commerce in maintaining and evaluating coal quality. They are not the preferred sample type for basic research, because it often is difficult to identify the material in terms of seam and site of origin. The problem is further compounded, as previously mentioned, by the fact that the sample may be a mixture of several coals from several sites assembled in proportions unknown to the researcher. Unfortunately, these samples are among the easiest to obtain with the result that much data has been generated on unidentifiable material.

In selecting samples, mine samples are preferred over outcrop samples in all cases, and *grab* samples should be avoided except for educational display purposes. A *clean coal sample* may be employed if there is no concern regarding the differential loss of certain coal components during the cleaning process and if the objective is to have any resultant data apply to beneficiated commercial products. *Raw samples* have the advantage of containing all material present in the seam, hence data generated from such samples provide a full base of information concerning the composition of the coal resource. *As-mined* samples are particularly useful in evaluating beneficiation potentials and in designing coal cleaning facilities. Because they contain nonseam material, they should be avoided for most basic research and many applied research programs.

In specifying or accepting crushed coal samples, it is important to know that the sample is a homogeneous mixture representative of the uncrushed sample. In addition, the identity of the uncrushed sample must be known. Was it a full seam channel, a lithobody channel, or a miscellaneous grab sample of lump coal. In addition, consideration must be given to the size consist of the crushed material. This is of major consequence if subsamples are to be extracted from the as-received material. Particle size determines the initial volume required to produce representative subsamples using conventional splitting techniques.[26,27] Research samples are often supplied in kilogram or 1 lb quantities of −20 mesh coal (Tyler sieve size). In some cases, this may require grinding to −60 mesh in order to generate representative subsamples. Care in grinding is required to prevent differential loss of the more friable components, for example, fusinite. The intentional use of particular size fractions appears undesirable, although it is standard practice in certain types of research and testing. If this is done, it would be helpful to recharacterize the sample in order to have knowledge of the composition of the coal to which the data apply.

1.7.4 Sample Condition

Sample condition is used here to refer to the extent to which the coal sample may have been chemically altered in place, either by exposure to atmospheric conditions or percolating ground water. The term also refers to any chemical alteration induced through handling and processing subsequent to extraction from the seam. In addition to simple exposure to the atmosphere, grinding of the coal during sample processing can generate elevated temperatures that will accelerate alternation or produce dehydration that otherwise might not occur. Other aspects of sample condition concern: the atmosphere in which the sample has been stored and received; the particle size involved during storage; and the temperature and duration of storage.

Recently collected, unweathered samples that have been stored at low temperatures (but above freezing) and in large particle sizes (i.e., $+\frac{1}{4}$ in.) in an inert atmosphere are the most desirable for many purposes. Whether these must be premium samples in the sense of always having been handled in an inert atmosphere should be considered on a case-by-case basis. As previously stated, conventional samples will prove adequate in most instances, provided they conform with the characteristics just described and have been processed with care.

Although the above applies for many kinds of samples used in both testing and research, reference has already been made to the fact that the pristine condition or lack thereof is of little concern in certain studies. Thus, although the sample history always should be known to the sample user, sample selection may or may not demand that the coal be in an unaltered condition and received in an inert atmosphere. A check list such as the one that follows is useful, however, in determining whether a given sample is appropriate for use and such a list can aid in developing specifications for samples being sought.

1. Date of sample collection.
2. State of weathering at time of collection.
3. Conditions associated with handling and processing.
4. Size consist during storage.
5. Temperature during storage.
6. Atmosphere during storage.
7. Size consist of sample being supplied.
8. Atmosphere in which sample is being supplied.

1.7.5   Samples for Industrial Use

Selecting coal samples for industrial purposes is frequently simplified by the fact that the identity and source of the samples are often predetermined. In such cases, the concerns are restricted to the adequacy of the sampling operations, the representativeness of the samples, and care and handling of the materials. To provide legal protection and/or to conform to procurement specifications, sampling and sample handling are usually conducted so as to conform with standardized and industrially accepted procedures as set forth in various ASTM or International Standards publications.[28] In coal exploration programs, the area to be evaluated will be predetermined, but judgment must be used in establishing sample spacing, sample handling, and the acceptability of various sample types (e.g., outcrop versus drill core). Drilling on a quarter mile grid pattern is normally regarded as maximum spacing, with smaller intervals being more desirable. Geostatistical techniques can be applied to reduce the number of holes required.

Applied research programs in industrial laboratories have greater flexibility in sample selection than that associated with coal exploration and quality control programs. Although selection should be based primarily on suitability for achieving the research objective, the selection of samples is customarily restricted to minable seams with acceptable reserves. Cleaned coal and full seam samples tend to be the preferred types. Although appearing practical, these biases are not always in the best interest of efficient and successful research.

Industrial coal testing programs (e.g., testing for coke strength) typically require large samples (sometimes a ton or more) and the source and sample identity usually are determined by ownership, prospective purchase, or actual purchase. The most important need in these cases is for an adequate characterization of the samples. Without this, the data are meaningful only for the sample tested.

1.7.6 Samples for Scientific Use

Sample selection and characterization are of prime importance in samples intended for both basic and applied research. Three forces sometimes operate to inhibit the rational selection of coal samples for scientific use. These forces are:

1. *Fame*:   the Pittsburgh, Illinois #6 seams, and the Wyodak coal products capture a sizable clientele on name alone (or on the basis that much work has been done using these coals);
2. *Size of reserve*:   the claim being that if there is not enough to mine, it is not useful for research; and
3. *Probable use*:   "this coal is wholly owned and will only be used for carbonization, so why study its combustion characteristics?".

The forces that *should* control sample selection are those outlined in the opening paragraphs of this section.

For scientific purposes, the merits of utilizing lithotype samples should be noted. In these naturally occurring mixtures, two features are of interest. First, the variety of compositions in the sample pool is greatly increased over that displayed by full seam samples. The reason for this is that the presence of multiple lithotypes in the full seam results in a dilution of the composition of each lithotype and a mixed composition characteristic of only that site. Second, predictability and the ability to generalize are both increased because lithotype composition is more constant than seam composition (the exinite–micrinite lithotype of the B seam of Lynch, Kentucky, has the same composition as the exinite–micrinite lithotype of the Clark seam in Alabama or the Harbour seam of Nova Scotia, but the full seam compositions are quite different in the three cases). It may be useful to recall the fact that coal seams are complex, stratified rock bodies and although their lateral changes may be predictable, no two coal seams are identical.

In recent years, two other kinds of samples have become available that are of interest for scientific purposes: maceral and mineral concentrates. The latter concentrates have become available as the result of the use of oxygen plasmas at low temperatures (LTA ashing techniques) to produce mineral residues that are free of the carbonaceous material that formerly inhibited x-ray and other standard mineralogical analysis techniques.[29,30] The mineral residue is largely unaltered although gypsum sometimes is produced in the process. The preparation of current day maceral concentrates has been the result of applying sophisticated separation technology to pulverized and demineralized coal.[17] With the further development of these techniques, samples of individual macerals may be available for selection and research.

1.7.7   Samples for Petrographic Characterization

Petrographic analyses are conducted in two principal modes: megascopic, meaning observation with the unaided eye, and microscopic, using several types of illumination, for example, transmitted white light, reflected white light, polarized light, and blue light. Megascopic analyses require block coal samples for which the orientation is known. These samples typically come from core or column samples. The resulting data find use in coal exploration, seam identification, seam correlation, and in various geologic applications. Analyses using microscopic observation conventionally employ the use of homogenized, crushed coal samples with a -20 mesh size consist.[31] The analyses can be performed on weathered or unweathered coal, although the latter is, of course, preferred. Coal blends can be detected and described using maceral and vitrinite reflectance analyses.

1.7.8   Samples for Palynological Characterization

Palynology (the study of pollen and spores) is employed in seam identification, seam correlation, and age determination. In geological studies, it also is used to determine the character of depositional environments. Samples should be selected from columns, cores, or seam faces and marked as to their vertical position in the seam. The sampling interval should be as small as feasible (5–10 cm if feasible, 25–30 cm maximum) as the objective is to develop a vertical profile of the pollen and spore content throughout the seam. Only a small sample is required (e.g., 5 g), and it may be in either lump or crushed form.

1.7.9   Samples for Chemical Characterization

Conventional chemical characterization generally employs -60 mesh samples of homogenized coal sample. These can be derived from core, column, channel, or other sample types but must be taken in a manner that ensures extraction of a representative sample. Sample specifications for each of the various analyses employed in characterization (proximate, ultimate, sulfur forms, etc.) can be found in their respective ASTM Standards.[28]

1.7.10   Samples for Physical Characterization

As with conventional chemical characterization, after sample identity and source have been determined, sample selection and processing is described in the several ASTM Standards.[14,15] Sample size and character vary considerably depending on whether the objective is to characterize grindability, hardness, or size. Lump coal is required for certain analyses such as permeability, whereas crushed coal is used in porosity and surface area studies. Discussion with the analyst is required to establish the constraints that will guide sample selection and handling.

1.7.11   Use of Sample Suites

Recognizing the vast variety of compositions available in the sample pool, it is useful to consider the merits of selecting research samples to form sample suites in which one or more compositional parameters can be varied systematically. This is possible with the aid of the various computerized coal data banks that are available for use. These include national, state, and university data systems. Such sample suites are useful in both basic and applied research, and they provide a practical and efficient means of evaluating the effect of a given compositional component on coal behavior.[32]

1.7.12   Special Research Samples

When samples having unusual compositions or qualities are required, it is usually necessary to enlist the aid of persons having a broad knowledge of the occurrence and characteristics of coals. Such individuals are typically coal geologists and/or coal petrologists, and they can be contacted through state geological surveys or universities having strong coal research programs. Individuals with the required expertise also can be found in federal agencies such as the U.S. Geological Survey and the U.S. Bureau of Mines.

The absence of a particular composition in a computerized data bank does not mean that a coal with that composition is nonexistent. Data banks tend to exclude coals with unusual compositions and high mineral contents as well as coals occurring in thin seams and unmined areas. The selection of a set of samples with abundant and varied mineral content may prove difficult but not impossible. Similarly, procurement of an intact 1 ft by 1 ft by 1 ft block of high-volatile A bituminous coal for permeability studies may be possible only from certain seams in certain areas. Assistances in these cases can be found if such is required.

1.7.13   Samples for Educational Purposes

Samples for display and teaching purposes are best if in lump form. The latter may be difficult to acquire in the case of low-volatile bituminous coal and difficult to retain in the case of low-rank lignite. In spite of this, the most commonly encountered display contains a set of coal lumps with one representing each ASTM class or rank. It is obviously impossible for a single lump to represent the variety of coal types found at a given level of coalification, but certain gross features such as color, luster, texture, and fracture pattern can be displayed. Another type of display and/or teaching collection can be assembled by selecting samples representing different lithotypes, including cannels and bogheads, either from one or several levels of coalification. Here, again, state geological surveys often can be of assistance, but several may be required as few states have a large spectrum of coal types and ranks available.

1.7.14   Handling and Restorage

Samples received from agencies or individuals concerned with supplying coal samples generally will arrive sealed in an inert atmosphere; those received from mining companies usually will not. If maintenance of the inert atmosphere is required by the research, the sample must be opened, subsamples obtained, and the residual sample resealed employing a glove box with the appropriate atmosphere. In extracting a subsample either inside or outside of an inert atmosphere, it must be realized that differential settling probably has occurred in the sample and any subsample extracted should be taken only after thorough remixing of the sample and then using an appropriate riffling or splitting device. The simple extraction of a teaspoon of sample from the top of the container (even in the case of −60 mesh coal) can negate all efforts previously made to supply a representative coal sample. Restorage in an inert atmosphere may not be required if the sample is to be consumed in a short time, but sealing in an airtight container is always advisable. It should be recognized that low-rank lignites and many coking coals are rapidly altered when exposed to the ambient atmosphere.

1.7.15   Need for Recharacterization

Any coal sample to be used in the development of data for scientific use must be documented as to identity, source, and sample history and should be fully characterized by conventional chemical and petrographic analyses. Samples known to be splits (subsamples) of larger samples should be checked upon receipt to determine what characterization data are lacking or to determine if recharacterization is required. The latter often can be assessed by analyzing the sample for moisture content and ash and volatile matter yields. If the resultant data do not conform with those supplied, recharacterization is indicated.

1.8   HAZARDS OF SELECTING COAL SAMPLES: A SUMMARY

The process of selecting coal samples offers many challenges that must be met and conquered if optimal samples are to be obtained. At the onset, coal must be recognized as a class of rocks as opposed to a single substance. Coal seams must be seen as laminated rock bodies which individual laminae may thicken, thin, and even vanish as the seam is traced laterally. The laminae are composed of distinctive mixtures of maceral and mineral material. Thus, seams are heterogeneous mixtures by virtue of their stratified nature and the variety of macerals and minerals present. Coal is a *particulate* rock in which the constituent grains vary greatly in size and shape and in chemical and physical properties. The maceral materials become transformed to form metamorphic series when subjected to slightly elevated temperatures through geologic time. Hence, a third dimension is added to coal's heterogeneity as the result of this coalification process, and lignites, through changes in their constituent macerals, can become bituminous coals and subsequently anthracites if subjected to appropriate temperatures for sufficient time. The precursor materials required for the production of coal stem from plants that themselves evolve through time, producing new organic substances and new mixtures. Coal seams have their origins in peat-forming swamps and marshes. These swamps and marshes are formed of several to many areally discrete, distinctively populated plant communities, each containing its own set of geochemical conditions and substance-altering microorganisms. It is in these environmentally distinct areas that the various mixtures of macerals and minerals have their origin. Hence, it is these individual ecosystems that control and limit the composition of the coal lithotypes.

All of the above have combined, through time, to produce a very large population of things to be sampled within the category of coal. The first and perhaps the greatest hazard to be overcome in sample selection may be the need to comprehend the nature and diversity of the sample population itself. This must include a knowledge of present-day occurrence and availability, which places an even greater burden on the selector.

In addition to their inherent compositions and properties, coal samples can be obtained in a variety of forms ranging from raw to cleaned, unweathered to weathered, column to channel, and so on. These present the selector with the need for another decision and the requirement that knowledge be acquired concerning the merits and demerits of all of the sample types. Moreover, samples may or may not be adequate, depending on how the sampling was conducted and how the samples were handled and processed. The method of sampling will determine the degree to which the sample is representative of the population being sampled, and subsequent handling and processing will determine the extent to which sample alteration occurs. Coals vary in their susceptibility to alteration upon removal from the seam, anthracites being quite stable and some coking coals and lignite being quite vulnerable.

Sample quantities and qualities will, of course, vary depending on the use intended for the sample. The identity and source of the sample often will be predetermined in the case of samples selected for industrial purposes. In selecting samples for scientific use, establishing the identity and source of samples is a primary objective. Samples scheduled for use by the organic chemist, the geologist, the petrologist, the mineralogist, or the physicist have their separate requirements that can best be determined by the sample user if the hazards to such determinations can be removed. Assistance in sample selection and procurement can usually be found in geological surveys at the state and local levels and at universities with strong coal-oriented programs.

Generalization from a body of data and correlation of data generated for a sample with data from other samples requires specific knowledge of the material to which the data apply. It is insufficient to know only that the sample comes from seam X, or North Dakota, or the Bear Mine, or is of subbituminous B rank. Collectively, this information is helpful but still insufficient. The best and most useful sample is one that has been fully characterized by means of:

1. Proximate and ultimate analyses,
2. Sulfur forms analyses,
3. Maceral and vitrinite reflectance analyses,
4. Calorific value determinations,
5. Determination of major and minor elements in the ash,
6. Trace element analyses,
7. Equilibrium moisture determination,
8. Determination of free swelling index or plasticity,
9. Mineralogical analyses, and
10. Physical properties analyses such as grindability, porosity, and surface area determinations.

Finally, coal sample selection involves identification of:

1. Qualities required in the sample,
2. Potential sources for the sample,
3. Appropriate sample type, and
4. condition of the sample at time of receipt along with those associated with sampling, handling, and processing.

One way to proceed is to identify qualities by first deciding on the level of coalification or rank that is most suitable for use. Once this is established, consideration can be given to chemical characteristics that are essential. This should be followed by determining the extent to which maceral content, mineral content, or physical properties should be specified. Identification of sample source involves specifying a seam by name and a site by geographic location.

Assistance at this point may be required. Designation of sample type and sample condition will be dictated by the research or experimental testing involved and usually is readily determined by the sample user.

The opportunites for selecting optimal samples for use in myriad cases are great and appropriate attention to the selection process can be of significant benefit to the user of the samples and to coal science generally.

REFERENCES

1. P. G. Gove, *Webster's Third New International Dictionary*, p. 422. G. & C. Merriam Co., New York, 1961.

2. ASTM Standard Method D2796-82: Standard definitions of terms relating to megascopic description of coal and coal seams and microscopical description and analysis of coal. *Annu. Book ASTM Stand.*, Vol. 05.05, pp. 378–381 (1986).

3. J. M. Schopf, A definition of coal. *Econ. Geol.* **51**, 521–527 (1956).

4. ASTM Standard Method D2798-79: Standard method for microscopical determination of the reflectance of the organic components in a polished specimen of coal. *Annu. Book ASTM Stand.* Vol. 05.05, pp. 388–391 (1986).

5. ASTM, Standard Method D3172-73, Standard method for proximate analysis of coal and coke. *Annu. Book ASTM Stand.* Vol. 05.05, pp. 401–402 (1986).

6. ASTM, Standard Method D388-84: Standard classification of coals by rank. *Annu. Book ASTM Stand.* Vol. 05.05, pp. 274–252 (1986).

7. ASTM, Standard Method D3176-84: Standard method for ultimate analysis of coal and coke. *Annu. Book ASTM Stand.* Vol. 05.05, pp. 414–417 (1986).

8. ASTM, Standard Method D2492-84: Standard test method for forms of sulfur in coal. *Annu. Book ASTM Stand.* Vol. 05.05, pp. 357–361 (1986).

9. ASTM, Standard Method D3682-78: Standard test method for major and minor elements in coal and coke ash by atomic absorption. *Annu. Book ASTM Stand.* Vol. 05.05; pp. 464–471 (1986).

10. H. J. Gluskoter, R. R. Ruch, W. G. Miller, R. A. Cahill, G. B. Dreher, and J. K. Kuhn, Trace elements in coal. Occurrence and distribution. *Circ.—Ill. State Geol. Surv.* **499** (1977).

11. P. C. Lindahl and R. B. Finkelman, Factors influencing major, minor and trace element variations in U.S. coals. *ACS Symp. Ser.* **301**, 61–69 (1986).

12. V. E. Swanson, J. H. Medlin, J. R. Hatch, S. L. Coleman, G. H. Wood, Jr., S. D. Woodruff, and R. T. Hildebrande, *Geol. Surv. Open-File Rep. (U.S.)* **76-468** (1976).

13. E. Stach, M. T. Mackowsky, M. Teichmüller, G. H. Taylor, D. Chandra, and R. Teichmüller, *Stach's Textbook of Coal Petrology*, pp. 341–344. Borntraeger, Berlin, 1982.

14. ASTM, Standard Method D409-71: Standard test method for grindability of coal by the Hardgrove-machine method. *Annu. Book ASTM Stand.* Vol. 05.05, pp. 253–258 (1986).

15. ASTM, Standard Method D291-60: Standard test method for cubic foot weight of crushed bituminous coal. *Annu. Book ASTM Stand.* Vol. 05.05, pp. 228–232 (1986).

16. ICCP, *International Handbook of Coal Petrography*, 2nd ed. Cent. Nat. Rech. Sci., Paris, 1963.

17. G. R. Dyrkacz and E. P. Horowitz, Separation of coal macerals. *Fuel* **61**, 3–12 (1982).

18. C. Kröeger, A. Pohl, and F. Kuthe, Uber die Isolierung der Steinkohlegefuge-bestandteile aus Glanz-und Mattkholen von Ruhrflözen. *Glueckauf* **93**, 122–135 (1957).

19. ASTM, Standard Method D2639-74: Standard test method for plastic properties of coal by the constant-torque Gieseler plastometer. *Annu. Book ASTM Stand.* Vol. 05.05, pp. 362–368 (1986).

20. ASTM, Standard Method D720-83: Standard test method for free-swelling index of coal. *Annu. Book ASTM Stand.* Vol. 05.05, pp. 278–284 (1986).

21. M. Teichmüller and R. Teichmüller, Geological aspects of coal metamorphism. In *Coal and Coal Bearing-Strata* (D. G. Murchison and T. S. Westoll, eds.), pp. 233–267. Oliver & Boyd, Edinburgh, 1968.

22. J. M. Schopf, Variable coalification: The processes involved in coal formation. *Econ. Geol.* **43**, 207–225 (1948).

23. H. N. M. Dormans, F. J. Huntgens, and D. W. Van Krevelen, Chemical structure and properties of coal. 20. Composition of individual macerals (vitrinites, fusinites, micrinites and exinites). *Fuel* **36**, 321 (1957).

24. N. Schapiro and R. J. Gray, Petrographic classification applicable to coals of all ranks. *Proc. Ill. Min. Inst.* **68**, 83–97 (1960).

25. M. Teichmüller and R. Teichmüller, Die stoffliche und strukturelle Metamorphose der Kohle. *Geol. Rundsch.* **42**, 265 (1954).

26. ASTM, Standard Method D2234-82: Standard methods for collection of a gross sample of coal. *Annu. Book ASTM Stand.* Vol. 05.05, pp. 336–352 (1986).

27. ASTM, Standard Method D2013-72: Standard method of preparing coal samples for analysis. *Annu. Book ASTM Stand.* Vol. 05.05, pp. 303–317 (1986).

28. ASTM, *Annu. Book ASTM Stand.* Vol. 05.05 (1986).

29. H. J. Gluskoter, Electronic low-temperature ashing of bituminous coal. *Fuel* **44**, 285–291 (1965).

30. C. P. Rao and H. J. Gluskoter, Occurrence and distribution of minerals in Illinois coals. *Circ.—Ill. State Geol. Surv.* **476**, 1–56 (1973).

31. ASTM, Standard Method D2799-72: Standard method for microscopical determination of volume percent of physical components of coal. *Annu. Book ASTM Stand.* Vol. 05.05, pp. 382–387 (1986).

32. P. H. Given, D. C. Cronauer, W. Spackman, H. L. Lovell, A. Davis, and B. Biswas, Dependence of coal liquefaction behavior on coal characteristics. 1. Vitrinite-rich samples. *Fuel* **54**, 34–39 (1975).

Commentary

Sample selection is the initial step in the utilization of coal, whether it be an experimental, industrial, or sample bank activity. Chapter 1, "Sample Selection," has established the framework. The next consideration, the sample selection having been made, is that of characterization. How well can a mass of coal be characterized through a small sample? Homogenization is essential, but in any inhomogeneous material, it behooves us to establish the natural limits of the homogenization process and hence maintain realistic expectations and assessments. The homogeneity–heterogeneity concept is addressed in Chapter 2. Whether it be the determination of the ash in a coal pile by way of a grab sample or the preparation of uniform samples for a coal sample bank, the theory and limits of homogenization are pertinent for an appreciation of the characterization process.

2

PREPARATION OF HOMOGENEOUS COAL SAMPLES: AN ANALYSIS OF THE HOMOGENEITY/HETEROGENEITY CONCEPT

P. M. Gy

Consulting Engineer (Sampling)
Cannes, France

Homogeneity is defined as the property of a batch of material made up of units having identical compositions. These units can be either individual particles (constitution homogeneity) or clusters of particles (distribution homogeneity). A sample of a certain number of whole units has the same average composition as the original batch from which it has been extracted. The obvious conclusion is that the sampling of a homogeneous material is an exact, that is, nonerror-generating, operation.

Heterogeneity is defined as the absence of homogeneity, the units making up the batch having different compositions. A sample of a certain number of units, each different from another, is quite unlikely to have the same average composition as the original batch. The sampling of a heterogeneous material is a random error-generating operation. A theory of sampling has to be closely linked to a theory of the homogeneity/heterogeneity concept, as sampling errors are always the consequence of one form or another of heterogeneity.[1]

In a coal sample bank, it is of the utmost importance that all standard samples representing a given quality of coal should have *identical* compositions or, more realistically since rigorous identity is an inaccessible limit, compositions *as identical as possible.*

Chapter 2 presents a theoretical analysis of the homogeneity/heterogeneity concept and derives practical recommendations for the preparation of standard samples.

51

1. There are various forms of homogeneity and as many forms of heterogeneity.
2. Constitution homogeneity and heterogeneity are properties of the population of individual particles.
3. Distribution homogeneity and heterogeneity are properties of the populations of clusters of particles, the cluster size being highly relevant.
4. Constitution and distribution heterogeneities can be mathematically related to one another and to the properties (mass and average grade) of the constituting units (particles or clusters of particles). They can be characterized by two parameters: constitution heterogeneity (CH) and distribution heterogeneity (DH). The properties of CH and DH are reviewed and analyzed.
5. Constitution homogeneity is characterized by CH = O and distribution homogeneity by DH = O.
6. Mixing, blending, and homogenizing reduce but never completely suppress the distribution heterogeneity. They have no effect on the constitution heterogeneity, an intrinsic property of the population of individual particles in a given state of comminution.
7. When a certain number of conditions are fulfilled, the sampling error, the difference between the composition of the sample and that of the original batch, is a random variable with a zero mean whose variance is proportional to the distribution heterogeneity. The relationship linking the sampling variance to CH, DH, and the other relevant sampling parameters is presented.
8. Practical experience has shown that in the handling of particles with different densities the system tends to segregate rather than to approach distribution homogeneity. Particle size and shape are also factors facilitating segregation and hindering homogenization.

2.1 INTRODUCTION

To eliminate analytical bias and more specifically to calibrate the instrumental methods of analysis, it is advisable to introduce one or several samples of reference materials in the series of samples to be analyzed. Such reference materials, whose composition has to be carefully and accurately estimated and certified, are available for only a few commodities such as metal-bearing concentrates. In the preparation of standard samples, certain conditions have to be fulfilled.

1. The standard sample should have a general composition similar to that of the samples to be analyzed when calibrating an analytical method. Hence, it is highly desirable to prepare a collection of standard samples covering a range of physical and chemical properties.

2. When a material of a given origin characterized by a certain number of properties has been selected, a primary sample, usually weighing a few hundred kilograms and truly representative of this material, must be collected. Its representativeness is assumed but must be confirmed subsequently.

3. The primary sample has to be divided into a large number of samples for distribution and the samples certified. This involves comminution, blending, and sampling.

4. The analysis of a set of these standard samples in a certain number of laboratories according to a certain procedure must be conducted to determine accurately (1) the average composition of the set of standard samples to be certified, and (2) the confidence interval of this composition. If the average composition is significantly different from that assumed in (2), either the primary sample was not truly representative or the samples had been altered during preparation (e.g., contamination by iron during comminution). If the confidence interval of the average composition as estimated by the same analysis on different samples is too large, the inference is that there is a deficiency in the preparative procedure (e.g., too coarse a top particle size).

If both the average composition and its confidence interval are in conformity with their expected values, then the set of standard samples may be certified. Some discussion of the preparation and certification of reference materials is available in the literature.[2,3]

5. Maintaining the stability of the standard samples requires a number of precautions. Points 1, 2, and 5 are considered in part in other chapters of this book. The scope of this chapter is to develop point 3 and more specifically to show, from a practical as well as from a theoretical standpoint, how top particle size, heterogeneity, and sampling error are closely related. Although the focus is on coal, the principles of homogeneity and heterogeneity are general, and the concepts are valid for any particulate material.

2.2 HETEROGENEITY AND SAMPLING ERROR—A QUALITATIVE APPROACH

Homogeneity is defined as the quality of a batch of material whose constituent units have the same composition. A sample made up of a certain number of whole units obviously has the same composition as the original batch. The important conclusion is that the sampling of a homogeneous material does not generate any sampling error; it is an *exact* process.

Heterogeneity, defined as the absence of homogeneity, is the quality of a batch of material whose constituent units usually have different compositions, differing from the average composition of the batch. It follows that a sample made up of several of these units generally does not have the same composition as the original batch. The second important conclusion is that the *sampling* of a heterogeneous material usually generates a sampling error; it is a *random* process. A third conclusion is easily deduced; sampling errors are the inevitable consequence of heterogeneity. From these very simple qualitative observations, it is apparent that a theory of sampling will have to be derived from a theory of heterogeneity.[1,4] The heterogeneity of a batch of material can be described by the relationship between the composition of an elementary volume of matter and its location in space or time. Three broad categories of relationships and therefore of heterogeneities may be postulated: (1) essentially continuous, (2) hybrid between continuous and discontinuous, and (3) essentially discontinuous.

Nature as well as industry provides a variety of examples of heterogeneous materials that can be represented by this model or one of its limits. The ash content of a coal along an axis running throughout a coal seam is a good example of the first category. It expresses the continuity of the geological phenomena. This continuity is the rule with most mineral deposits. It is for the study of such functions that geostatistitics had to be devised as an independent science.

Consider coal of the first category after it has been mined, crushed, screened, processed, and fed to a pile by way of a 1 km long belt conveyor. The ash content of the coal along the belt, characterized by the superposition of continuous and discontinuous fluctuations, is a good example of the second category. The continuous fluctuations are derived from the original geological continuity. They are very rarely completely suppressed. The discontinuous fluctuations are to be ascribed to a discrete set of fragments redistributed more or less at random along the belt. Particulate materials, as opposed to continuous media such as compact solids or liquids, always generate a discontinuous term, random in nature, in the overall fluctuation function.

Suppose that a 200 kg sample is removed from a 1 m long section of the feed stream. The residual continuity perceptible at the scale of the 1 km long belt is rarely, if at all, perceptible at the scale of the 1 m section and the sample is a near-perfect example of the third category of heterogeneity. The phenomena governing the continuous component of the overall heterogeneity are of geological origin. Those governing the discontinuous, quasi-random component are of human origin and can be regarded as completely independent of the former. For this reason, both components of the overall heterogeneity are simply additive. The study of the homogeneity/heterogeneity concept is built on two independent and complementary models, the continuous model describing the heterogeneity of a continuous set, and the discrete model describing the heterogeneity of a discrete set. The development of the continuous model and its properties have been treated elsewhere.[1,4] This chapter concerns the discontinuous or discrete model directly involved in the preparation of homogeneous coal samples.

2.3   HETEROGENEITY OF A DISCRETE SET—NOTATIONS

Any batch of particulate material, any lot L of coal such as the primary samples used as a raw material to prepare small standard samples, can be regarded either as a set L_F of individual fragments F_i or as a set L_G of complementary clusters or groups of fragments G. The properties of the set L_F of individual particles will describe the constitution of the material. Those of the set L_G will describe the distribution of the fragments throughout the lot at a certain observational scale that is characterized by the cluster size. From the standpoint of the mathematical model, there is no difference between L_F and L_G. Parts of our demonstration can therefore be carried out on a set L_U of unspecified units U that can be either individual fragments F or clusters of fragments G.

We shall use the following notations:

L = lot of particulate material whose heterogeneity is being studied.

M_L = mass of L.

A_L = mass of critical component in L (by critical component is meant the component of particular interest, e.g., ash in coal).

a_L = critical content of L. By definition

$$a_L = \frac{A_L}{M_L}.$$

The lot will be referred to as L, L_F, L_G, or L_U when regarded as a whole, a set of fragments F, a set of clusters G, or a set of nonspecified units U, respectively.

U_m is an unspecified unit belonging to L, with $m = 1, 2, \ldots N_U$ where

N_U = number of units U_m in L_u,

M_m = mass of unit U_m,

A_m = mass of critical component in U_m,

a_m = critical content of U_m. By definition $a_m = A_m/M_m$.

F_i is defined as a fragment of L with $i = 1, 2, \ldots N_F$.

N_F = number of fragments F_i in L_F,

M_i = mass of F_i,

A_i = mass of critical component in F_i,

a_i = critical content of F_i with the value A_i/M_i.

G_n is a cluster of fragments belonging to L, with $n = 1, 2, \ldots N_G$.

N_G = number of clusters in L_G,

M_n = mass of G_n,

A_n = mass of critical component in G_n,
$a_n = A_n/M_n$.

F_{nj} is a fragment belonging to the cluster G_n, with $j = 1, 2, \ldots N_n$.

N_n = number of fragments in cluster G_n,
M_{nj} = mass of F_{nj},
A_{nj} = mass of critical component in F_{nj},
$a_{nj} = A_{nj}/M_{nj}$.

Mass conservation leads to the following obvious relationships:

$$M_L = \sum_n M_n = \sum_n \sum_j M_{nj} = \sum_i M_i,$$

$$A_L = \sum_n \sum_j A_{nj} = \sum_i A_i,$$

$$a_L = \frac{A_L}{M_L} = \frac{\sum_n A_n}{\sum_n M_n} = \frac{\sum_n \sum_j A_{nj}}{\sum_n \sum_j M_{nj}} = \frac{\sum_i A_i}{\sum_i M_i},$$

$$N_F = \sum_n N_n$$

For reasons that will appear more clearly in the following sections, we shall also define the average unit $\bar{U}_m$ of a certain lot L_U. The average unit $\bar{U}_m$ of the set L_U is defined by the two following properties:

1. The mass $\bar{M}_m$ of $\bar{U}_m$ is the mean of the masses M_m of U_m.
2. The mass $\bar{A}_m$ in $\bar{U}_m$ is the mean of A_m in U_m.

$$\bar{M}_m = \frac{1}{N_U} \sum_m M_m = \frac{M_L}{N_U} \text{ and } \bar{A}_m = \frac{1}{N_U} \sum_m A_m = \frac{A_L}{N_U} = \frac{a_L M_L}{N_U}$$

$\bar{F}_i$ is defined as the average fragment of L_F. Thus, $\bar{F}_i$ has the following characteristics:

$\bar{M}_i$ = the mass of $\bar{F}_i$,
$\bar{A}_i$ = the mass of critical component in $\bar{F}_i$,
$\bar{a}_i$ = the critical content of $\bar{F}_i$.

$$\bar{M}_j = \frac{M_L}{N_F} \qquad \bar{A}_i = \frac{A_L}{N_F} \qquad \bar{a}_i = \frac{\bar{A}_i}{\bar{M}_i} = \frac{A_L}{M_L} = a_L$$

$\bar{F}_{nj}$ is the average fragment of the cluster G_n. By designation,

$\bar{M}_{nj}$ = mass of $\bar{F}_{nj}$,

$\bar{A}_{nj}$ = mass of critical component in $\bar{F}_{nj}$,

$\bar{a}_{nj}$ = critical content of $\bar{F}_{nj}$

$$\bar{M}_{nj} = \frac{M_n}{N_n} \quad \bar{A}_{nj} = \frac{A_n}{N_n} \quad \bar{a}_{nj} = \frac{\bar{A}_{nj}}{\bar{M}_{nj}} = \frac{A_n}{M_n} = a_n$$

$\bar{G}_n$ similarly is the average cluster of fragments of L_G.

$\bar{M}_n$ = mass of $\bar{G}_n$,

$\bar{A}_n$ = mass of critical component in $\bar{G}_n$,

$\bar{a}_n$ = critical content of $\bar{G}_n$.

$$\bar{M}_n = \frac{M_L}{N_G} \quad \bar{A}_n = \frac{A_L}{N_G} \quad \bar{a}_n = \frac{\bar{A}_n}{\bar{M}_n} = \frac{A_L}{M_L} = a_L$$

Insofar as the sampling of the lot L_U by selection of a certain number of whole units U_m is concerned, these units being regarded as indestructible, the unit U_m is completely defined by any two of the three parameters M_m, A_m, and a_m since $A_m = a_m M_m$.

The purpose of the approach to the concept of heterogeneity of a discrete set of units U_m is to quantify this heterogeneity in terms of M_m, A_m, and a_m and to derive mathematical relationships between variance of the sampling error and the heterogeneity.

2.4 HETEROGENEITY OF A DISCRETE SET— QUANTIFICATION OF HETEROGENEITY

2.4.1 Heterogeneity Carried by Unit U_m of Lot L_U

The heterogeneity of a certain lot L_U necessarily involves all units U_m making up the lot. The quantification of this heterogeneity will require two steps, namely,

1. Definition of the heterogeneity h_m carried by unit U_m as a function of the descriptive parameters M_m and a_m, and
2. Definition of the overall heterogeneity H_L of lot L as a function of the individual elementary heterogeneities h_m.

A certain lot L_U is said to be homogeneous when all units U_m have the same composition, that is, the same critical content a_L as the lot L. The lot L_U is said to be heterogeneous when at least two units have a critical content different from a_L. From these definitions, it appears that the elementary heterogeneity carried by unit U_m should be proportional to the difference $(a_m - a_L)$. This is consistent with the fact that when $a_m = a_L$ there is no heterogeneity.

It is reasonable to assume that the perturbation associated with the heterogeneity carried by a certain unit should be proportional to its mass M_m. This perturbation will be defined as

$$D_m = (a_m - a_L)M_m$$

D_m has the meaning of a mass of critical component. However, it is more convenient to use relative, dimensionless magnitudes rather than absolute ones such as D_m. To convert D_m into a dimensionless quantity, the mass $\bar{A}_m$ of critical component in the average unit $\bar{U}_m$ of the lot L_U is used. The heterogeneity h_m associated with the unit U_m is the ratio

$$h_m = \frac{D_m}{\bar{A}_m} = \frac{(a_m - a_L)M_m}{\bar{a}_m \bar{M}_m} = N_U \frac{(a_m - a_L)M_m}{a_L M_L}$$

From this general definition the following specific definitions are adduced.

2.4.2   Heterogeneity Carried by a Fragment F_i of the Lot L_F

$$h_i = \frac{D_i}{\bar{A}_i} = \frac{(a_i - a_L)M_i}{\bar{a}_m \bar{M}_m} = N_F \frac{(a_i - a_L)M_i}{a_L M_L}$$

or, for the fragment F_{nj} of the cluster G_n:

$$h_{nj} = \frac{D_{nj}}{\bar{A}_i} = \frac{(a_{nj} - a_L)M_{nj}}{\bar{a}_i \bar{M}_i} = N_F \frac{(a_{nj} - a_L)M_{nj}}{a_L M_L}$$

2.4.3   Heterogeneity Carried by a Cluster G_n of the Lot L_G

$$h_n = \frac{D_n}{\bar{A}_n} = \frac{(a_n - a_L)M_n}{\bar{a}_n \bar{M}_n} = N_G \frac{(a_n - a_L)M_n}{a_L M_L}$$

2.4.4   Constitution Heterogeneity CH_L of the Lot L

The constitution of the lot L is characterized by the properties of the set L_F made up of N_F fragments. The constitution heterogeneity CH_L of the lot L_F will be defined as the variance of the population of N_F values of h_i. The mean $\bar{h}_i$ of this population is zero:

$$\bar{h}_i = \frac{1}{N_F} \sum_i h_i = \frac{1}{a_L M_L} \left[\sum_i a_i M_i - a_L \sum_i M_i \right] = 0$$

$$CH_L = \sigma^2(h_i) = \frac{1}{N_F} \sum_i h_i^2 = N_F \sum_i \frac{(a_i - a_L)^2 M_i^2}{a_L^2 M_L^2}$$

The constitution heterogeneity CH_L is an intrinsic property of the set of fragments L_F in their present state of comminution. Blending has no effect on this form of heterogeneity.

2.4.5   Distribution Heterogeneity DH_L of the Lot L

At the scale of the clusters G_n that must be defined in size, shape, and orientation, the distribution of the lot L is characterized by the properties of the set L_G made up of N_G clusters. The distribution heterogeneity DH_L of the lot L_G is defined as the variance of the population of N_G values of h_n.

$$DH_L = \sigma^2(h_n) = \frac{1}{N_G} \sum_n h_n^2 = N_G \sum_n \frac{(a_n - a_L)^2 M_n^2}{a_L^2 M_L^2}$$

Through their effect on the distribution of the fragments throughout the domain occupied by the lot L, blending and mixing on the one hand and segregation on the other are likely to modify the distribution heterogeneity DH_L. Homogenization is therefore the art of reducing the value of DH_L.

2.4.6   Observation Module

Only two assumptions, based on unspecified clusters of fragments, have been made until now, namely,

1. The clusters are complementary and can be expressed by

$$L_G = \sum_n G_n,$$

2. The cluster G_n is made up of N_n fragments, and the N_G values of N_n are of the same order of magnitude.

More specificity is now introduced. Suppose that the volume V_L occupied by the lot L is divided into a large number N_G of identical modules V_n, and the cluster G_n is defined as the set of fragments whose center of gravity falls within V_n. The fact that there should be no gap between the modules restricts the possible shape of the module to a limited number of models. Two sorts of modules may be invoked:

1. An isotropic module that would naturally be a cube, and
2. An anisotropic module that would be a prism with a triangular, square, rectangular, or hexagonal section. In the field of gravity, a certain amount of vertical segregation may be assumed and these prisms would have a vertical axis. They would extend throughout the volume V_L occupied by the lot L.

The choice of the observation module to be taken into account in the definition of the distribution heterogeneity in each particular case is a matter of convenience. This point will be developed in Section 2.5.7.

2.4.7   Generalization of the Notion of Distribution Heterogeneity

For the sake of simplicity, DH_L has been defined as the variance of a finite population of N_G complementary clusters, but it would be more satisfactory to define the distribution heterogeneity as the variance of an infinite population of clusters derived from a shift in module. An isotropic module will be used, but the results could easily be adapted to an anisotropic one. The perfect isotropic module is a sphere S of radius r, volume V_r and center C. It is assumed that V_r remains small in comparison with the volume V_L occupied by the lot L, and larger than the volume of the coarsest fragment in the lot L. The cluster G will be defined as the set of fragments whose center of gravity falls within the sphere S. The center C of S will be allowed to take all possible positions in the volume V_L. By shifting C throughout V_L we shall generate an infinite population of clusters G whose variance will be the distribution heterogeneity of the lot. The following notation is adopted:

$P(x, y, z)$ = a certain point of V_L,

$S_r(P)$ = the module S of radius r centered on point P,

$G_r(P)$ = the cluster of the fragments whose center of gravity falls within the boundaries of $S_r(P)$,

$M_r(P)$ = the mass of $G_r(P)$,

$A_r(P)$ = the mass of critical component in $G_r(P)$,

$a_r(P)$ = the critical content of $G_r(P)$. By definition $a_r(P) = A_r(P)/M_r(P)$,

$\bar{A}_r(P)$ = the mass of critical component in the average cluster.

$$\bar{A}_r(P) = \frac{1}{V_L} \iiint_{V_L} A_r(x, y, z)\, dx\, dy\, dz$$

which can easily be reduced to

$$\bar{A}_r(P) = \frac{V_r}{V_L} A_L = \frac{V_r}{V_L} a_L M_L$$

$h_r(P)$ = the heterogeneity carried by the cluster $G_r(P)$. Its definition is directly derived from that of h_n (Section 2.4.3).

$$h_r(P) = \frac{\{a_r(P) - a_L\}M_r(P)}{\bar{A}_r(P)} = \frac{V_L}{V_r}\left[\frac{\{a_r(P) - a_L\}M_r(P)}{a_L M_L}\right]$$

$\bar{h}_r(P)$ = the mean of $h_r(P)$ over the volume V_L. It can be shown that $\bar{h}_r(P) = 0$. IH_L = the intrinsic distribution heterogeneity of the lot L, at the scale of a module of radius r, defined as the variance of $h_r(P)$ over the volume V_L. For a given lot L and a given radius r, IH_L is determined in a unique way. As the mean of $h_r(P)$ is zero, the variance is equal to the mean square.

$$IH_L = \sigma^2\{h_r(P)\} = \frac{1}{V_L}\iiint_{V_L} h_r^2(x, y, z)\,dx\,dy\,dz$$

$$IH_L = \frac{V_L}{V_r^2}\iiint_{V_L} \frac{\{a_r(P) - a_L\}^2 M^2}{a_L^2 M_L^2}\,dx\,dy\,dz$$

It can easily be shown that DH_L is a random variable (the present breaking up of the lot L into N_G clusters G_n is but one of a practically infinite number of possibilities) whose mean is simply IH_L.

2.4.8 Relationship Between the Constitution and Distribution Heterogeneities

If we write

$$h_{nj} = (h_{nj} - h_n) + h_n$$

and square this value of h_{nj}, remembering the following properties,

$$\sum_j (h_{nj} - h_n) = 0 \text{ and } \sum_n h_n = 0$$

it readily follows that

$$\sum_i h_i^2 = \sum_n \sum_j h_{nj}^2 = \sum_n \sum_j (h_{nj} - h_n)^2 + \sum_n N_n h_n^2$$

This is the decomposition of $\sum_i h_i^2$ into a sum of squares, a technique borrowed from the analysis of variance. If we assume that all values of N_n are of the same order of magnitude, we obtain, after dividing by N_F,

$$CH_L = \frac{1}{N_F} \sum_n \sum_j (h_{nj} - h_n^2) + DH_L$$

It has been shown elsewhere[1] that

$$\overline{CH_n} = \frac{1}{N_F} \sum_n \sum_j (h_{nj} - h_n)^2$$

is in fact the average constitution heterogeneity of the clusters G_n. Then, $CH_L = \overline{CH_n} + DH_L$, hence, $CH_L \geq DH_L \geq 0$ and $CH_L \geq IH_L \geq 0$.

2.4.9 Provisional Summing Up

We have introduced the notion of heterogeneity carried by each unit making up the lot from which we have derived a general, quantitative definition of the heterogeneity of the lot. According to the nature of the unit considered, we have been able to distinguish two very different forms of heterogeneity, namely,

1. The constitution heterogeneity CH_L of the lot L, observed at the scale of individual fragments. This particular form of heterogeneity is an intrinsic property of the material in its present state of comminution. It is therefore independent of the spatial distribution of the fragments; mixing, blending, and homogenization have no effect on it.

2. The distribution heterogeneity DH_L of the lot L, observed at the scale of clusters of fragments. This second form of heterogeneity depends on two factors: (1) the degree of blending or segregation of the fragments throughout the lot L, and (2) the size, shape, and orientation of the observation module, that is, the definition of the cluster of fragments taken into consideration.

2.5 PROPERTIES OF THE CONSTITUTION AND DISTRIBUTION HETEROGENEITIES

2.5.1 Constitution Homogeneity of the Lot *L*

The constitution of the lot L is said to be homogeneous when

$$CH_L = 0$$

CH_L has been defined as the variance of h_i. This variance will be zero if and only if

$$h_i = 0 \text{ for all } i, \text{ with } h_i = N_F \frac{(a_i - a_L)M_i}{a_L M_L}$$

The solution $M_i = 0$ for all i being ruled out, there remains the solution $a_i = a_L$ for all i. The constitution of the lot L is homogeneous if, and only if, all fragments have the same critical content. This apparently obvious statement means that our definitions are not inconsistent. From Section 2.4.8 we know that

$$CH_L = \overline{CH_n} + IH_L$$

Where $\overline{CH_n}$ and IH_L are non-negative quantities. Their sum is zero, which requires

$$\overline{CH_n} = 0 \text{ and } IH_L = 0$$

This property can be expressed in the following way: Constitution homogeneity implies distribution homogeneity. There can be *no* distribution heterogeneity without constitution heterogeneity. This point is purely academic as constitution homogeneity is an inaccessible limit never observed in practice.

2.5.2 Rigorous Distribution Homogeneity of the Lot L

The distribution of the fragments throughout the volume occupied by the lot L, or the distribution of the lot L is said to be homogeneous when

$$IH_L = m(DH_L) = 0 \text{ hence } DH_L = 0$$

DH_L has been defined as the variance of h_n. This variance will be zero if and only if

$$h_n = 0 \text{ for all } n, \text{ with } h_n = N_G \frac{(a_n - a_L)M_n}{a_L M_L}$$

The solution $M_n = 0$ for all n being ruled out, there remains the solution

$$a_n = a_L \text{ for all } n, \text{ with } a_n = \frac{\sum_j a_{nj} M_{nj}}{\sum_j M_{nj}}$$

We must observe that sets of values of a_{nj} and M_{nj} are data of the problem and the only possibility for a_n to be equal to a_L for the N_G clusters G_n is

$$a_{nj} = a_L \text{ for all } n \text{ and } j$$

We already know this solution from Section 2.5.1. The constitution of the lot L is homogeneous and we know that constitution homogeneity entails distribution homogeneity. If this condition is not fulfilled, the changes for a_n to be equal to a_L for all n are, even after thorough mixing, infinitely smaller than those of a foursome of card players receiving the thirteen cards of each color, respectively, and this is very small indeed. The following very important consequence follows. Under natural conditions, the distribution heterogeneity DH_L is never zero. In other words a rigorous distribution homogeneity must be regarded as an inaccessible limit. Thus we can state that

$$DH_L > 0 \text{ and } IH_L > 0$$

2.5.3 Modular Distribution Homogeneity of the Lot *L*

The model that naturally comes to mind when thinking of homogeneity in the mineral world is the modular homogeneity of a perfect crystal. We examine how such a form of homogeneity fits in with our definition of DH_L and IH_L. A perfect crystal lattice consists of a space-filling repeat module. We shall assume that the crystal is made up of four components A, B, C, D, of equal mass and present in equal proportions, and that A is the critical component. Then the critical content a_L of the lot L is $a_L = 0.25$. With a particulate material, we can always imagine that we have achieved the same crystal-like modular distribution through some artifice such as hand positioning of each particle according to its identity. Figure 2.1 shows a cross section of such a distribution. The $4 \times 4 \times 4$

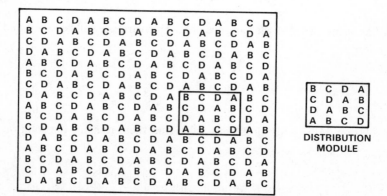

Figure 2.1 Cross section of a crystal-like modular distribution. The distribution module, a $4 \times 4 \times 4$ cube, is represented by 4×4 square.

distribution module is represented by its two-dimensional cross section, a 4×4 square.

Recall the definition of DH_L

$$DH_L = N_G \sum_n \frac{(a_n - a_L)^2 M_n^2}{a_L^2 M_L^2}$$

All observation modules of a given size have equal masses M_n with

$$M_n = \frac{M_L}{N_G}$$

we obtain the simple expression

$$DH_L = \frac{1}{N_G} \sum_n \frac{(a_n - a_L)^2}{a_L^2}$$

We observe that the composition of a cluster is completely defined when the following data are specified: (1) the identity of its corner element, and (2) the edge of the cube used as the observation module.

There are only four possible corner elements (A, B, C, D) with a $4 \times 4 \times 4$ modular distribution, the consequence being that for a given edge r there are but four possible compositions. These we shall denote by G_A, G_B, G_C, G_D after their corner element; these four compositions occur in equal proportions. For a population of N_G clusters, we observe

1. $N_G/4$ clusters of the G_A type (critical content a_A),
2. $N_G/4$ clusters of the G_B type (critical content a_B),
3. $N_G/4$ clusters of the G_C type (critical content a_C),
4. $N_G/4$ clusters of the G_D type (critical content a_D).

We readily derive

$$IH_L = DH_L = \frac{1}{4} \left[\frac{(a_A - a_L)^2 + (a_B - a_L)^2 + (a_C - a_L)^2 + (a_D - a_L)^2}{a_L^2} \right]$$

This result is valid independently of the size of the observation module, but the critical contents a_A, a_B, a_C, and a_D are functions of the edge r of this module. Several cases must be distinguished.

r = 1.   The observation module is a $1 \times 1 \times 1$ cube, that is, it contains a single element. The population L_G of clusters G_n and the population L_F of individual particles F_i are one and the same. The consequence is that in this case the distribution heterogeneity is equal to the constitution heterogeneity.

$$IH_{LI} = DH_{LI} = CH_L$$

It is easy to calculate that

$$a_A = 1 \quad a_B = 0 \quad A_C = 0 \quad a_D = 0 \quad a_L = 0.25$$

$$IH_{LI} = DH_{LI} = CH_L = 3 \text{ (dimensionless)}$$

r = 4k *(with k Integer)*.   The observation module is made up of k^3 (whole number) distribution modules whose composition is identical with that of the lot *L*. We obviously have $a_A = a_B = a_C = a_D = a_L = 0.25$

$$IH_{Lr} = DH_{Lr} = 0 \text{ whenever } r = 4k \text{ (with } k \text{ integer)}$$

This case is represented in Figure 2.2 for $r = 4$ (or $k = 1$). The observation module is a $4 \times 4 \times 4$ cube that for the sake of clarity has been shown broken up into its four cross sections *A, B, C,* and *D*. One can easily check that the average composition of the four cross sections, as well as that of the observation module that is the sum of these four cross sections, is identical with that of the lot.

r = 4k + k' *(with k Integer and k' = 1, 2, 3)*.   The edge of the observation module is *not* an entire multiple of that of the distribution module. Furthermore, it can be shown that the observation module is *never* made up of a whole

CROSS-SECTION D
CROSS-SECTION C
CROSS-SECTION B
CROSS-SECTION A

4x4x4 OBSERVATION MODULE. EACH CROSS-SECTION
IS DENOTED AFTER ITS LOWER-LEFT ELEMENT

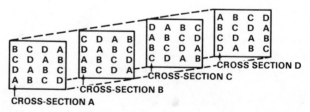

Figure 2.2 Observation module identical with (or multiple of) the distribution module. $r = 4k$ (with k integer); in the present case, $r = 4$ (i.e., $k = 1$).

number of distribution modules, with the consequence that *no* observation module ever has the same composition as the lot L.

$$a_A \neq a_L \quad a_B \neq a_L \quad a_C \neq a_L \quad a_D \neq a_L$$

$IH_{Lr} > 0$ and $DH_{Lr} > 0$ whenever $r = 4k + k'$ (with k integer and $k' = 1, 2, 3$).

This case is represented in Figure 2.3 for $r = 3$ (or $k = 0$ and $k' = 3$). The observation module is a $3 \times 3 \times 3$ cube, shown broken up into its three cross sections. There are four possible compositions corresponding to four cluster types G_A, G_B, G_C, G_D, after their corner element. Figure 2.3 shows clusters G_A and G_B. The other two can be derived from the former by cyclical permutation.

According to its position, the 27-element observation module will have one of the compositions given in Table 2.1.

From these figures, we easily calculate

$$IH_{L3} = DH_{L3} = 4.115 \times 10^{-3}$$

Value of IH_{Lr} as a Function of the Edge r of the Observation Module.
These values are given in Table 2.2.

Table 2.2 shows that for each value of k' (1, 2, 3) the value of IH_{Lr} is a decreasing function of the edge r of the observation module. As the value of k increases, IH_{Lr} tends toward zero. Observed through a magnifying lens, as we have done in our examples, the modular structure shows its local heterogeneity, but observed from far away it is the modular homogeneity that stands out.

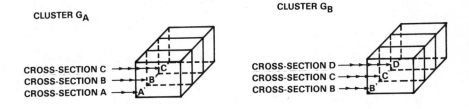

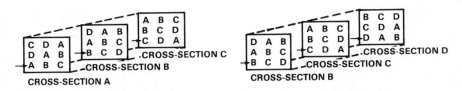

Figure 2.3   A $3 \times 3 \times 3$ observation module; $r = 3$ (i.e., $k = 0$ and $k' = 3$).

TABLE 2.1 Composition of the four possible clusters of 27 Elements of 4 × 4 × 4 Distribution Module and 3 × 3 × 3 Observation Module

Corner element	A	B	C	D
Cluster type	G_A	G_B	G_C	G_D
Nb of elements A	6	7	7	7
Nb of elements B	7	6	7	7
Nb of elements C	7	7	6	7
Nb of elements D	7	7	7	6
Nb of $A+B+C+D$	27	27	27	27
$a_n = A/(A+B+C+D)$	0.222 ...	0.259 ...	0.259 ...	0.259 ...

TABLE 2.2 Distribution Heterogeneity H_{Lr} as a Function of the Edge r of the Observation Module

r	IH_{Lr}	r	IH_{Lr}	r	IH_{Lr}
1	3	21	350×10^{-10}	41	632×10^{-12}
2	2500×10^{-4}	22	$1411 \times$	42	$2915 \times$
3	$41 \times$	23	$203 \times$	43	$475 \times$
4	0	24	0	44	0
5	192×10^{-6}	25	184×10^{-10}	45	361×10^{-12}
6	$343 \times$	26	$518 \times$	46	$1689 \times$
7	$25 \times$	27	$77 \times$	47	$278 \times$
8	0	28	0	48	0
9	565×10^{-8}	29	50×10^{-10}	49	217×10^{-12}
10	$1600 \times$	30	$219 \times$	50	$1024 \times$
11	$169 \times$	31	$34 \times$	51	$170 \times$
12	0	32	0	52	0
13	62×10^{-8}	33	23×10^{-10}	53	135×10^{-12}
14	$212 \times$	34	$104 \times$	54	$645 \times$
15	$26 \times$	35	$16 \times$	55	$108 \times$
16	0	36	0	56	0
17	1243×10^{-10}	37	1169×10^{-12}	57	87×10^{-12}
18	$4704 \times$	38	$5314 \times$	58	$420 \times$
19	$638 \times$	39	$853 \times$	59	$71 \times$
20	0	40	0	60	0

Important Conclusions

1. The modular distribution is the most homogeneous distribution that we can imagine.
2. Even though it can be imagined, there is no mechanical way of achieving such a modular distribution.
3. Even assuming that the stage of modular homogeneity has been achieved, the distribution heterogeneity is never identically zero.

This confirms the conclusion reached in the preceding section that the distribution heterogeneity IH_L is always larger than zero.

We shall show subsequently that the sampling variance is proportional to the value of IH_L. The fact that we are not capable of reducing IH_L to zero leads to the consequence that the sampling variance is never zero. *Sampling is always an error-generating process.*

2.5.4 Natural Distribution Homogeneity of the Lot L

Even if the distribution heterogeneity IH_L cannot be completely suppressed under ideal conditions such as hand positioning of the elements to achieve a modular distribution, it is of the utmost importance from a practical as well as from a theoretical standpoint to investigate whether and how the distribution heterogeneity can be minimized and to determine the minimal value of the distribution heterogeneity that can be practicably achieved. Homogenization is approached by means of mixers, stirrers, and the like. How do they work?

When analyzing the mixing process, as implemented for instance in a cube or Vee-type mixer, we observe that its theoretical model is represented by an aggregate of particles in which the coordinates of the center of gravity of any fragment F_i are independent of its *identity* characterized by its physical properties and more particularly its size, density, and shape. Let

M_i = the mass of the fragment F_i, a function of its size, density, and shape,

a_i = its critical content, and

$P_i(x, y, z)$ = the position of its center of gravity.

Assuming that the independence has been actually achieved, the resulting distribution can be characterized in the following equivalent ways:

1. There is no correlation between the position $P_i(x, y, z)$ of F_i and its physical properties M_i and a_i. M_i is proportional to the cube of its diameter d_i, to its density λ_i, and to a shape factor f_i.
2. The mixing process has suppressed the segregation that existed or was suspected as a consequence of the selective effect of gravity on size and density. This suppression was not instantaneous—the longer the mixing, the better.

3. There is no correlation between the identities of neighboring fragments.
4. The fragments are distributed at random throughout the volume of the lot.

The state of homogeneity we have just defined and that is characterized by a random distribution is what we shall call the state of natural distribution homogeneity of the lot. We must be conscious of the fact that we can do no better with natural means. Intelligent positioning alone could overcome this natural limit. What are the properties of this state of natural distribution homogeneity? How can we determine the nonzero residual distribution heterogeneity?

2.5.5   Evaluation of the Residual Distribution Heterogeneity of the Lot L

The residual distribution heterogeneity is that fraction of the distribution heterogeneity IH_L of the lot L that must be regarded as irreducible, that homogenization is incapable of suppressing, and that characterizes the state of natural distribution homogeneity. For the sake of simplification, we shall assume that the lot L is a sample of N_F fragments F_i that have been extracted at random from an infinite lot L_∞ having the same composition as the lot L. We shall denote by σ^2 the variance of h_i (heterogeneity carried by the fragment F_i, defined in Section 2.4.2) in this infinite population. One of the consequences of the random distribution assumed in the state of natural distribution homogeneity is the independence of neighboring fragments. For this reason, any cluster G_n of N_n neighboring fragments can be regarded as a sample of N_n fragments extracted one by one and at random from the infinite lot L_∞. It is a known result of mathematical statistics (analysis of variance) that under such conditions and for a given set L_G of clusters G_n,

1. the quantity $\sum_n \sum_j h_{nj}^2/\sigma^2 = N_F CH_L/\sigma^2$ follows a chi-squared distribution with $N_F - 1$ degrees of freedom,

2. the quantity $\sum_n \sum_j (h_{nj} - h_n)^2/\sigma^2 = N_F \overline{CH_n}/\sigma^2$ follows a chi-squared distribution with $N_F - N_G$ degrees of freedom,

3. the quantity $\sum_n N_h h_n^2/\sigma^2 = N_F DH_L/\sigma^2$ follows a chi-squared distribution with $N_G - 1$ degrees of freedom.

As a consequence of these properties, we can define three unbiased, independent estimators of the variance σ^2 of the mother distribution. These are

1. $\dfrac{N_F}{N_F - 1}\, CH_L$

2. $\dfrac{N_F}{N_F - N_G} \overline{CH_n}$

3. $\dfrac{N_F}{N_G - 1} IH_L$

We let RH_L = the residual distribution heterogeneity, defined as the value of the intrinsic distribution heterogeneity IH_L, observed when the state of natural distribution homogeneity is achieved.

By equating the first and the third estimators of σ^2, we obtain

$$RH_L = \text{natural minimum of } IH_L = \frac{N_G - 1}{N_F - 1} CH_L$$

$\bar{N}_n$ = the average number of fragments in the cluster G_n.

$$\bar{N}_n = \frac{N_F}{N_G}$$

γ = a *cluster-size factor* the physical meaning to be particularized later,

$$\gamma = \frac{N_F - N_G}{N_G - 1} \quad \text{or} \quad \frac{N_G - 1}{N_F - 1} = \frac{1}{1 + \gamma}$$

It is nearly always the case that

$$N_F \gg N_G \gg 1$$

and hence,

$$\gamma = \frac{N_F}{N_G} = \bar{N}_n.$$

The order of magnitude of γ is the average number of fragments in a cluster G_n. We note that $\gamma = 0$ if $N_G = N_F$, that is, when each cluster is made up of one and only one fragment. The consequences of this property will be reviewed in Section 2.5.10. In terms of γ, the residual distribution heterogeneity can be written

$$RH_L = \frac{1}{1 + \gamma} CH_L$$

To summarize Section 2.5.5, a material is said to be in a state of natural distribution homogeneity when the equality

$$IH_L = RH_L = \frac{1}{1 + \gamma} CH_L$$

is satisfied for a given observation module whose shape, size, and orientation are highly relevant. This will be developed in the Section 2.5.6.

2.5.6 Various Forms of Natural Distribution Homogeneity

From now on, we shall drop the qualifier *natural* and simply speak of distribution homogeneity, implicitly excluding all forms of artificial homogeneity such as the modular homogeneity that cannot be achieved by natural (i.e., mechanical) means. We have already shown that there were two main forms of heterogeneity, the constitution and distribution heterogeneities. We now show that distribution homogeneity can take several forms, having different properties. This multiplicity of the forms of homogeneity and heterogeneity explains why the words *homogeneous* and *heterogeneous* are so often misleading and justifies a thorough analysis.

Three-dimensional Distribution Homogeneity. This is the only true distribution homogeneity. It is the form that one attempts to achieve with mixers and stirrers. The condition of distribution homogeneity stated in Section 2.5.5 is fulfilled when using an isotropic module such as a sphere (theoretical definition) or a cube (experimental estimation). The degenerate forms that will be defined now are anisotropic hybrids between homogeneous and heterogeneous distributions.

Two-dimensional Distribution Homogeneity. It is well known that in a gravitational field some vertical segregation occurs and affects either certain size classes or certain density classes. For this reason, the state of three-dimensional distribution homogeneity, which has just been defined, is difficult to achieve with multisize materials or with pulps containing materials of different densities. Even if this homogeneity is attained, it must be regarded as highly unstable. Vertical segregation, taking place in a population of fragments three-dimensionally homogeneous, generates a two-dimensional distribution homogeneity. In mineral processing techniques, this fairly well describes the distribution that can be observed in a batch laboratory jig. This state is characterized by a near-perfect vertical segregation and a reasonably good horizontal homogeneity.

To match the anisotropy of the distribution, we shall use an anisotropic observation module such as a vertical cylinder (theoretical definition) and a vertical prism with triangular, square, rectangular, or hexagonal section (experimental estimation). It is advisable to restrict the application to flat heaps of material with a near constant thickness for this model to be efficient. The observation module must involve the totality of the lot thickness. We suppress the third dimension by using a two-dimensional model in which the material making up the lot is projected on and condensed in the horizontal plane. In the original three-dimensional space, the condition of distribution homogeneity stated in Section 2.5.5 is fulfilled when using an anisotropic module (cylinder or prism). In the two-dimensional horizontal projection, the condition of distribution homogeneity is fulfilled when using an isotropic two-dimensional module (circle, triangle, square, rectangle, hexagon).

One-dimensional Distribution Homogeneity. This is the form of homogeneity achieved by bed-blending systems where the pile is made up of a very large number of layers (usually several hundreds to a few thousands), deposited by a stacker moving at uniform speed along the pile and delivering a uniform flow rate. The reader (especially if he happens to be a manufacturer of bed-blending equipment) should not forget that homogeneity is achieved in one dimension only, that of the horizontal axis of the pile, the distribution remaining heterogeneous in a vertical or oblique plane (such as the plane generated by messiter-type reclaimers). This is obvious from watching the reclaiming of a multisize aggregate.

To match the anisotropy of the distribution, we must use an anisotropic observation module such as a vertical or oblique slice of material of constant thickness involving the totality of the pile cross section. For this reason, the reclaimer of a bed-blending system will achieve a true blending if, and only if, it reclaims the totality of the cross section of the pile simultaneously. This is practically (but never rigorously) achieved by systems such as windscreen wiper, messiter, disc, screw-cage, or drum-type reclaimers. It is not achieved with bucket-wheel reclaimers. A bed-blending theory, derived from the sampling theory is developed in Gy.[5]

In the original three-dimensional space, the condition of distribution homogeneity is fulfilled when using a constant-thickness slice module. In the one-dimensional model according to which the material is projected on and condensed in a horizontal axis, the condition of homogeneity is fulfilled when using an isotropic one-dimensional module such as a segment of given length.

Revolution Distribution Homogeneity. This form of homogeneity corresponds with a symmetry of the distribution toward a vertical axis. It is achieved or nearly achieved in cylindrical or conical piles fed vertically, such as at the discharge of a low-velocity belt conveyor, when the grain-size contrast of the material is not significant. It is also the form of homogeneity tentatively achieved during the first step of the old coning and quartering sampling method. Apart from this symmetry toward the vertical axis, the distribution must be regarded as heterogeneous. We must use a particular module made up of a sector of the pile of a given angle to match this particular form of homogeneity. The condition of distribution homogeneity is fulfilled when using such a sectorial module.

Three-dimensional Distribution Heterogeneity. When none of the forms of homogeneity can be observed in the lot L, its distribution must be regarded as heterogeneous in the three dimensions of space. In the presence of gravity, three-dimensional heterogeneity is the rule; distribution homogeneity even of a degenerate kind is the exception. Regardless of the module shape, size, and orientation, the distribution heterogeneity IH_L is characterized by

$$IH_L > RH_L = \frac{1}{1 + \gamma} CH_L > 0$$

Experimental Check of Distribution Homogeneity. From a theoretical standpoint, the experimental check will consist of comparing the intrinsic distribution heterogeneity IH_L that characterizes the actual state of heterogeneity of the lot L with the residual distribution heterogeneity RH_L that characterizes the state of natural distribution homogeneity of the same lot L. Both estimations involve the same observation module adapted to the form of heterogeneity to be checked. From a practical standpoint, this test would require the splitting of the lot L into N_G clusters G_n of comparable (if not equal) bulk without disturbing the original essentially unstable fragment distribution to be checked. The clusters would then be weighed and analyzed and the value of the confidence interval of IH_L would be derived from the value of DH_L experimentally estimated. There remains the estimate of RH_L and we shall see in Section 2.8.2 how this may be done.

2.5.7 Notion of Degrees of Homogeneity

The state of three-dimensional distribution homogeneity is characterized by three degrees of homogeneity. Vertical segregation entails a loss of one degree of homogeneity and results in a two-dimensional distribution homogeneity. In contrast, the state of three-dimensional distribution heterogeneity is characterized by zero degree of homogeneity. Bed-blending secures one (but only one) degree of homogeneity and results in a one-dimensional distribution homogeneity.

2.5.8 Maximal Distribution Heterogeneity

Let MH_L = the maximal distribution heterogeneity of the lot L. We have shown (Section 2.4.8) that

$$IH_L \leqq CH_L$$

Then, obviously,

$$MH_L = CH_L$$

If and how IH_L can reach its maximal value MH_L remains to be explored. We remind the reader of the relationship

$$CH_L = \overline{CH_n} + IH_L$$

When IH_L equals CH_L, the term $\overline{CH_n}$, which is equal to the average constitution heterogeneity of the clusters G_n, is zero, and

$$\overline{CH_n} = \frac{1}{N_G} \sum_n CH_n = 0$$

Now the CH_n are sums of squares, that is, non-negative quantities, so that the only solution is $CH_n = 0$ for all n.

All clusters must have a homogeneous constitution, which means that all fragments making up a given cluster have identical composition, thus $a_{nj} = a_n$ for all n and j. From a practical standpoint, this condition is fulfilled if, and only if, the following conditions are simultaneously fulfilled:

1. The material making up the lot L is completely liberated in the sense given to the verb *liberate* by the mineral processing industry, that is, each fragment of L is made of a pure mineral.
2. The various minerals present in the lot L are perfectly segregated.
3. The fragments are grouped in clusters according to their mineral composition.

Such a distribution can be observed in the bed of a batch laboratory jig concentrating a completely liberated ore.

2.5.9 Definition of the Segregation Factor ξ

From the results obtained in the preceding sections, the distribution heterogeneity IH_L ranges from 0 to CH_L but we must distinguish the natural range, which alone is of interest to us, and the artificial range, which is interesting from a purely academic standpoint.

Natural Range of IH_L. It is in the natural range that we shall find all the materials with which we are concerned. From Sections 2.5.4 to 2.5.8, this range is defined by

$$MH_L \geq IH_L \geq RH_L$$

$$CH_L \geq IH_L \geq \frac{1}{1+\gamma} CH_L$$

For the sake of convenience we write

$$\frac{1+\gamma}{1+\gamma} CH_L \geq IH_L \geq \frac{1}{1+\gamma} CH_L$$

We now introduce ξ, a segregation factor (dimensionless) characterizing the degree of segregation of the minerals present in the lot L. It is defined as

$$IH_L = \frac{1 + \gamma\xi}{1 + \gamma} CH_L$$

or

$$\xi = \frac{(1 + \gamma)CH_L - 1}{\gamma}$$

The range of ξ is specified by

$$1 + \gamma \geqq 1 + \gamma\xi \geqq 1$$

or

$$1 \geqq \xi \geqq 0$$

This defines the actual fluctuation range of the segregation factor.

For $\xi = 1$, the distribution heterogeneity is maximal. The state of perfect segregation is achieved. We shall never observe this state in practice.

$1 > \xi > 0$ is the natural range of the distribution heterogeneity.

For $\xi = 0$, the distribution heterogeneity is minimal. The state of natural distribution homogeneity is achieved but never observed practically.

Artificial Range of IH_L. This artificial range describes what we shall call the state of hyperhomogeneity that can be achieved only by artificial means involving an intelligent positioning of the elements (as opposed to random positioning).

The artificial range of DH_L is defined by

$$\frac{1}{1 + \gamma} CH_L > IH_L \geqq 0$$

It is characterized by negative values of the segregation factor

$$0 > \xi \geqq -\frac{1}{\gamma}$$

with

$0 > \xi > -\dfrac{1}{\gamma}$, state of hyperhomogeneity achieved by intelligent positioning,

$\xi = -\dfrac{1}{\gamma}$, state of rigorous hyperhomogeneity inaccessible even by intelligent positioning of the elements. This state of rigorous distribution homogeneity would be achieved only with materials having a rigorous constitution homogeneity, an inaccessible limit.

2.5.10 Influence of the Observation Scale on the Value of IH_L

The observation scale is characterized by the grouping factor γ defined in terms of the number N_F of fragments and the number N_G of clusters in the lot L. For a given lot L and a given distribution (in the natural range) characterized by a certain value of the segregation factor ξ, Table 2.3 shows that IH_L is a decreasing function of the grouping factor γ (which means that the larger the observation scale, the smaller the apparent distribution heterogeneity), or an increasing function of the number N_G of clusters (observation modules) in the lot.

2.5.11 General Expression of the Distribution Heterogeneity IH_L

The distribution heterogeneity IH_L can be represented as follows:

$$IH_L = \frac{1 + \gamma\xi}{1 + \gamma} \, CH_L = (1 + \gamma\xi)(N_G - 1) \frac{N_F}{N_F - 1} \sum_i \frac{(a_i - a_L)^2 M_i^2}{a_L^2 M_L^2}$$

This expression specifies the role played by each factor and by all relevant data of the problem. It will be useful when developing a relationship between the sampling variance and the distribution heterogeneity of the material being sampled. This point will be considered subsequently.

TABLE 2.3   Influence of the Observation Scale on the Distribution Heterogeneity DH_L Expressed in Terms of the Average Number of Fragments in the Observation Module N_n

$N_n = 1$ Each observation module contains one fragment and one fragment only		$1 < N_n < N_F$		$N_n = N_F$ There is only one observation module and it contains the N_F fragments
N_F	↓	N_G	↓	1
CH_L	↓	IH_L	↓	0
0	↑	γ	↑	$+\infty$

2.5.12 Influence of the State of Comminution on the Value of CH_L

According to its definition, the constitution heterogeneity CH_L is an intrinsic property of the population of fragments making up the lot L under observation. This population can be altered either by comminution or by agglomeration.

Agglomeration. Consider a lot L made up of N_F fragments (set L_F) and of N_G clusters (set L_G). Assume now that all fragments making up a cluster are agglomerated so as to become a new fragment (set $L_{F'}$). From the properties of the constitution and distribution heterogeneities reviewed in the preceding sections, we know that

$$CH_{L'} = IH_L \leq CH_L$$

The conclusion is that agglomeration can only reduce the constitution heterogeneity of the material under investigation.

Comminution. By inverse reasoning, we easily deduce that comminution can only increase the constitution heterogeneity of the material.

2.5.13 Influence of the State of Comminution on the Value of RH_L

According to its definition (Section 2.5.5), the residual distribution heterogeneity RH_L is

$$RH_L = \frac{N_G - 1}{N_F - 1} CH_L = (N_G - 1) \frac{N_F}{N_F - 1} \sum_i h_i^2$$

with

$$h_i = \frac{(a_i - a_L)M_i}{a_L M_L}$$

As N_F is always a very large number, we may write

$$RH_L = (N_G - 1) \sum_i h_i^2$$

We already know that the observation scale, represented in the formula above by the number N_G of clusters, is highly relevant. For comparability we must keep the observation scale, N_G, constant. After comminution, the residual heterogeneity RH_L varies in proportion to the sum

$$\sum_i h_i^2$$

Suppose that the lot L_F made up of N_F fragments F_i $(a_i = M_i)$ has been comminuted and has become the lot $L_{F'}$, made of $N_{F'}$ fragments $F_{i'}(a_{i'} = M_{i'})$, and assume that

$$N_{F'} = kN_F \text{ with } k > 1$$

The number of terms in the sum is multiplied by k, but it is easy to show that each term of the sum, proportional as it is to the square of M_i, becomes M_i'. Each term is divided on average by k^2, the overall result of comminution being that the sum, and with it the residual heterogeneity RH_L, is reduced approximately k times. This accounts approximately for the fact that minerals being nearer their state of liberation, the value of $(a_i - a_L)$ will increase slightly on average. We therefore reach the conclusion that though increasing the constitution heterogeneity CH_L, comminution always tends to reduce the residual distribution heterogeneity RH_L. This result is very important from a theoretical as well as from a practical standpoint. The sampling variance that we always want to reduce is proportional to RH_L after homogenization, provided that the sampling is done correctly.

2.6   CORRELATION BETWEEN THE POSITION OF A FRAGMENT AND ITS IDENTITY

2.6.1   Notion of Correlation

This notion is familiar to those acquainted with mathematical statistics. We define again the natural range of the intrinsic distribution heterogeneity IH_L (Section 2.5.9)

$$MH_L \geq IH_L \geq RH_L$$

and fix our attention on its limits RH_L and MH_L.

1. RH_L, the residual distribution heterogeneity, is the incompressible minimum that defines the state of natural distribution homogeneity. This state is characterized by the following properties.

   a. The fragment distribution is random.

   b. There is a complete absence of order in the fragment positioning.

   c. There is independence, that is, absence of any kind of relation between the position of a fragment and its identity.

2. MH_L, the maximal distribution heterogeneity, defines the state of perfect segregation of the components of the material. This state is characterized by the following properties.

  a. The fragment distribution is arranged according to a certain pattern.

  b. There is a strict order in the fragment positioning.

  c. There is a complete dependence between the position of a fragment, especially along a vertical axis, and its identity.

From the inequality defining its natural range, we know that the distribution heterogeneity IH_L lies somewhere between those limits in a state intermediate between total disorder and complete order, between independence and absolute dependence, between absence of any kind of relation and a completely dependent relationship.

Bartlett[6] defines a stochastic process as a "physical process that has some random element involved in its structure." We would say similarly that a stochastic process is a physical process that has some random element and some nonrandom element involved in its structure. According to this definition, homogenization and gravity segregation are stochastic processes. The force of gravity accounts for the nonrandom element, acting differentially and selectively on the different fragments making up the lot. The probabilistic nature of the homogenizing process, as well as a random residue present in all physical processes involving a large number of factors, accounts for the random element and tends to oppose the segregating action of gravity.

There is a correlation, a hybrid between relationship and absence of relation, between the position of a fragment and its identity. If this correlation were linear, it could be characterized by the coefficient ρ of linear correlation. Then the state of natural distribution homogeneity would be represented by

$$IH_L = RH_L \text{ and } \rho = 0$$

The state of maximum distribution heterogeneity would be represented by

$$IH_L = MH_L \text{ and } \rho = 1$$

All intermediate states, including all those encountered in actual practice are represented by

$$MH_L > IH_L > RH_L \text{ with } 1 > \rho > 0$$

2.6.2 Notion of Identity of a Fragment

As far as sampling and homogenizing are concerned, the identity of a fragment F_i is completely defined by two parameters, its mass M_i and its critical content a_i. The mean and variance of the sampling error on one hand, and the constitution and distribution heterogeneities of the material on the other, can be expressed in terms of a_i and M_i. With regard to the action of the forces of gravity, the mass M_i must be specified. Denote by

d_i—the diameter or dimension of the fragment F_i defined as the opening of the imaginary square mesh that would just pass F_i (cm),

v_i—the volume of F_i (cm^3),

λ_i—the density of F_i (g/cm^3),

f_i—a shape factor, defined as a coefficient of cubicity of F_i (dimensionless). It is the ratio of the volume of F_i to that of a cube with edge d_i.

$$f_i = \frac{v_i}{d_i^3}$$

or

$$v_i = f_i d_i^3$$

According to these definitions, the mass M_i (grams) is

$$M_i = \lambda_i V_i = \lambda_i f_i d_i^3$$

2.6.3   How Gravity Interferes with the Fragment Distribution

Between homogeneity and segregation, between order and disorder, there is no natural factor working in favor of homogeneity. On the contrary the force of gravity as well as other minor forces that will be reviewed in the Section 2.6.4 can work only in favor of segregation. The three factors involved in the definition of the mass M_i are highly pertinent to the differential, selective action of gravity.

Density.   Mineral processing engineers know very well that associated with shocks, shaking, or vibrations such as those frequently occurring in handling or processing devices, gravity can effect a stratification of the components according to their density. All gravity concentration processes take advantage of this property to separate dense from light minerals. Clearly, it is much easier to separate minerals than to homogenize them. It can be added that the state of homogeneity must be regarded as unstable in the presence of gravitational forces.

Size.   Size segregation occurs in much the same way as density segregation but to a lesser extent.

Shape. Segregation by shape may occur but usually not in stationary materials. It usually does so as an interaction between density, size, and shape rather than directly.

2.6.4   Possible Interference of Other Forces

Gravity is undoubtedly the most effective of all forces that can interfere with the fragment distribution and frustrate homogeneity, but it is not the only one. There are also centrifugal, magnetic, and electrostatic forces and drafts.

Centrifugal forces.   Centrifugal forces act in much the same way as gravity.

Magnetic Forces.   A certain number of materials such as magnetite, ferrosilicon and even steel powder abraded from rods and balls in rod-and-ball mills have a magnetic permeability that is often used to separate them from nonmagnetic minerals. In the natural magnetic field, magnetic forces may be disregarded in comparison with gravitational forces. Nevertheless as far as coal is concerned, we must observe that magnetic suspensoids are often used in heavy media separation, and concentrated coals may occasionally retain a small amount of residual magnetite, usually in the form of tiny particles concealed in the anfractuosites of coarser coal lumps. These magnetic particles may retain some remnant magnetism and there is a risk, however small, of magnetic aggregation of magnetic materials after grinding and pulverizing coal lumps. Perhaps coals concentrated by means of magnetic heavy media should not be used, and steel equipment for grinding and pulverizing coal should be used with caution or avoided when preparing coal standard samples. Magnetic aggregates would probably prove very difficult to disperse evenly throughout the entire volume of the primary sample prior to its splitting.

Electrostatic Forces.   In dealing with very dry, finely pulverized materials, as is always the case when preparing samples for analysis, electrostatic forces are liable to cause differential interactions between particles of different minerals or between mineral fragments and pieces of equipment they may contact.

Drafts.   Most sample preparation laboratories are equipped with dust-collecting systems with suction openings near all pieces of equipment. When handling dry, fine materials, the drafts created by the air suction (fans are nearly always generously dimensioned) may affect the various minerals present in the material in a selective way. Drafts may also upset fragment distribution when blowing across a stream of dry material falling at the discharge of a conveying device.

Multistage sampling installations usually extend vertically for gravity transfer of successive samples from one piece of equipment to the next. A number of free falls are involved. When the material contains dry fines, such free falls are likely to generate dust. Completely enclosed circuits are often used to avoid spreading. In the absence of dust collectors, ascending drafts are created by a chimney effect that sometimes perturb the fragment distribution. They may even practically dedust samples at the level of sample cutter openings where the drafts concentrate. When, in addition, dust-collecting systems have been installed, further complications may arise. Drafts are always harmful in sampling and homogenizing operations such as those involved in the preparation of standard samples. Dust-collecting systems must be used with the utmost discretion, the problem being to choose between the protection of the environment and that of the samples. It is our opinion that in recent years the emphasis has been put on the former to the detriment of the latter.

2.6.5 Correlation Between the Mass and Grade of a Fragment in a Given Population

The fragment F_i has a mass M_i, where $M_i = \lambda_i f_i d_i^3$. The grade of the fragment F_i is characterized by its critical content a_i, the proportion of the critical component in F_i. Consider two extremes, namely, F_i a fragment of pure valuable mineral or a fragment of pure gangue mineral. Valuable and gangue minerals nearly always have different densities λ_i and quite often different shapes characterized by different shape factors f_i. Moreover, they usually have different hardnesses, which result in different size distributions after comminution. All these factors contribute to the generation of a correlation between the mass M_i and the grade a_i.

2.6.6 Prospects of Achieving the State of Natural Distribution Homogeneity

1. The identity of a fragment F_i is characterized by two parameters, its mass M_i and its grade a_i.
2. The heterogeneity h_i carried by F_i is a function of a_i and M_i (Section 2.4.2).
3. M_i is mathematically related and a_i is rather closely correlated with the fragment shape f_i, density λ_i, and diameter d_i, as is the heterogeneity h_i.
4. Alterations in the fragment distribution may occur as the result of movements induced by external forces.
5. External forces involved in all handling and processing operations, principally the omnipresent force of gravity, are closely related with the fragment shape, density, and diameter and correlated with its grade and heterogeneity. They tend therefore to act in a differential way on different minerals.
6. Nature by gravity and humans by handling and processing operations are, under normal industrial conditions, operating in favor of segregation. Homogeneity always tends to be destroyed even if it has been achieved.
7. The definition of conditions under which segregation can be avoided and homogeneity achieved is required. This is the object of the Section 2.7.

2.7 HOMOGENIZATION

2.7.1 Aim to Be Achieved

The target is the achievement of natural distribution homogeneity. Intelligent positioning (excluded in the present instance as it is in almost all industrial processes) alone is capable of crossing the borderline separating natural from artificial homogeneity. We have used the qualifier *natural* as opposed to *artificial* to describe the form of distribution homogeneity that occurs when fragments are distributed at random throughout the lot. There is, however, nothing natural in the homogenizing effect that nature does its best to negate and make the state of homogeneity unstable.

2.7.2 Logical Analysis of the Homogenizing Process

The first step of our approach must be to seek to transform a nonrandom fragment distribution into a random one. We can think of two models, a physical model—the Brownian motion—and a mathematical model—the random walk. In both models, the moving element describes a broken line made up of straight segments articulated in such a way that it starts its motion in a random direction, then bounces in another direction independent of the previous one. Assume that such a process can be applied to all fragments F_i making up a certain lot L to be homogenized. This is repeated a certain number of times N_i for each fragment F_i. The position of F_i tends to become independent of its initial position as N_i increases, as does that of all fragments of L, with the consequence that the position of the fragment F_i tends to become independent of its identity, and the distribution tends toward the state of natural homogeneity characterized by this independence. The problem is whether and how it is possible to implement such a process in the real world. This must be done through mechanical means; we must exclude the application of forces produced by magnetic or electric fields whose action is known to be very selective and likely to generate segregation rather than homogeneity.

In the previous paragraph, we have used the verb *bounce* to describe the abrupt change in direction of a fragment trajectory, because such an alteration of the fragment course can only occur as the result of a collision whether with another object or an interface. To comply with the model, the direction of the course after the collision should be independent of the course before the collision. This is hardly possible. Both courses are governed by the direction of the resultant of all forces applied to the fragment. Gravity is one of these, and the condition of independence required in the model is unachievable. Both the Brownian motion and the random walk, however, provide us with a model that can be approached mechanically and that consists of multiplying the collisions between fragments and moving parts of the equipment to induce rebounds in all directions. This will be treated in Section 2.7.4. If gravity cannot be overcome, can its relative influence be minimized?

2.7.3 Influence of Comminution on the Relative Importance of Gravity

The two main forces contributing to define the trajectory of a fragment are the force of gravity and the resistance of the surrounding fluid (air or water). The force of gravity is proportional to the mass of the fragment F_i, hence, to the cube of its diameter d_i. The resistance of the fluid, at least in the relevant Newtonian range, is proportional to the square of the diameter d_i. We may therefore surmise that the finer the material to be homogenized, the smaller the relative influence of the force of gravity and the easier it should be to attain a state of natural distribution homogeneity. We already know from Section 2.5.13 that comminution tends to reduce the value of the residual distribution heterogeneity involved in the state of natural homogeneity and to which the sampling variance is proportional. Comminution is therefore beneficial on two accounts: It reduces the natural limit of the homogenizing process, namely, the residual distribution heterogeneity, and it facilitates the attainment of this natural limit.

2.7.4 Homogenizing Processes

Particular Case of Bed-Blending.   Bed-blending is often correctly referred to as prehomogenization but sometimes as homogenization, a misuse of the term because of a misinterpretation of the action in bed-blending systems. We noted in Section 2.5.7 that the stacking phase of bed-blending tends to achieve a state of one-dimensional distribution homogeneity along the pile axis, and that across the section of the pile the distribution remains heterogeneous. Multisize aggregates such as the raw feed to cement plants are always segregated. Bed-blending must therefore not be regarded as a true homogenizing process. A theory of bed-blending derived from the sampling theory is presented in Gy.[5]

Industrial Homogenizing Processes.   In industrial facilities, the purpose of homogenization always is to feed a processing plant with a material as uniform or as homogeneous as possible. The cement industry provides the best examples available in the mineral industries. We know of no true homogenizing process capable of handling coarse materials (say, more than a few millimeters) on an industrial scale. The only materials to which homogenization is applicable are powders or pulps of finely ground materials. It is not the purpose here to consider industrial homogenization. The reader interested in this subject may refer to Schofield.[7]

Laboratory Homogenizing Processes—General.   On the laboratory scale, the purpose of homogenization is always to reduce sampling errors, whether the samples are used for comparative tests or for chemical analyses. When preparing coal standard samples the purpose is to obtain samples as identical as possible within a confidence interval as narrow as possible.

Laboratory Homogenization of Medium-size Materials. By medium size we mean materials having a top particle size not exceeding a few millimeters. Homogenizing can be achieved by means of a tumbling or cascading effect, in batch blenders with capacities not exceeding a few hundred kilograms. They can be cylindroconical, cubic, or Vee-shaped. Materials contained in standard drums can be homogenized by end-over-end or axial rotation.

Laboratory Homogenization of Dry, Fine Materials. The equipment reviewed in the preceding paragraph can be used with dry, fine materials as well.

Laboratory Homogenization of Pulps of Finely Ground Materials. This can be and usually is done by turbulent stirring in cylindrical or conical tanks, with or without baffles, using propeller or paddle-type mixers. The higher the concentration of the pulp, the easier and more stable the homogenization. The capacity of such mixers can reach several cubic meters. Another efficient way of homogenizing pulps is to pump them in a closed circuit with or without stirring.

Combination of Comminution and Homogenization. We have pointed out, in Section 2.7.3, that a reduction of the top particle diameter is always beneficial. We must now observe that fine grinding nearly always involves ball mills, working either in batch (laboratory-scale) or in continuous (industrial-scale) mode, and that the tumbling of the load and balls in the shell of the mill achieves comminution and homogenization at the same time. We can therefore surmise that the ball mill, achieving two purposes in a single operation, will prove very useful in a sampling and homogenizing laboratory as it does in a cement plant.

2.7.5   Splitting

The batch homogenizers reviewed in the preceding section may nearly achieve a state of three-dimensional homogeneity in situ. This property is to be maintained in the actual sampling. From a theoretical standpoint, if we assume the distribution to be truly homogeneous, any form of grab sampling should be acceptable since any fraction of a homogeneous batch can be regarded as a representative sample of this batch. Practically, the state of three-dimensional homogeneity is always precarious and unstable, except perhaps with very finely ground materials. It is risky to place too great a reliance on such a form of homogeneity. We think it advisable to transform a static three-dimensional homogeneity, perhaps of poor quality, into a dynamic one-dimensional flow of nearly uniform quality by extracting the material from the mixer and feeding it to a convenient splitting system. The most convenient is the revolving feeder sectorial splitter, such as the one schematized in section 24.6.1 of Gy.[1] These splitters, separating clusters of fragments, absorb residual fluctuations because they distribute hundreds of increments in each of the samples they prepare. This approximates ideal samples that could be extracted from a material characterized by the state of natural distribution homogeneity.

2.7.6 Practical Implementation

The implementation of the whole grinding, homogenizing, and splitting process to prepare a large number of standard samples must be specifically adapted to the problem to be solved. The reader has been presented with the relevant information that can be derived from the theories of heterogeneity and sampling under a very general form, but the success of an operation requires experience and know-how that can hardly be condensed in the pages of this chapter.

2.8 DISPERSION OF THE ASH CONTENT OF A SERIES OF COAL STANDARD SAMPLES

We shall assume in this section that

1. A 250 kg primary sample (lot L) has been retained to represent a certain quality of coal.
2. Lot L has been ground to a certain top particle size d that for our calculations will be given values ranging from 250 to 50 μm.
3. Lot L, after grinding, has been homogenized and the state of natural distribution homogeneity has been attained.
4. The homogenized material has been fed to a multistage splitter and eventually split into 2500 100 g samples. We shall assume that the homogeneous distribution previously obtained has not been significantly altered.
5. We want to estimate the confidence interval of the true ash content a_S of this series of samples. In other words, our purpose is to estimate the sampling variance isolated from the analysis variance.

2.8.1 Relationship Between the Sampling Variance and the Distribution Heterogeneity

This general relationship was derived by Gy in 1953 (see Section 20.5.4.3)[1] with notations slightly different from those of the present work.

$$\sigma^2(SE) = \frac{1-P}{P} \frac{1}{N_G} IH_L$$

with

$$SE = \frac{a_S - a_L}{a_L}$$

where

SE = the sampling error

P = the sampling ratio, or ratio of the mass of the sample to that of the lot. In the present instance, the sample mass is 100 g and the lot mass is 250 kg. The sampling ratio is

$$P = 4 \times 10^{-4}$$

N_G = the number of samples involved. In the present case,

$$N_G = 2500 \text{ and } PN_G = 1$$

IH_L = the intrinsic distribution heterogeneity of the material at the moment of its splitting. The above formula obviously reduces to

$$\sigma^2(SE) = IH_L$$

The sampling variance is equal to the intrinsic distribution heterogeneity.

2.8.2   Sampling Error Involved When Sampling a Naturally Homogeneous Distribution

When sampling a homogenized material, the total sampling error SE reduces to what we have called the fundamental error FE, the incompressible lower limit of the sampling error. Similarly the residual distribution heterogeneity is the incompressible lower limit of the intrinsic distribution heterogeneity. The state of natural distribution homogeneity assumed here is characterized by

$$IH_L = RH_L \text{ or } \sigma^2(SE) = \sigma^2(FE) = RH_L = (N_G - 1)_i \sum_i h_i^2$$

The fundamental variance is equal to the residual distribution heterogeneity. This remarkable property fully justifies our choice of h_i

$$h_i = N_F \frac{(a_i - a_L)M_i}{a_L M_L}$$

to characterize the heterogeneity carried by the fragment F_i (see Sections 2.4.1 and 2.4.2).

2.8.3   Estimation of the Variance of the Fundamental Error

Recall that this error is called fundamental for the simple reason that it is the only component of the total sampling error that we can never suppress, even theoretically. The expression of the fundamental variance, as given in the preceding section, involves a sum$_i$ extended to the N_F fragments F_i making up the lot L. When dealing with finely ground materials as assumed here, N_F is always much too large to be even roughly estimated. This difficulty, more apparent than real, has been overcome and the reader will find a full proof in Section 20.4 of Gy.[1] The fact is that after a mathematical treatment involving a few minor approximations, the expression of the fundamental variance can be transformed into the following very simple expression.

$$\sigma^2(FE) = \left[\frac{1}{M_S} - \frac{1}{M_L}\right] clfgd^3$$

where

M_S = the sample mass (grams)

M_L = the lot mass (grams)

c = a mineralogical composition factor defined as

$$c = \frac{1 - a_L}{a_L} \left[(1 - a_L)\lambda_a + a_L\lambda_c\right]$$

a_L = the ash content of the lot L (decimal value),

λ_a = the ash density (g/cm^3),

λ_c = the coal density (g/cm^3),

l = a liberation factor (dimensionless) expressing the state of liberation of the mineralogical components of the material. The liberation factor is equal to 1 when all mineralogical components have been completely liberated and to 0 when the material constitution is perfectly homogeneous.

$$1 \geq l \geq 0$$

In the present instance it is safe to assume that $l = 1$.

f = a shape factor. We have already defined the shape factor f_i of the fragment F_i (Section 2.6.2). The factor f must be regarded as the average shape factor of all fragments making up the lot L. We have shown that, with a very limited number of exceptions, the value of f was always very near $f = 0.50$.

g = a size factor that can be computed from the size analysis of the material. The study of a very large number of factors shows that for natural, uncalibrated material such as those resulting from grinding in a ball mill, the value of g is always very near $g = 0.25$

d = the top particle diameter in centimeters, more precisely defined as the opening of the square mesh that would retain a 5 percent oversize.

2.8.4   Computation of the Fundamental Variance

We shall compute this variance for a certain number of values of the ash content a_L (involved in the definition of c) and of the top particle diameter d.

Value of c as a Function of a_L.   For the computation of c, we have used the following densities:

$$\lambda_a = 2.2 \text{ g/cm}^3$$
$$\lambda_c = 1.4 \text{ g/cm}^3$$

These are average densities applicable to French coals. With coals of other origins and more specifically with American coals, the value of ash and coal densities can be adjusted, but this should not alter the result significantly. Table 2.4 gives the value of c for values of a_L ranging from 1 to 80 percent, covering high purity coals as well as raw coals, concentrated coals, and even tailings of coal washing plants.

Fundamental Variance of a Population of 100 Gram Samples. With M_s, M_L, l, f, and g values given, the fundamental variance is

$$\sigma^2(FE) = 0.001251 \ cd^3$$

TABLE 2.4   Mineralogical Composition Factor c as a Function of the Ash Content a_L

Ash Content a_L (%)	Mineralogical Composition Factor c (g/cm³)
1	217.0
2	107.0
4	52.0
6	33.7
8	24.6
10	19.1
15	11.8
20	8.16
25	6.00
30	4.57
35	3.57
40	2.82
50	1.80
60	1.15
70	0.70
80	0.39

The 99 percent probability confidence interval of the ash content is

$$I_{99} = 2.576a_L\sigma(FE)$$

The 99.9 percent probability confidence interval of the ash content is

$$I_{99.9} = 3.291a_L\sigma(FE)$$

Tables 2.5 through 2.10 give the value of variance, standard deviation, and 99 percent and 99.9 percent probabilities confidence intervals for values of a_L ranging from 1 to 80 percent (same table), and for values of d ranging from 250 to 50 μm (Tables 2.5 to 2.10). When selecting a confidence interval, it should be emphasized that a risk of 0.1 percent is not negligible when dealing with a series of 2500 standard samples. In each series, an average of 2.5 samples should have a content falling outside the 99.9 percent confidence interval. From a practical standpoint we shall retain

$$a_S = a_L \pm I_{99.9}$$

which means that

$$\text{Prob}(\{a_S + I_{99.9}\} > a_L) + \text{Prob}(\{a_S - I_{99.9}\} < a_L) = 0.1 \text{ percent}$$

TABLE 2.5 Population of 100-Gram Samples—
$d_{95} = 250$ Microns—Fundamental Variance,
Standard Deviation, Confidence Intervals

a_L % ash	$\sigma^2(FE)$ 10^{-6}	$\sigma(FE)$ 10^{-3}	I_{99} % ash	$I_{99.9}$ % ash
1	4.238	2.059	0.0053	0.0068
2	2.090	1.446	0.0075	0.0095
4	1.016	1.008	0.0104	0.0133
6	0.658	0.811	0.0125	0.0160
8	0.480	0.693	0.0143	0.0182
10	0.373	0.611	0.0157	0.0201
15	0.230	0.480	0.0186	0.0237
20	0.159	0.399	0.0206	0.0263
25	0.117	0.342	0.0220	0.0282
30	0.0893	0.299	0.0231	0.0295
35	0.0697	0.264	0.0238	0.0304
40	0.0551	0.235	0.0242	0.0309
50	0.0352	0.188	0.0241	0.0308
60	0.0225	0.150	0.0232	0.0296
70	0.0137	0.117	0.0211	0.0270
80	0.0076	0.087	0.0180	0.0230

TABLE 2.6 Population of 100-Gram Samples—$d_{95} = 200$ microns—Fundamental Variance, Standard deviation, Confidence Intervals

a_L % ash	$\sigma^2(FE)$ 10^{-6}	$\sigma(FE)$ 10^{-3}	I_{99} % ash	$I_{99.9}$ % ash
1	2.170	1.473	0.0038	0.0048
2	1.070	1.034	0.0053	0.0068
4	0.520	0.721	0.0074	0.0095
6	0.337	0.581	0.0090	0.0115
8	0.246	0.496	0.0102	0.0131
10	0.191	0.437	0.0113	0.0144
15	0.118	0.344	0.0133	0.0170
20	0.0816	0.286	0.0147	0.0188
25	0.0600	0.245	0.0158	0.0202
30	0.0457	0.214	0.0165	0.0211
35	0.0357	0.189	0.0170	0.0218
40	0.0282	0.168	0.0173	0.0221
50	0.0180	0.134	0.0173	0.0221
60	0.0115	0.107	0.0166	0.0212
70	0.0070	0.084	0.0151	0.0193
80	0.0039	0.062	0.0129	0.0164

TABLE 2.7 Population of 100-Gram Samples—$d_{95} = 150$ Microns—Fundamental Variance, Standard Deviation, Conference Intervals

a_L % ash	$\sigma^2(FE)$ 10^{-6}	$\sigma(FE)$ 10^{-3}	I_{99} % ash	$I_{99.9}$ % ash
1	0.9155	0.9568	0.0025	0.0031
2	0.4514	0.6719	0.0035	0.0044
4	0.2194	0.4684	0.0048	0.0062
6	0.1422	0.3771	0.0058	0.0074
8	0.1038	0.3222	0.0066	0.0085
10	0.0806	0.2839	0.0073	0.0093
15	0.0498	0.2231	0.0086	0.0110
20	0.0344	0.1855	0.0096	0.0122
25	0.0253	0.1591	0.0102	0.0131
30	0.0193	0.1389	0.0107	0.0137
35	0.0151	0.1227	0.0111	0.0141
40	0.0119	0.1091	0.0112	0.0144
50	0.00759	0.0871	0.0112	0.0143
60	0.00485	0.0697	0.0108	0.0138
70	0.00297	0.0545	0.0098	0.0125
80	0.00164	0.0406	0.0084	0.0107

TABLE 2.8 Population of 100-Gram Samples—
$d_{95} = 100$ **Microns—Fundamental Variance,**
Standard Deviation, Confidence Intervals

a_L % ash	$\sigma^2(FE)$ 10^{-6}	$\sigma(FE)$ 10^{-3}	I_{99} % ash	$I_{99.9}$ % ash
1	0.27125	0.5208	0.00134	0.00171
2	0.13375	0.3657	0.00188	0.00241
4	0.06500	0.2550	0.00263	0.00336
6	0.04212	0.2052	0.00317	0.00405
8	0.03075	0.1754	0.00361	0.00462
10	0.02386	0.1545	0.00398	0.00508
15	0.01475	0.1214	0.00469	0.00600
20	0.01020	0.1010	0.00520	0.00665
25	0.00750	0.0866	0.00558	0.00712
30	0.00571	0.0756	0.00584	0.00746
35	0.00446	0.0668	0.00602	0.00770
40	0.00352	0.0594	0.00612	0.00782
50	0.00225	0.0474	0.00611	0.00781
60	0.00144	0.0379	0.00586	0.00749
70	0.00088	0.0296	0.00534	0.00683
80	0.00049	0.0221	0.00455	0.00581

TABLE 2.9 Population of 100-Gram Samples—
$d_{95} = 75$ **Microns—Fundamental Variance, Standard**
Deviation, Confidence Intervals

a_L % ash	$\sigma^2(FE)$ 10^{-6}	$\sigma(FE)$ 10^{-3}	I_{99} % ash	$I_{99.9}$ % ash
1	0.11443	0.3383	0.00087	0.00111
2	0.05643	0.2375	0.00122	0.00156
4	0.02741	0.1656	0.00171	0.00218
6	0.01777	0.1333	0.00206	0.00263
8	0.01297	0.1139	0.00235	0.00300
10	0.01007	0.1004	0.00258	0.00330
15	0.00622	0.0789	0.00305	0.00389
20	0.00430	0.0656	0.00338	0.00432
25	0.00316	0.0562	0.00362	0.00463
30	0.00241	0.0491	0.00379	0.00485
35	0.00188	0.0434	0.00391	0.00500
40	0.00149	0.0386	0.00397	0.00508
50	0.00095	0.0308	0.00397	0.00507
60	0.00061	0.0246	0.00381	0.00486
70	0.00037	0.0193	0.00347	0.00444
80	0.00021	0.0143	0.00296	0.00378

TABLE 2.10 Population of 100-Gram Samples—$d_{95} = 50$ Microns—Fundamental Variance, Standard Deviation, Confidence Intervals

a_L %ash	$\sigma^2(FE)$ 10^{-6}	$\sigma(FE)$ 10^{-3}	I_{99} % ash	$I_{99.9}$ % ash
1	0.03390	0.1841	0.00047	0.00061
2	0.01672	0.1283	0.00067	0.00085
4	0.00812	0.09014	0.00093	0.00119
6	0.00526	0.07256	0.00112	0.00143
8	0.00384	0.06200	0.00128	0.00163
10	0.00298	0.05463	0.00141	0.00180
15	0.00184	0.04294	0.00166	0.00212
20	0.00128	0.03571	0.00184	0.00235
25	0.00094	0.03062	0.00197	0.00252
30	0.00071	0.02672	0.00207	0.00264
35	0.00056	0.02363	0.00213	0.00272
40	0.00044	0.02099	0.00216	0.00276
50	0.00028	0.01677	0.00216	0.00276
60	0.00018	0.01340	0.00207	0.00265
70	0.00011	0.01048	0.00189	0.00241
80	0.00006	0.00781	0.00161	0.00206

Fundamental Variance of a Population of 1-Gram Assay Portions. When selecting the top particle size below which the lot L must be pulverized prior to its splitting into a series of standard samples, it must be kept in mind that a standard sample is always to be used as received, especially without any further comminution. What has to be taken into consideration here is therefore the confidence interval associated with the assay portion that the analyst extracts from the standard sample.

According to AFNOR French standards, 1 to 2 gram assay portions are to be used to carry out the ash analysis of a coal sample. The fundamental variance is inversely proportional to the sample mass and is 100 times larger with 1 g assay portions than with 100 g samples. The standard deviation and the confidence intervals are multiplied by 10.

We have condensed in Table 2.11 the values of the 99.9 percent probability confidence intervals associated with 1-g assay portions, for values of the ash content between 1 and 80 percent ash and for values of d_{95} between 250 and 50 μm.

TABLE 2.11 Population of 1-Gram Assay Portions—99.9 Percent Probability Confidence Intervals Expressed in Percent Ash

a_L % ash	$d_{95}=250$ μm (%)	$d_{95}=200$ μm (%)	$d_{95}=150$ μm (%)	$d_{95}=100$ μm (%)	$d_{95}=75$ μm (%)	$d_{95}=50$ μm (%)
1	0.068	0.048	0.031	0.0171	0.0111	0.0661
2	0.095	0.068	0.044	0.0241	0.0156	0.0085
4	0.133	0.095	0.062	0.0336	0.0218	0.0119
6	0.160	0.115	0.074	0.0405	0.0263	0.0143
8	0.182	0.131	0.085	0.0462	0.0300	0.0163
10	0.201	0.144	0.093	0.0508	0.0330	0.0180
15	0.237	0.170	0.110	0.0600	0.0389	0.0212
20	0.263	0.188	0.122	0.0765	0.0432	0.0235
25	0.282	0.202	0.131	0.0712	0.0463	0.0252
30	0.295	0.212	0.137	0.0746	0.0485	0.0264
35	0.304	0.218	0.141	0.0770	0.0500	0.0272
40	0.309	0.221	0.144	0.0782	0.0508	0.0276
50	0.308	0.221	0.143	0.0781	0.0507	0.0276
60	0.296	0.212	0.138	0.0749	0.0486	0.0265
70	0.270	0.193	0.125	0.0683	0.0444	0.0241
80	0.230	0.164	0.107	0.0581	0.0378	0.0206

2.9 DISPERSION OF THE KAOLINITE CONTENT OF A SERIES OF 10 MILLIGRAM ASSAY PORTIONS

As an example of the application of some of the previous considerations, suppose we wish to determine the kaolinite content of a coal. Should the sample be -20 mesh or -100 mesh? It is assumed that the kaolinite content is about 4 percent and the samples are furnished in 5 g ampules. If a few milligrams are needed for each measurement, how many samples are required and from how many ampules? Pure kaolinite has a density of 2.2 g/cm³. The corresponding composition factor c can easily be calculated from its definition in Section 2.8.3, giving 56 g/cm³. The samples, as we have assumed, come in 5 g quantities and were extracted from a much heavier, practically infinite batch (any mass above 1 kg would satisfy this requirement). A sample application of the formula for $\sigma^2(FE)$ gives a value of the intrinsic variance, using the following data:

$c = 56$ g/cm³ (estimate)
$I = 1$ (dimensionless)
$f = 0.5$ (dimensionless)
$g = 0.25$ (dimensionless)

From this it follows that the sampling constant $C = 7$ g/cm³.

2.9.1 First Example: The Coal is -20 Mesh

$d_1 = 0.084$ cm (20 mesh) and $d_1^3 = 6 \times 10^{-4}$ cm³
$M_s = 5$ grams
$M_L = \infty$

$$\sigma^2(FE) = \left[\frac{1}{M_S} - \frac{1}{M_L} \right] cd^3$$

or, more simply, when M_L is large enough,

$$\sigma^2(\text{FE}) = \frac{cd^3}{M_s} = 2.9 \times 10^{-2}$$

Thus, the standard deviation of the relative fundamental error associated with a 5 g sample of coal ground to 20 mesh is of the order of 3 percent. For reference coal samples, it is advisable to choose the smallest possible risk. The 99.9 percent confidence interval (assuming the distribution to be normal) is ± 3.291 $\sigma(FE) = 9.5$ percent. Assuming that maximum care has been taken in the sampling operation (see Gy[1]), the 5 g samples will have a kaolinite content expressed by the confidence interval 4 percent $\pm$ .38 percent. This is a very poor precision indeed, especially with reference materials.

2.9.2 Second Example: The Coal Batch is Ground to 100 Mesh

$d_2 = 0.0149$ cm (100 mesh) and $d_2^3 = 3.3 \times 10^{-6}$ cm^3

$M_s = 5$ g ($M_s = 0.2$)

$M_L = \infty$ ($1/M_L = 0$)

$\sigma^2(FE) = (0.2 - 0) \times 56 \times 1 \times 0.5 \times .0, 25 \times 3.3 \times 10^{-6}$

$\sigma = 2.15 \times 10^{-3}$

Thus, the standard deviation of the relative fundamental error associated with a 5 g sample of 100 mesh coal is about 0.2 percent. The 99.9 percent confidence interval is ± 0.71 percent and hence the kaolinite level is expressed as 4 percent ± 0.028 percent. It is questionable whether that precision is sufficient for a reference sample.

2.9.3 Third Example: The Assay Portion is 10 Milligrams and the Maximum Particle Size is 20 Mesh

As compared with the first example, the fundamental variance is multiplied by the mass ratio $5/0.01 = 500$. Accordingly, the standard deviation and confidence interval becomes 4 ± 8.4 percent kaolinite. This means that, first, the distribution is no longer normal, for the number of 20 mesh fragments in a 10 mg sample is much too small, and, second, even for a routine analysis such a lack of precision is ridiculous.

2.9.4 Fourth Example: Posing the Problem

The correct way to pose the analytical problem is: "At what size should the coal be pulverized to obtain a certain 99.9 percent confidence interval I_o for assay portions with M_o mass?"

From I_o the tolerated standard deviation σ_0 is calculated as

$$\sigma_o = \frac{I_o}{3.291}$$

Then the equation

$$d_o^3 = \frac{M_o \sigma_o}{c}$$

with $c = 7$ g/cm^3 is solved. For reference samples, $\sigma_o = 10^{-3}$ would seem to be appropriate. The mass M_o may be imposed by the assaying method. If it is supposed that it is 10 mg, then it is evident that $d_o = 1.13 \times 10^{-3}$ cm. From a theoretical standpoint, the whole batch from which the 10 mg sample is extracted should be pulverized to about $10\,\mu$m. It is obvious that 10 mg is an inappropriately small sample size.

2.10 CONCLUSIONS

The notions of homogeneity and heterogeneity are complex and multiform and most people use the words *homogeneous* and *heterogeneous* loosely, unaware of their implications. These adjectives can be applied to the constitution of the material or to the distribution of its fragments. Both the constitution and the distribution heterogeneities are given mathematical definitions and their expressions are related to each other. The purpose of homogenization is to reduce the distribution heterogeneity. By nature it is incapable of reducing the constitution heterogeneity that is an intrinsic property of the population of fragments making up the material to be homogenized. Even though it can be reduced, the distribution heterogeneity can never be completely suppressed. Its incompressible lower limit is called the residual distribution heterogeneity. The corresponding state is called the natural distribution homogeneity and is characterized by a random distribution of the fragments. It is the best that can be achieved by homogenization. Due to the presence of gravity, it is always easier to segregate the components of an aggregate than to mix them homogeneously. When the sample mass is small in comparison with the mass of the original batch, the variance of the relative sampling error is equal to the distribution heterogeneity. When the latter is minimized by homogenization and reduced to its residual value, the sampling error is minimized also and reduced to the incompressible fundamental error that, thanks to the results of the theory of sampling, can easily be estimated. There is one conclusion that stands out from this theoretical analysis; the finer the material to be homogenized, the more stable the state of homogeneity attained and the smaller the residual heterogeneity and the sampling variance. It is always advisable to pulverize the material as finely as possible when preparing standard samples. This is of course on the basis of statistical consideration alone and does not take into account other restraints dictated by end use and storage requirements.

REFERENCES

1. P. M. Gy, *Sampling of Particulate Materials—Theory and Practice*, 2nd ed. Elsevier, Amsterdam, 1982; *Heterogeneite—Echantillonnage—Homogenéisation: Un ensemble coherent de théories* [*Heterogeneity—Sampling—Blending, A Consistent Set of Theories*]. Masson, Paris, 1988.

2. J. Bastin et al., Utilisation de matériaux de référence pour analyse commerciale de grande précision—Exemple des minérais concentres de zinc [Use of reference materials for commercial analysis of high accuracy. Example—zinc ores]. Part I. Use

of results from a standardization circuit. Part II. Reference materials elaboration. *Analusis* **9**(9), 417–428 (1981); **10**(6), 253–265 (1982).

3. B. Lister, Inter-laboratory surveys of ore analysis and the development of analytical standards. In *Sampling and Analysis for the Mineral Industry*, pp. 39–56 (contains 47 references). Inst. Min. Metall., London, 1982.

4. P. M. Gy, *Théorie et pratique de l'Echantillonnage des matières morcelées.* L'Echantillonnage, 14, av. J. de Noailles, Cannes, France, 1975.

5. P. M. Gy, A new theory of bed-blending derived from the theory of sampling— Development and full-scale experimental check. *Int. J. Miner. Process.* **8**, 201–238 (1981).

6. M. S. Bartlett, *Stochastic Processes.* Cambridge Univ. London and New York, 1960.

7. C. G. Schofield, *Homogenization/Blending System Design and Control for Minerals Processing.* Trans-Tech Publications, Rockport MA, 1980.

Commentary

Having considered homogeneity from a theoretical concept, we now focus on the process by which coal is homogenized, that is, grinding. The overview Chapter 3 furnishes covers the process of coal grinding from first principles to industrial machinery. As the authors note, coal size reduction is an essential part of virtually every process for which coal is utilized. Whether one is concerned with sample banks, power production, coke, synthetic fuels, or any of the other myriad of coal applications, it is useful to have an understanding of the coal size reduction process. Chapter 3 considers coal reduction in a manner that encompasses the essentials of the field, starting with the basic relationships and continuing through to the engineering by which the comminution is effected.

3

COAL SIZE REDUCTION

R. S. C. Rogers
BP America.
also
Adjunct Associate Professor
The Pennsylvania State University

and

L. G. Austin
Mineral Engineering Department
The Pennsylvania State University

3.1 INTRODUCTION

Coal size reduction is integral to every aspect of coal processing, beginning with the mining of the coal. Run-of-mine (ROM) coal generally falls into two major groups, that from underground mining (continuous mining machines) and that from strip mining. Continuous miners produce a finer product and strip mining a coarser product that is crushed to produce the desired size. Subsequent to mining, coals must often be beneficiated, that is, cleaned of deleterious mineral impurities. This is achieved by creating particles, by size reduction, that are predominantly composed of impurities and then removing these impurities from the rest of the material. Further size reduction depends on the proposed use or uses for the coal. For example, for pulverized coal power plants, the coal is ground or pulverized to a size distribution that measures, as a rough general rule, 80 weight percent smaller than 200 mesh (74 μm) before it is blown into the boiler furnace as a coal-air suspension. In fluidized-bed uses, the coal is typically crushed to less than 3.2 mm; and for Lurgi pressure gasifiers the desired size is 6–30 mm. In the newly developed high-concentration coal–water slurries for direct boiler combustion, the slurries contain about 70 weight percent of coal ground typically 70 to 80 weight percent smaller than 200 mesh; for coal–water slurries for pipeline transportation the coal particle size is reduced to 100 percent less than 14 mesh (1.2 mm).

Coal size reduction is also important in the testing of coals. One important test (ASTM D409-71) provides an index of the grindability of a coal known as the Hardgrove Grindability Index (HGI); the test involves the grinding of a coal sample in a specially designed grindability testing machine. Another test (ASTM D441-45) gives a measure of the friability of coals to breakage during handling and is based on the size reduction of a sample tumbled for a preset time in a horizontal rotating cylinder. Other tests that provide physical data for a coal relevant to size reduction include the Drop Shatter Test (ASTM D440-49) and the non-ASTM Bond Work Index test.[1] Additionally, many analytical tests for coal call for size reduction as a part of sample preparation (see, for example, ASTM D2013-72, Standard Method for Preparing Coal Samples for Analysis).

Since all large-scale uses of coal require size reduction, the electrical energy used for comminution and the capital cost of the milling equipment are substantial. These considerations, and the increased value of coal in the last decade, has led to a renewed interest in the efficiencies of coal comminution processes. Attempts have been made to relate these efficiencies to concepts arising from fracture mechanics, but due to the complexity of the fracture process, these attempts have been only qualitatively successful. Thus, there is very active research in the systems engineering of coal size reduction processes, with the objective of applying modern process engineering and design methods that use computer simulation models of equipment and process flow sheets.

This chapter acquaints the reader with coal size reduction, beginning with a discussion of fracture mechanics as it pertains to coal and proceeding to a description of recent advances in the process engineering of size reduction. Coal properties related to size reduction are discussed, and the observed patterns of change of breakage properties with coal type are reviewed. Finally, a summary description of coal size reduction applications is presented, and recommendations are made for further research.

3.2 FRACTURE MECHANICS

3.2.1 Brittle Fracture: Stress, Strain, and Energy

Coal is geologically a sedimentary rock with distinct bedding planes containing a mixture of petrographic constituents. Yet, because of its chemical nature, it also has some of the characteristics of a polymeric material. In addition, it is extensively precracked to give a macropore structure and also has a substantial internal volume of microporosity. Add to this the mineral constituents dispersed in the coal in a variety of forms and sizes, and it can be readily appreciated that coal is a complex and heterogeneous material. However, the evidence is that coal breaks by brittle fracture in the industrially important size reduction machines and therefore the theory of brittle fracture is sufficient to illustrate the important features of the fracture of coal.

To produce size reduction a solid must be fractured, and it must be stressed by the application of force in order to produce fracture. The force can be a tensile (stretching), compressive, or shear force. The three-dimensional stress pattern in the solid represents the tensile, compressive, or shear force per unit area acting in any direction through any point in the solid. Corresponding to *stress* there is the change in dimensions of the solid referred to as *strain*. For the case of a simple one-dimensional tensile stress, stress is defined as $\sigma = F/A$, where F is the force, and strain, ε, is defined as the fractional change in the length of the solid, so that $\varepsilon = x/L_0$, where x is the extension of an original length of L_o.

Materials can be divided into two broad categories, elastic and ductile. If the solid is an elastic material it can be stressed, producing elongation. If not stretched too far, the material returns to its original shape when the stress is removed; otherwise catastrophic failure occurs and the solid fractures at a stress termed the *tensile strength.* If the solid is a ductile material, it will undergo a partially irreversible stretching before failure occurs. Figure 3.1 shows the characteristics of elastic and ductile materials. The failure under stress of an elastic material is termed brittle fracture; that for a ductile material is called nonbrittle fracture.

Elastic materials fail at small strain so $\sigma \approx \sigma_o$, and the stress–strain relation up to where brittle fracture occurs is the empirical Hooke's law, $\sigma = Y\varepsilon$, where Y is Young's modulus. For a perfect crystal, Y depends on the orientation of the stress with respect to the crystal axis. Most brittle solids are polycrystalline with a random arrangement of crystallites, so Y is an effective isotropic elastic constant. The integral of the stress times the strain is the work done on the solid to take the solid from zero external stress to a stressed state by slowly increasing F up to a final stress σ. This reversible *strain energy* is stored in the solid and, using Hooke's law, is $\sigma^2/2Y$ per unit volume. If the solid is immediately loaded to σ, the work done is $\varepsilon A\sigma L$, which is σ^2/Y per unit volume. Half of this work is strain energy and the other half will accelerate the solid and cause it to oscillate until frictional damping converts the kinetic energy to heat. Similarly, if a solid suddenly stretches at a constant σ, the work done per unit volume is σ^2/Y and only half is reversible strain energy.

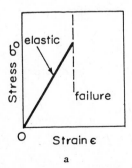
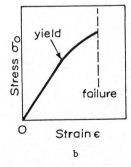

a   b

Figure 3.1   Illustration of stress–strain curves for simple one-dimensional tensile tests, σ_o = force/original cross section, strain $\varepsilon = x/L_o$.

3.2.2   Ideal Strength, Stress Concentration, and the Griffith Crack Theory

In principle, an ideal solid will fail when the applied stretching force just exceeds the ideal strength of the solid. The concept of *ideal strength* can be illustrated by considering an ideal solid made up of planes of molecules or atoms, subjected to simple one-dimensional tension. The tension stretches the bonds between the molecules, as illustrated in Figure 3.2, where the arrows indicate intermolecular attractive–repulsive forces. In the stretched state, any molecule still has an equilibrium balance of forces on it. However, when an external tension is applied that exceeds the maximum attractive force that the solid can exert on the layer of surface molecules, this tension causes an unbalance of forces and acceleration of one plane of molecules from another. The solid then catastrophically disintegrates at all planes in the solid. Assuming Hooke's law to apply up to the maximum attractive force, the strain energy per unit volume E is $\sigma^2/2Y$. The area produced by disintegration of the solid is $2N$ per unit volume of solid, where N is the number of planes per unit length; $N = 1/a$, where a is the interplanar spacing. Thus the ideal tensile strength T_I is given by

$$T_I = -\sigma_{\text{ideal failure}} \approx \sqrt{\frac{4Y\gamma}{a}} \tag{3.1}$$

where γ is the surface energy, defined as the reversible work to create a unit area of surface from the unstressed solid.

Real solids are not ideal. They contain many minute flaws (cracks, holes, etc.) that can cause a *stress concentration* within the solid. The concept of stress concentration can be illustrated by considering a planar solid containing a small hole, under a uniform externally applied tensile stress of S in the x direction and zero in the y direction. The solution of the differential stress–strain equations for the case of a small elliptical hole gives a maximum stress within the solid of

$$\sigma_{\text{max}} = S(1 + 2a/b) \tag{3.2}$$

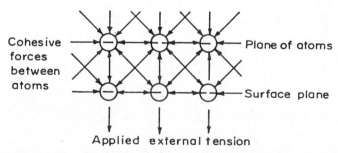

Figure 3.2   Illustration of forces between molecules in a solid.

where a is the ellipse axis in the y direction and b in the x direction. For an elliptical hole with its long axis perpendicular to the stress direction, a is greater than b. Stress concentration can be very high if $a \gg b$, that is, $\sigma_{max}/S \gg 1$.

In his well-known crack theory, Griffith[3,4] argued that real solids contain flaws that approximately correspond to the three-dimensional equivalent of the elliptical holes discussed above and that these points of weakness, called *stress-activated flaws*, initiate cracks at stress levels much below ideal. He made four basic assumptions: (1) that stress concentration occurs at the tip of the flaw; (2) that the solid is stressed to where the intermolecular bonds at the tip are stretched to breaking point; (3) that the stress state is reproduced at the tip for an infinitesimal expansion of the flaw; and (4) that energy for expanding the flaw as a propagating crack is available because the solid cannot immediately relax from its externally applied stressed or strained state. By performing a simplified energy balance for the case of a small elliptical hole in a large thin plate, Griffith showed that the overall stress required to initiate a crack at the tips of the ellipse must have a critical value σ_c, which in turn defines the *critical tensile strength T_o* of the solid:

$$T_o = -\sigma_c = \sqrt{\frac{2\gamma Y}{\pi c}} \tag{3.3}$$

where c is the long half axis of the ellipse, that is, half the crack length.

Comparing Equation 3.3 with Equation 3.1, values of a are no more than a few angstroms, so a flaw with a half length of hundreds of angstroms can give orders of magnitude reduction in tensile strength T_o from the ideal strength. As the crack progresses after initiation, extra strain energy becomes available to accelerate the crack tip. Unless this energy is relieved, the system is unstable and the crack rapidly expands, accelerating to high velocities. The strength is lower than ideal because the bulk stress does not have to be sufficient to break all the bonding forces at once, since only the bonds around the crack tip are breaking at any instant in time. In addition, Equation 3.3 is valid for a single flaw, whereas the presence of many flaws close together will give further reductions in strength.

Obviously, pure compressive stress does not cause a flaw to open and will not cause crack propagation, so tensile stress is necessary for brittle fracture. It might be thought that tensile stress will not exist under conditions of simple one-dimensional compression. However, a more detailed analysis considering all possible orientations of flaws shows that tensile stresses are produced at the tip of an ellipse at a suitable orientation even under conditions of bulk compression. The result for a planar system that would give a tensile strength of T_o under one-dimensional tension, with the crack axis perpendicular to the stress, gives a compressive strength of $8T_o$. This indicates that compressive strengths of brittle materials are about an order of magnitude higher than tensile strengths.

Under combined stress conditions the crack will propagate in a direction perpendicular to the local tensile stress conditions and may run into a region of compression that prevents further crack growth. Solutions of the stress–strain equations for simple compression of discs, cylinders in the "Brazilian" radial mode of testing, and spheres show that tensile stresses are present, with maximum values along the loaded radial axis. Even for cubes and cylinders loaded along the axis, friction between the loading plate and the sample leads to nonuniform compressive stress and regions of tensile stress. Thus, compressive loading of irregularly shaped lumps or particles will certainly produce local regions of tensile stress and, hence, brittle fracture.

3.2.3 Qualitative Applications of Fracture Theory

It is concluded from studies of coal fracture that coal normally breaks by brittle tensile fracture, originating from a stress-activated flaw lying in a region of local tensile stress, when the local value of tensile stress exceeds Griffith's critical value. Thus, the magnitude of the force necessary to cause fracture depends on the bulk geometry of the specimen, how force is applied, and the number and size distribution of flaws. In addition, it is well established that the rapidly moving crack tip will usually branch and rebranch to form a tree of cracks. This appears to be due to the high dynamic stresses produced at the tip of a propagating crack that will produce many stress-activated flaws in the path of the crack. If a large force is applied dynamically to a solid instead of being applied by slow increments of additional force, the solid will undergo rapid strain at the point of impact. The strain will spread through the solid and the resulting stress wave can cause fracture to initiate at many stress-activated flaws in the path of the stress wave. There appears to be no fundamental theory that can be used to predict the set of fragment sizes produced by the branching crack.

It is known that different particle sizes break at different rates in size reduction devices, and theory states that the following factors are involved. Since fracture originates from stress-activated flaws in the solid, small particles will require higher levels of stress to cause fracture because there is a smaller probability of finding a large flaw in a small volume. Put another way, the continued process of fracture removes weak flaws leaving the solid stronger. In addition, the geometry of stress application does not favor the breakage of small particles. Many particles must be captured for subjection to stress to break a unit mass of small particles. However, when a small particle is stressed to a level where fracture occurs, the high stress level required is likely to lead to many branching cracks and a more explosive disintegration. This results in greater surface production. Because coal does have some degree of viscoelastic behavior like an organic polymer, it is also possible that the high stresses necessary to fracture very small particles of coal cause plastic behavior in the coal. In this case, the critical stress equations must be modified to allow for absorption of energy in plasticity, and Griffith's critical value is greater than his relation would predict.

Large particles, on the other hand, may not break readily in the size reduction machine because the forces required to stress them to the fracture point may never be reached. Obviously, even if a given force applied to a small particle of

1 mm diameter produces a local tensile stress of 10^3 psi, the same force applied under identical conditions to a larger particle of 10 mm diameter may produce a local tensile stress of no more than 10 psi because of the larger area and volume involved. In addition, the geometry of capture of large particles for stress application may not be favorable.

There has been a great deal of misconception in the grinding literature concerning *grinding energy*. The previous discussions show that a strong solid must be raised to a higher state of stress for fracture to proceed, especially from applied compressive forces. Once the fracture has initiated, only a fraction of the local stored strain energy around the propagating cracks is used to break bonds (the γ term). The fragments of solid are removed from external stress when the solid disintegrates, and the rest of the strain energy stored in the solid is converted to heat and sound. Experiments on mills show that the fraction of the electric power input to the mill that is used directly to break bonding forces is very small (< 1 percent), usually less than the errors involved in the measurement of the energy balance. Rittinger's law,[5] that the "energy of size reduction is proportional to the new surface produced," has no correct theoretical base.

Under certain circumstances, the addition of more power to a size reduction device causes a directly proportional increase in the number of breakage actions per unit time. In these circumstances, doubling the power doubles the breakage rate, and the energy per ton of product remains approximately constant. Energy per ton equals the power to mill divided by the tons per hour of production. This is a simple and useful concept that does not rely on any fundamental fracture relation. It is also clear from the discussion that small particles break slower than larger particles, so that finer grinding requires more mill energy. Finer grinding also produces more surface area, but there is no valid reason why the mill energy per unit of surface produced should be exactly constant for even finer grinding.

In summary, the theory of fracture implies that the terms *weak* and *strong* should be used with respect to the ease of breakage of materials, rather than the terms *soft* and *hard*. This usage is recommended because the flaw structure is involved in breakage as well as in the basic strength indicated by the surface energy. There is virtually no correlation, for example, between ease of grindability and indentation hardness results on coal. The concept of branching cracks states that any normal fracture process will produce a natural set of products, including some fine powder; it is not possible to fracture solids at high rates without producing fines. The function of a size reduction device is to stress many particles to the fracture point at high rates, so that the capacity in tons per hour for a given capital outlay is high. At the same time, the efficiency of transfer of input energy to stress–strain energy should be as high as possible, so that the specific grinding energy is as low as possible. Under the best conditions, the fraction of input energy converted to surface energy is small, giving very low grinding efficiencies based on the minimum thermodynamic energy to create new surface. To a large extent, however, this is irrelevant because it is necessary to stress particles to a high energy state to produce fracture, and the energy inevitably is converted to heat.

To describe the rates of breakage of coal in an industrial size reduction device from fundamental reasoning, it is necessary to know:

- How each particle is stressed by the forces produced in the device,
- How frequently the particles are stressed,
- The tensile regions produced in each particle,
- What fraction of the stress actions produce tensile stresses exceeding the failure point for the particular flaws in the solid, and
- The distribution of fragments produced by the branching tree of cracks.

When one considers the complexity of describing the dynamic applied forces in size reduction machines, the complexity of the distribution of flaw sizes in solids, and the difficulty of solving stress–strain equations for irregular shapes and glancing blows, it is not surprising that the designer has to rely on other means for predicting the results of size reduction.

3.3 COAL PROPERTIES RELATED TO SIZE REDUCTION

3.3.1 The Emprical Nature of Tests of Coal Strength

Coal is obviously not an ideal solid for experimentation because of its variability and complexity, so much of the work on the mechanical properties of coal has been use oriented and empirical. As a rock that is mined, there are data on the compressive strengths; as a material that is handled dry and wet, there are hardness and abrasiveness tests; as a fuel that is ground prior to utilization, there are tests of grindability. Theories for the design of picks for cutting coal out of a seam involve tensile strengths and coefficients of friction.

In spite of the importance of coal, available information on the variation of physical properties with coal rank is fragmentary for many of the properties. The most useful general reference is the book by Evans and Pomeroy that reports results for British coals.[6] The articles by Rad[7] and McClung and Geer[8] contain between them almost 350 references. A growing body of information exists concerning the comminutive properties of coal. In most cases, these properties are defined by an appropriate empirical test that tries to simulate the breakage action of the large-scale machine. The properties are known to change in a consistent manner with the degree of coalification (coal rank), although other factors often cause a variability that tends to obscure the basic pattern of change. The purpose of this section is to demonstrate patterns of change as the coal increases in rank, using as an index coal characteristics related to rank (e.g., percent carbon content and DMMF volatile matter). It is important to recognize, however, that virtually all information pertaining to the mechanical and comminutive properties of coal is empirical. There is very little fundamental understanding of the relationship between these properties of coal and its physical and chemical structure.

3.3.2 Grindability

One of the most important mechanical properties of coal from an industrial point of view is its grindability. Grindability is a loose term meaning the ease with which the coal can be comminuted by mechanical action. The equipment expense and power requirements for coarse crushing of coal are not high, so there has been little incentive to develop a grindability test for coarse crushing. On the other hand, there has been a great deal of work done on tests of fine pulverizing of coal.

The most widely used standard grindability test for coal is the Hardgrove test.[9,10] This consists of grinding 50 g of 16 × 30 mesh (U.S. standard, 1190–595 μm) coal for sixty revolutions in a carefully standardized ball-and-race mill (see Figure 3.3). The sample must be carefully produced from larger lumps by repeated crushing in jaw crushers at low size reduction ratios, so that as much as possible of the original coal ends up in the 16 × 30 mesh fraction. The sample is also air-dried to a stable moisture content. The coal is removed after the test and sieved at 200 mesh to determine the weight w passing through the sieve. The

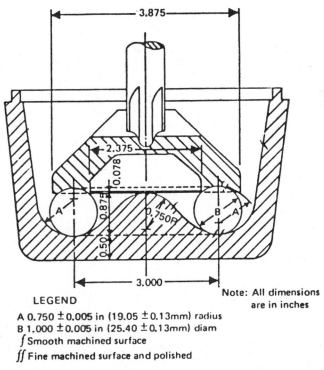

LEGEND

A 0.750 ± 0.005 in (19.05 ± 0.13mm) radius
B 1.000 ± 0.005 in (25.40 ± 0.13mm) diam
∫ Smooth machined surface
∫∫ Fine machined surface and polished

Note: All dimensions are in inches

Figure 3.3 Grinding elements of the Hardgrove machine.

Hardgrove Grindability Index (HGI) is then determined from

$$HGI = 13.6 + 6.93w \qquad (3.4)$$

This equation was originally proposed to relate the energy of comminution to the fresh surface area produced, but in practice the HGI is used simply as an empirical correlation number. In addition, current practice is to use Equation 3.4 in the form

$$HGI = k_1 + k_2 w \qquad (3.4a)$$

where k_1 and k_2 for a particular machine are calculated by tests on four coals of known HGI, supplied as calibrating materials by the National Bureau of Standards.

Figure 3.4 shows the variation of HGI with coal rank for British and American coals. As with all correlations of coal properties with rank defined by a simple number, the correlation consists of a rather wide band. With grindability, this is to be expected because of the influence of mineral matter[11,12] in addition to differences in petrographic make-up.[13] The test result can be sensitive to moisture content,[14] and oven-drying the coal can cause significant changes in HGI.[15] Trends similar to the one in Figure 3.4 have been observed for HGI plotted against rank as related to DMMF volatile matter,[16] and percent carbon.[17] Empirical equations have been presented for estimating HGI from coal characteristics relating to coal rank.[17,18]

It is interesting to compare the pattern seen in Figure 3.4 to those for some of the more precisely defined mechanical properties of coal. For example, a

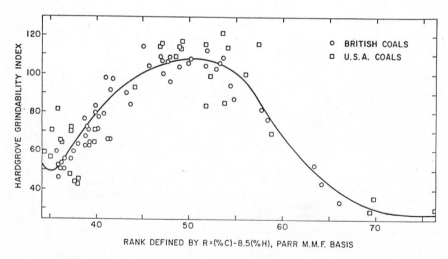

Figure 3.4   Hardgrove grindability of coal as a function of coal rank.

comparison of bulk moduli of elasticity for different coals as a function of vitrinite carbon content has shown that the bulk modulus increases with coal rank.[19] Since the stress required to cause deformation and, ultimately, to cause fracture is (for constant strain) proportional to the value of the bulk modulus, the result would indicate that a higher rank coal must be subjected to greater stress to cause fracture. This conclusion is qualitatively in agreement with the trend in Figure 3.4. However, direct correlation of elastic constants with grindability is difficult. Higher-rank coals can be highly nonisotropic[20] and elastic constants can vary depending on the magnitude of the stress applied during testing.[21]

Table 3.1 gives data for nine British coals relating average compressive and tensile strengths to coal rank.[22] If one assumes that low strength gives high HGI and vice versa, these data display the same pattern as seen in Figure 3.4. However, again, direct correlation of strength to grindability is difficult. It has been shown that, for the same coal, compressive strength decreases with increasing size of specimen and can vary over a wide range even for a single size of specimen.[23] Strength can also vary with petrographic composition[24] and for a given coal can vary with orientation of the specimen with respect to the load and even with the temperature at which tests are performed.[25]

TABLE 3.1   Tensile and Compressive Strengths of Coal as a Function of Coal Rank

Volatiles (% DAF)	Compressive Strength (lb/in²)		Tensile Strength (lb/in²)		
	(i)	(ii)	(i)	(ii)	(iii)
6	6630 ± 510	6340 ± 300	230 ± 20	350 ± 20	280 ± 20
12	2680 ± 80	2110 ± 160	100 ± 10	140 ± 10	100 ± 10
22	1420 ± 90	2110 ± 180	70 ± 10	110 ± 10	90 ± 10
30	3110 ± 90	1660 ± 90	80 ± 10	140 ± 20	80 ± 10
32	7320 ± 510	4860 ± 200	160 ± 10	590 ± 30	400 ± 40
35	5570 ± 400	3950 ± 210	160 ± 20	450 ± 40	330 ± 40
38	4950 ± 350	4140 ± 120	90 ± 10	250 ± 20	270 ± 20
38	5450 ± 270	3970 ± 130	110 ± 10	360 ± 30	240 ± 20
38	5570 ± 320	4020 ± 280	130 ± 10	420 ± 30	290 ± 30

Stress applied:
(i) Perpendicular
(ii) Parallel to
    bedding planes

Stress applied:
(i) Perpendicular to bedding
    planes (90,0,0)
(ii) Parallel to bedding planes
    and main cleat (0,0,90)
(iii) Parallel to bedding planes
    and perpendicular to main
    cleat (0,90,0)

Other mechanical properties have been examined in relation to coal rank. For example, the *hardness* of a coal is usually determined by using the Rockwell, Brinell, or Vickers hardness testers, all of which measure resistance to indentation under static loads. These tests have all been applied to coal as a function of rank, and the test results displayed the same pattern.[26] That is, hardness increases with rank, reaching a maximum at about 80 percent carbon, decreases to a minimum at about 90 percent carbon, then increases again. The reverse is true with respect to the HGI of coal.

The energy E required to reduce coal from a feed size consisting of 80 percent-by-weight less than a size x_F to a product 80 percent-by-weight less than size x_P in a tumbling ball mill (see below) can be estimated from

$$E = WI \left(\frac{10}{\sqrt{x_P}} - \frac{10}{\sqrt{x_F}} \right) \tag{3.5}$$

where x_F and x_P are microns and the value of 10 is $\sqrt{100\,\mu m}$. WI is an empirical grindability index called the Bond Work Index, determined by a standardized test in a 1-ft diameter tumbling ball mill,[1] with units of kWh/short ton. The Bond Work Index was derived by empirically comparing the results from the test mill with results from an 8 ft diameter production tumbling ball mill; another empirical law is that to convert to a mill diameter of D, WI should be multiplied by $(8\,\text{ft}/D\,\text{ft})^{0.2}$. A number of correction factors are necessary, depending on the milling conditions being used. The test is considerably more time-consuming than the Hardgrove test, and it has not been widely used in the coal industry. An approximate relation between the two tests is

$$WI = 435/(\text{HGI})^{0.91}, \text{ kWh/short ton} \tag{3.6}$$

so that WI increases as HGI decreases. This relationship is not very accurate, especially for coals with HGI between 50 and 65.

The surface energy for slow propagation of a crack in hard coal has been measured by a wedge technique, giving $(0.8)(10^4)$ ergs/cm^2 for the direction perpendicular to the bedding plane and $(1.4)(10^4)$ ergs/cm^2 parallel to the bedding plane.[27] However, it seems likely that the figure is meaningless for coal because of the development of an extensive microcrack system around the crack tip as it progresses.[28]

As seen, many of the mechanical and comminutive properties of coal change in a consistent manner with coal rank and thus with each other. This leads to qualitative descriptions of coal such as: a strong, hard coal is a higher-rank coal and is expected to be difficult to grind; a weak, soft coal is a lower-rank coal and is expected to be easier to grind. Unfortunately, since coal is a complex material and results can differ based on the test methodology employed, even these qualitative statements are not always true.

3.3.3 Primary Progeny Fragment Distributions

In recent years there have been substantial advances in the description of breakage from a process engineering point of view, as discussed in Section 3.4. This involves the concept of the *primary progeny fragment distribution*. When particles of size x_j are subjected to breakage action in a size reduction device, the branching tree of cracks that occurs on brittle failure produces a set of progeny fragments. These fragments can mix into the bulk of the powder or slurry in a machine, and in turn may eventually be reselected for further breakage. However, the mean set of progeny fragments produced by primary breakage before refracture occurs can be determined experimentally and is expressed as the fraction of broken material that falls below the size x, where $0 \leqslant x \leqslant x_j$. For coals, it is usually found that this fraction is a function of relative size x/x_j. Thus, it appears that a given coal, when stressed to fracture, produces a mass-relative size distribution of fragments that is independent of the original particle size.

It is also found that this *primary progeny fragment distribution* can be represented by the empirical equation

$$B(x/x_j) = \Phi \left(\frac{x}{x_j}\right)^{\gamma} + (1 - \Phi) \left(\frac{x}{x_j}\right)^{\beta} \qquad (3.7)$$

as illustrated in Figure 3.5. Thus the parameters Φ, γ, and β are the characteristic parameters describing the effective primary fracture distribution. Figure 3.6 shows the plot of B values versus relative size for breakage of a range of air-dried

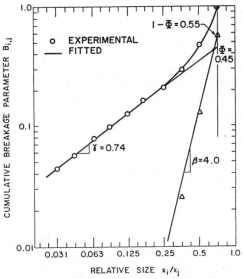

Figure 3.5 Experimental B values for Pittsburgh #8 coal ground in a ball mill.

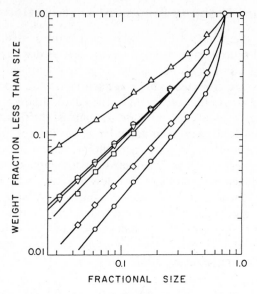

Figure 3.6 Primary progeny fragment distributions for various coals ground in the Hardgrove grindability machine.

	HGI
○ Buck Mountain: Anthracite	35
◇ Adaville 1: Subbituminous A	36
□ Illinois No. 6: High Volatile C	58
▽ Coteau-Glen Harold: Lignite	63
⬡ Pittsburgh Seam: High Volatile Bituminous A	69
△ Helena Mine: High Volatile Bituminous A	101

coals in the Hardgrove grindability tester. It is seen that primary breakage of *softer* (weaker) coals gives a larger proportion of the mass in the fine size fractions of the product, suggesting that the tree of cracks has more branching for the weaker coals.

3.4 PROCESS ENGINEERING

3.4.1 Design Criteria

The process engineering of size reduction involves the design of crushing and grinding systems where the size characteristics of the feed and product and the tonnage per hour of the product have been specified. The operating systems must be stable and controllable, to meet a range of product specifications if necessary. They should produce the product tonnage in as efficient a manner as possible, with a minimum of capital expenditure, energy costs, and maintenance/labor costs. Thus, in the design of any size reduction system, the following questions are of concern:

1. What type of crusher or grinder is best, or most satisfactory, for reducing a given feed size and type of coal to a desired product size distribution?
2. How big does the machine have to be for a given throughput rate?
3. How much electrical energy (or its equivalent) is required per ton of product?
4. How does the size distribution vary with change in throughput rate, and is a control scheme necessary or desirable?
5. Can the size distribution be readily varied if desired?
6. What is the optimum way to operate a machine in a given system?

For the most part, good system design evolved by trial and error, starting from common-sense applications of the concepts of fracture. An example of this is the Bond Method[1] for the design of ball mill circuits. However, for devices that reduce large tonnages of material, there is considerable impetus for accurate process design rules and for techniques for system optimization. As in other industrial unit operations, it is invaluable to construct mathematical models of size reduction to aid in its understanding and system design. In the last decades, considerable advances have been made in this respect, and models have been developed for a variety of size reduction devices.[29–50] These models are forming the basis for an improved approach to the process engineering of coal size reduction and the concepts involved in their construction are described in the following sections.

3.4.2 Definitions and General Concepts

The rate at which a material breaks in a machine depends on its particle size as well as the strength characteristics of the material. Normally, for any given mechanical action there will be particle sizes that are too big for efficient breakage because the action is not powerful enough, and particles that are too small for efficient breakage because the statistics of applying the action are not favorable. It is also apparent that the specific energy (kWh/ton) used for size reduction increases for breakage of finer and finer sizes, because (1) it becomes more and more difficult to apply stress efficiently to millions of tiny particles, and (2) the basic strength of brittle particles increases because large flaws (which can be stress-activated to fracture at low stress) become broken out as grinding proceeds to finer sizes. It is necessary, then, to analyze the breakage of each size range. It has been found convenient to use a $\sqrt{2}$ screen sequence to define the size ranges (e.g., *size* is defined as 16×20 mesh [$1180\,\mu m \times 850\,\mu m$], 20×30 mesh [$850\,\mu m \times 600\,\mu m$], etc.) because material in one of these size intervals appears to behave like a uniform material, to a sufficient approximation. Since a geometric progression never reaches zero, it is necessary to define a *sink* interval containing all material less than the smallest size measured. Thus a feed size range can be split into *n* intervals, numbered 1 for the top size interval to *n* for the sink interval.

Using this basis, the size distribution from breaking a given *size* in one pass through the device is called the *progeny fragment distribution*, and is conveniently represented in the form "d_{ij} equals the weight fraction of larger size *j* transferred by breakage to smaller size *i*." The set of numbers d_{ij} is called the *transfer* matrix and can be used to write a generalized mass-size balance equation for size reduction devices[51]

$$p_i = \sum_{j=1}^{i} d_{ij}f_j, \qquad 1 \leqslant i \leqslant n \tag{3.8}$$

where the f_i and p_i give the weight fraction of size *i* material in the feed and product, respectively. The assumption implicit in this equation is that there is no briquetting or agglomeration of particles, so that small feed does not appear in larger product.

In principle, d_{ij} values can be determined experimentally for any coal passed through any device at any given operating conditions for the device. In practice, however, the cost and time required for such testing of industrial-scale machines is prohibitive. Thus, the function of a simulation model is to calculate the d_{ij} for operating conditions of importance.

With this method of treating breakage balances, the *primary progeny fragment distribution* is expressed as $B_{i,j}$, where $B_{i,j}$ is the weight fraction of material broken from size *j* that is broken to less than size *i*, upon primary breakage, and Equation 3.7 becomes

$$B_{i,j} = \Phi \left(\frac{x_{i-1}}{x_j}\right)^\gamma + (1 - \Phi) \left(\frac{x_{i-1}}{x_j}\right)^\beta \tag{3.7a}$$

It has been found that this form of the primary *B* values applies for many brittle materials and machine types. Some typical values for coals have been plotted in Figure 3.6. The slope of the finer end of the *B* plot is characteristic of the material and appears to be the same for all breaking sizes. For coals, the *B* values are size normalizable, that is, the curves of Figure 3.6 for any type of coal fall on top of one another for different breaking sizes. Thus $B_{i,j} = B_{i+1,j+1} = B_{i+2,j+2}$, and so on.

Other elements of a simulation model depend on whether the machine is a *once-through* or a *retention* device. Once-through devices are those that subject a feed to essentially a single breakage action and pass out the products immediately. Thus, size reduction is not a strong function of feed rate, unless the device becomes choked with material at too high feed rates. Crushers can generally be classified as once-through devices. Retention devices, on the other hand, contain a reservoir of powder being acted on by the breakage action and the fineness of the product depends on how long the material is retained. That is, the product becomes finer as the feed rate decreases. Machines for pulverizing and grinding are generally retention devices.

The most popular models for once-through devices[39,42,52-54] incorporate the concept that each size interval of feed material may or may not be broken as it passes through the device, depending on its size with respect to how close the crushing surfaces approach each other. This physically real situation is mathematically treated using a series of *by-pass* values, a_i, which give the weight fraction of size i feed broken upon passage through the device. One model[42,54] defines these as *primary by-pass* values and introduces *secondary by-pass* values to account for the fact that products of primary fracture have a different probability of by-passing out of the device.

Models for retention devices are constructed using concepts very similar to those of chemical reactor theory. That is, the device is considered equivalent to a *reactor* that accepts the feed size distribution and *reacts* it to a product size distribution, and a size-breakage rate balance is performed on the *reactor*. Thus for retention devices the concept of *specific rate of breakage*, S_i, is used. This concept can be illustrated by considering a mass W of powder in a retention device, of which a weight fraction w_i is of size i. The specific rate of breakage S_i is defined by "rate of breakage of size i to smaller sizes $= S_i w_i W$." S_i has the units of time^{-1} and is comparable to a first-order rate constant in chemical reaction kinetics. The size reduction of a feed of size i in a batch retention device is comparable to a homogeneous, first-order *rate of reaction* experiment, and if S_i is constant, gives

$$-d(w_i W)/dt = S_i w_i W \qquad (3.9)$$

and

$$w_i(t) = w_i(0) \exp{(-S_i t)} \qquad (3.10)$$

Figure 3.7 shows a typical result for a coal. This first-order relation is observed so frequently that it can be called *normal* breakage, whereas nonfirst-order kinetics indicate some abnormal feature.

A nonfirst-order effect of particular importance is the *slowing-down* effect[55] that is observed for very fine dry grinding, or fine wet grinding at relatively high slurry densities. The physical mechanisms for the decrease of breakage rates of all sizes as the powder bed accumulates fine material are not clearly understood but are lumped under the vague term of *cushioning*. It appears that the presence of a substantial fraction of fine material in the powder bed allows the bed to absorb impact or crushing force with diminished fracture of the individual particles. The effect is so pronounced that dry size reduction almost ceases if the bed becomes too fine. Therefore, fine industrial dry grinding machines are designed to keep the bed stripped from fines by gas sweeping.

The concepts of S and B plus the time that material spends in a retention device are generally sufficient for the mathematical description of such devices

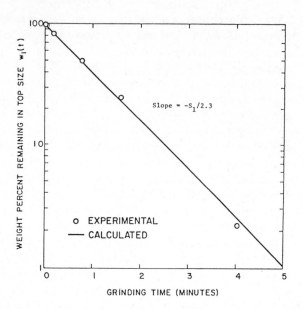

Figure 3.7   Typical first order plot (dry grinding of 16 × 20 mesh anthracite in a 0.6 m diameter ball mill).

operated in the batch mode. However, for continuous retention devices, the additional concept of *residence time distribution,* $\varphi(t, \tau)$, is required to account for the observation that some fraction of the feed may stay in the device for a short time, and other fractions for longer times. The mean value of the residence time distribution, the *mean residence time,* τ, is defined by W/F where F is the solids feed rate into the device.

Finally, as in many reactor systems, the use of several stages of size reduction combined with recycle can be advantageous. It is not uncommon to pass the product coming out of a size reduction device through a size *classifier,* which splits the product into two streams, one containing coarser (oversize) particles and the other finer particles. The coarser stream is recycled back to the feed. The process of selective size separation is known as *classification* and is treated mathematically using the concept of *size selectivity,* s_i. The size selectivity value s_i gives the fraction of size i feed to the classifier, which reports to the coarser stream. The general reason for employing a classifier in this manner, that is, for closed circuit operation, is to remove particles that are already fine enough and prevent energy being wasted on further size reduction.

3.4.3   Examples of Simulation Models

Examples are given here of mathematical simulation models for the process engineering of coal size reduction.

Double-Roll Crushers.   Probably the most commonly used crusher in the coal industry is the double roll crusher. The operation of a double-roll crusher is schematically illustrated in Figure 3.8. Material of a controlled tip size is fed to two rolls rotating toward each other, where it is fractured and passes through the minimum distance between the rolls (the gap). At least one roll is spring-loaded to permit the passage of tramp material without damage to the unit, and the distance between the crushing surfaces is usually adjustable. Double-roll crushers in the coal industry typically employ toothed rolls (the teeth are designed to mesh) and can reduce ROM coal with a maximum top size of 0.9 m to a product with a top size in the range of 360 to 20 mm. A typical industrial scale unit for reducing ROM coal would be a 0.83 m × 1.83 m (roll diameter × roll length) toothed double-roll crusher having a capacity of 1100 tph (metric) and a power requirement of 75 kW.[56]

The simulation model for double-roll crushers presented here has been derived, developed, and validated in a number of papers.[42,43,54,57–59] The basic assumptions of the model are:

1. The breakage of material of each size interval ($\sqrt{2}$ screen sequence) occurs independently of other material.
2. Provided the rolls are large in diameter compared to the feed sizes and the gap, the product size distribution of breaking a single size depends on the ratio of the size to the gap.
3. The description of the overall breakage of a single $\sqrt{2}$ feed size is made of:
   a. The primary progeny fragment distribution, $B_{i,j}$,
   b. The fraction of the feed that passes through the gap without breaking (primary by-pass), a_j,
   c. The fraction of primary fragments of a given size that by-pass without further fracture (secondary by-pass), a'_i,
   d. An algorithm to describe repeated fracture and by-pass until all feed material has passed through the gap.
4. The $B_{i,j}$ are normalized.

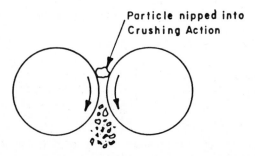

Figure 3.8   Schematic illustration of double-roll crusher operation.

On the basis of these assumptions, the algorithm for computing the progeny fragment distribution is

$$
d_{ij} = \begin{cases}
1 - a_i, & i = j \\
(1 - a_i')\left[b_{i,j}a_j + \displaystyle\sum_{\substack{l=j+1 \\ i>j+2}}^{i-1} b_{il}a_l d_{lj}/(1 - a_l') \right], & i > j
\end{cases}
\tag{3.11}
$$

where $b_{ij} = B_{i,j} - B_{i+1,j}$.

A methodology for estimating $B_{i,j}$ and the by-pass parameters from laboratory-scale crusher tests has been described and demonstrated, and systematic studies have been performed to investigate factors that could influence their values.[57] It has been found that for a given feed material, the $B_{i,j}$ and by-pass parameters are not affected by crusher feed rate (an expected result, as long as the crusher is not fed at too high rates), crusher roll speeds, crusher roll surface configurations, or crusher size. It was found that the $B_{i,j}$ and by-pass values are material dependent (see below). Figure 3.9 gives a typical comparison of experimental and simulated product size distributions for the crushing of a coal in an industrial-scale double-roll crusher.

The additional facets of the process engineering analysis of double-roll crushers are: (1) the maximum force required on the rolls, (2) the maximum power required, and (3) crusher capacity. These are generally based on the

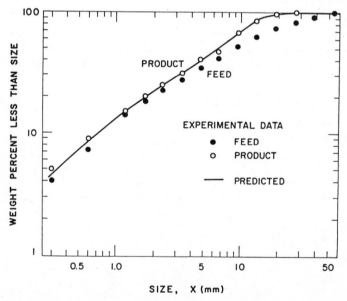

Figure 3.9 Actual and predicted product from a Kennedy Van Saun 3020 Double Corrugated Roll Crusher with gap setting of 13.5 mm and roll speeds of 170 and 85 rpm, respectively.

empirical experience of the manufacturer, although Gaudin[60] has described the concept of a capacity calculation based on the ribbon of packed solid that can be pulled through the gap by the roll rotation.

Wet Ball Mills.   The tumbling ball mill is the most widely used device for fine grinding of brittle materials on an industrial scale. Because of its simplicity, it is mechanically reliable, a very important consideration in continuous process streams. It is available in sizes ranging from small laboratory mills to industrial mills of 5 m diameter by 15 m long, or even larger, and drawing 4500 kW or more. The current use of wet ball mills for coal size reduction is to prepare coal–water slurries for direct combustion, although in time they may also find routine application for preparing slurry feed for coal gasifiers.[61]

Figure 3.10 shows a typical ball mill used for the wet grinding of coal. It consists of a horizontal steel cylinder with end walls, which is partially filled with steel balls. The cylinder rotates on trunnion bearings at each end, and the feed of coal–water slurry enters through the trunnion at one end and leaves through the trunnion at the other end, normally discharging into a sump. As the cylinder rotates, it carries the balls upward. They tumble in the mill and break lumps and particles of coal by impact between balls. The cylinder is of thick steel to withstand the stresses created by the tumbling balls. In addition, abrasive and corrosive action in the mill causes wear of the internal mill surface and the balls. The cylinder has internal steel liners several inches thick, which can be replaced after some years of operation.

Although the wet ball milling of coal–water slurries is a relatively recent development, ball milling historically has been one of the most important unit operations in mineral engineering. Thus, the construction of mathematical simulation models of ball mills and ball mill circuits has reached a fairly advanced stage, and their process engineering application is the subject of a

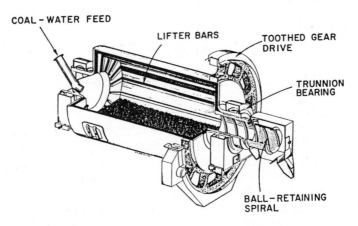

Figure 3.10 A wet overflow ball mill at rest.

recent textbook by Austin et al.[62] As discussed in this text, ball mills are retention devices; therefore, the basic elements that comprise the calculation of the transfer parameters d_{ij} are the primary progeny fragment distribution $B_{i,j}$, the specific rates of breakage S_i, and an expression of the time of grinding. For batch mills, the expression would simply be the grinding time t, and the d_{ij} for first-order grinding and constant $B_{i,j}$ would be

$$d_{ij} = \begin{cases} E_i, & i = j \\ \displaystyle\sum_{k=j}^{i-1} c_{ik}c_{jk}(E_k - E_i), & i > j \end{cases} \tag{3.12}$$

where $E_i = e^{-S_i t}$ and

$$c_{ij} = \begin{cases} \displaystyle\sum_{k=1}^{j-1} c_{ik}c_{jk}, & i < j \\ 1, & i = j \\ \dfrac{1}{S_i - S_j} \displaystyle\sum_{k=j}^{i-1} S_k b_{ik}c_{jk}, & i > j \end{cases}$$

In a manner analogous to the treatment of batch versus continuous chemical reactors, the product from a continuous ball mill at steady state is computed as the batch grinding product integrated over the residence time distribution of the continuous mill.

$$p_i = \int_0^\infty p_{i\text{BATCH}}\varphi(t, \tau)dt$$

Thus for batch grinding as described above, the d_{ij} for a continuous mill would be Equation 3.12 but with

$$E_i = \int_0^\infty e^{-S_i t}\varphi(t, \tau)dt \tag{3.13}$$

Although not necessary, it is convenient to have a mathematical expression for $\varphi(t, \tau)$ for use in Equation 3.13. For a discussion of residence time distribution models for ball mills, see the paper by Rogers and Austin.[63]

Factors that affect the power requirements, product size, and tonnage per hour of a wet ball mill for a given size distribution of feed include mill size, mill operating conditions, and grindability characteristics of the mill feed. Austin et al. describe the methodology that incorporates these factors into process engineering simulations. This methodology has been demonstrated to give accurate process engineering information, including product size distributions, as illustrated in Figure 3.11.

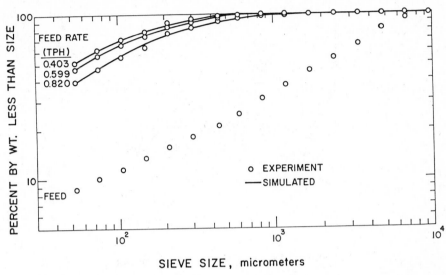

Figure 3.11   Product size distributions of Pittsburgh #8 coal ground in a Kennedy Van Saun 0.9 m × 1.5 m wet ball mill.

Additional factors can influence the performance of wet ball mills for coal grinding, particularly for the production of high density coal–water fuels. The most important of these is slurry rheology. It has been found[64–66] that ball milling at a high concentration of solids in the water creates rheological conditions that can cause breakage to become nonfirst order, gives broader primary progeny fragment distributions (more fines) than would normally occur at lower solids densities, and gives lower mill power requirements. Under these circumstances, mill simulation models can still be used for process design calculations, but the models must generally be more detailed than the ones presented here.[67]

Air-swept Ball Mills.   The most widely used equipment for dry grinding coal to the sizes prescribed for dry pulverized coal firing of furnaces, boilers, and kilns are the air-swept vertical types of mill (see below). There are some advantages to the use of air-swept ball mills for coal grinding, such as lower maintenance costs and the ability to handle strong or abrasive coals, and several manufacturers supply such systems. Figure 3.12 shows a typical air-swept ball mill for the dry grinding of coal. The characteristic feature of air-swept ball mills is that, unlike wet ball mills, there is no flow of bulk powder out of the mill since the ground coal is removed entirely in suspension in the air stream. The product coal–air suspension can be used immediately for direct firing of a furnace or boiler, or it can be passed through a cyclone and filter system to recover and store the coal. Air sweeping is used to dry the coal with hot entering air and to

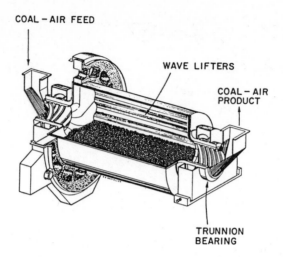

Figure 3.12 An air-swept ball mill at rest.

keep the size distribution in the mill relatively free of fines to prevent nonfirst-order grinding due to the slowing-down effect. Such mills have large feed and exit trunnions to reduce pressure drop, since high pressure drop means high energy use for fan power to move the air through the mills. They are usually operated at relatively low ball loadings, for example, 25 volume percent of the mill versus 35–40 percent for wet ball mills. The air flow rate is adjusted to give an optimum hold-up of coal for the feed rate being used. These mills are normally operated in closed circuit with an air classifier. Product size specification is obtained by adjustment of the air classifier with return of the coarser product to the mill feed. Air-swept ball mills can also be designed as double-ended, where air and coal enter and exit at both ends of the mill. This is accomplished by use of a partition plate in the trunnion at each end to separate the entering and exit streams. Double-ended mills employ two air classifiers, one for each end, and offer the advantage of reducing fuel distribution problems in large power plants. Air-swept ball mills are available in sizes ranging from pilot plant to industrial mills of 4 m diameter by 6 m long and drawing 1700 kW of power.

A model has been developed to aid in the process engineering of air-swept ball mills[68] that is similar in many respects to the one given earlier for wet ball mills. Since the air-swept ball mill is a retention device, the model incorporates the concepts of specific rates of breakage, primary progeny fragment distribution, and residence time distribution. Tests of an air-swept mill have shown that to a reasonable approximation the mill is fully mixed, so that the residence time distribution can be estimated as that for a fully mixed reactor. Additional assumptions concern the effect of air sweeping. These are: the powder hold-up in the mill is uniform along the mill length; a constant fraction η of the powder hold-up mass W is exposed to the air stream per mill revolution; material of size i exposed to the air stream has a probability c_i of being retained in the mill and hence a probability $1 - c_i$ of being swept into the product. On the basis of these assumptions, confirmed experimentally, the progeny fragment distribution for an air-swept ball mill can be calculated from

$$d_{ij} = \begin{cases} \dfrac{\eta\omega(1 - c_i)}{\eta\omega(1 - c_i) + S_i}, & i = j \\[4mm] \dfrac{S_j}{\eta\omega(1 - c_j) + S_j} \displaystyle\sum_{k=j+1}^{i} d_{ik}b_{jk}, & i > j \end{cases} \tag{3.14}$$

where ω is the mill rotational speed. For an estimated value of W the mill product rate F is computed from

$$F = W\eta\omega \sum_{i=1}^{n} p_i/(1 - c_i) \tag{3.15}$$

Factors affecting the power requirements, product size, and tonnage per hour of an air-swept ball mill for a given size distribution of feed are similar to those that apply to the wet ball mill.[68] Note Figure 3.13.

Vertical Mills.   Figure 3.14 shows one type of vertical mill, known as a ball-and-race mill. Coal falls by gravity down a central pipe onto a rotating table. The centrifugal action of the table throws the coal outward so that it passes

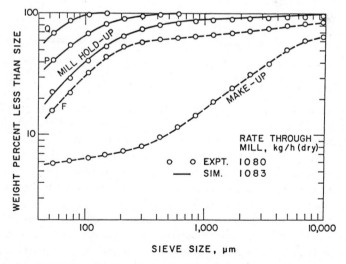

Figure 3.13   Comparison of experimental and simulated results for the grinding of a Belle Ayre-Wyoming coal in a Kennedy Van Saun 1.1 m × 1.5 m air-swept ball mill in closed circuit with a 0.7 m diameter twin cone classifier.

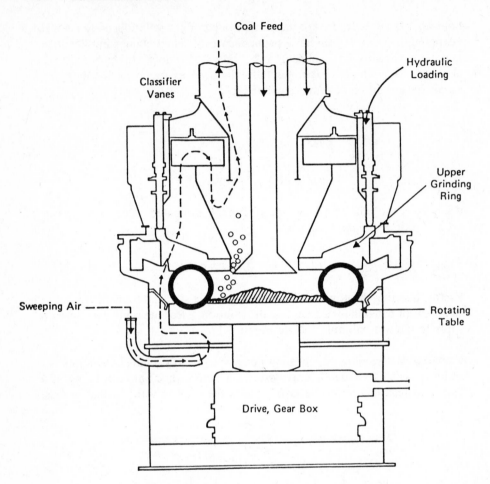

Figure 3.14 An industrial ball-and-race mill.

under heavily loaded balls rolling in a rotating race, subjecting coal in the race to high crushing forces. The ground coal passes over the rim of the rotating table and is swept up by air flowing at high velocity through the annulus between the rim and the mill housing. As the air expands in the space above the grinding zone, larger particles fall back onto the table; the remaining coal–air suspension enters an air classifier, usually an integral part of the mill housing as shown, where the finer particles pass out in the air stream and the coarser particles return to the table to repeat the cycle. For coal that contains large lumps of a strong dense mineral such as pyrite, vertical mills can be designed so that relatively large crushed lumps will fall through the annulus to give a reject stream leaving at the mill base. As with air-swept ball mills, heated sweeping air is used to dry the coal and keep the size distribution in the grinding zone relatively free of fines. The slowing-down effect is aggravated by too much free moisture, so the operation can be adversely affected by wet, sticky coal.

There are a number of variations of this type of mill offered by different manufacturers. The important differences in design insofar as the grinding action is concerned are illustrated in Figure 3.15. The table carrying the bottom grinding element rotates in all cases, and the major force between the roll or ball grinding element and the bottom grinding element is produced by hydraulic or mechanical action, not by the weight of the roll. Vertical mills range in size from pilot plant to industrial scale; the largest ball-and-race mill is 7.85 m tall, has a race 3.3 m in diameter, uses ten 0.98 m hollow grinding balls under 89 kN of crushing load per ball, and has a power requirement of 570 kW.

A simulation model has been developed to air in the process engineering of vertical mills and, to date, has been demonstrated for ball-and-race and roller-race mills,[69-72] that is, mills whose grinding elements are configured as Figure 3.15(e) and (f), respectively. Since there is a measurable hold-up of powder (and thus a measurable retention time) in the grinding zone, this model treats vertical mills as retention devices. The concepts of specific rates of breakage, primary progeny fragment distribution, and residence time distribution by regarding the mill as a fully mixed reactor are incorporated. The falling of particles out of the air stream and back to the table is described by a set of internal classification numbers c_i, where c_i is defined as the weight fraction of size i material in the stream returned to the table. The mass rate of material passing out of the grinding zone is denoted F'. Using these concepts and definitions, the model is constructed as a steady-state mass size balance equation for a mill without an air classifier. The solution gives a progeny fragment distribution for vertical mill grinding that can be written as

$$d_{ij} = \begin{cases} \dfrac{1 - c_i}{1 - c_i + (A_i/F')}, & i = j \\[4mm] \dfrac{(A_j/F')}{1 - c_j + (A_j/F')} \displaystyle\sum_{k=j+1}^{i} d_{ik}b_{kj}, & i > j \end{cases} \tag{3.16}$$

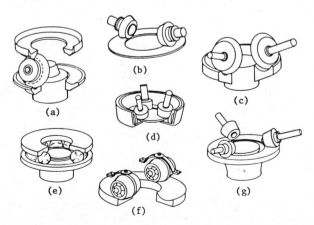

Figure 3.15 Assorted ball-and-race and roller-race designs.

where $A_i = S_i W$ and the product rate F from the mill is related to F' by

$$F'/F = \sum_{i=1}^{n} p_i/(1 - c_i) \tag{3.17}$$

This type of mill has the added feature that the breakage values depend on the amount of powder in the race. A higher flow rate of solid through the mill gives a thicker bed, more power is required to drive the balls or rollers through the bed, and breakage is consequently greater. The inclusion of the external classifier is accomplished as described in the next section.

Factors affecting the power requirements, product size, and tonnage per hour of a vertical mill for a given size distribution of feed include race diameter, mill operating conditions, that is, crushing load and rotational speed, grindability characteristics of the feed, and rate of air sweeping. The methodology for incorporating these factors into process engineering simulations is described in Austin et al.[70-72] and has been successfully demonstrated, as illustrated in Figure 3.16.

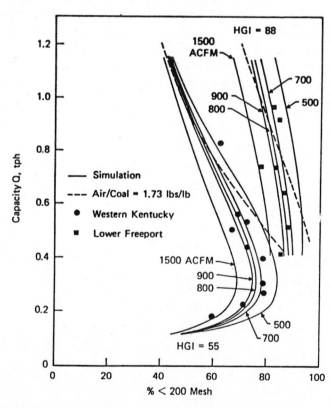

Figure 3.16 Comparison of simulated and experimental capacities and product fineness for a Babcock and Wilcox E1.7 ball-and-race mill.

Closed Circuit Operations.   Simulation models based on the transfer parameter concept are readily extended to treat closed-circuit size reduction operations. Consider the three common closed-circuit configurations in Figure 3.17. In all cases g_i and q_i give the weight fraction of size i material in the circuit feed and product, respectively. The s_i are size selectivity parameters for the classifiers (or screens), with s_{1i} and s_{2i} being values for the pre- and post-classifiers of the combined circuit, respectively.

Rogers[51] has presented a general algorithm that can be used for simulating open circuit plus these three closed circuits for known d_{ij}, s_i, and g_i. This algorithm is

$$q_i = (1 - s_{1i})g_i + (1 - s_{2i}) \sum_{j=1}^{i} s_{1j}g_i e_{ij}, \quad 1 \leqslant i \leqslant n \qquad (3.18)$$

where

$$e_{ij} = \begin{cases} d_{ii}/(1 - s_{2i}d_{ii}), & i = j \\ \dfrac{d_{ij}/(1 - s_{2j}d_{jj}) + \displaystyle\sum_{\substack{k=j+1 \\ i>j+1}}^{i-1} d_{ij}s_{2k}e_{kj}}{1 - s_{2i}d_{ii}}, & i > j \end{cases} \qquad (3.19)$$

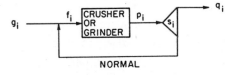

NORMAL

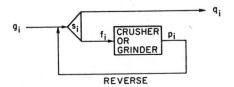

REVERSE

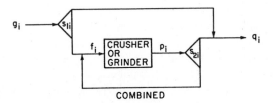

COMBINED

Figure 3.17 Three possible configurations for closed-circuit crushing and grinding devices.

Simulations are performed by letting g_i be the vector of circuit feed size distribution values, open or closed circuit, and by supplying the algorithm with pre- and postclassifier selectivity values as follows:

	s_{1i}	s_{2i}
Open circuit	All equal to 1	All equal to 0
Normal closed circuit	All equal to 1	All as input
Reverse closed circuit	Equal to postclassifier values	All as input
Combined closed circuit	All as input	All as input

3.5   EFFECT OF COAL TYPE AND TYPE OF BREAKAGE ACTION ON COAL BREAKAGE PROPERTIES

In Section 3.2.2 it was explained that a particle subjected to compression or shear force will contain regions of tensile stress, which initiate brittle fracture from stress-activated flaws. There are five main methods by which stress is applied to particles in machines used for size reduction:

1. Direction compression of the particle between hard surfaces,
2. Compression by relatively light impacts from grinding media tumbling in a bed of powder,
3. Impactive compression by hammers traveling at high speeds,
4. Shear of the particle between surfaces moving at high relative speeds, and
5. Impactive compression of particle on particle, produced by carrying the particles in a highly turbulent gas stream.

In reality, all machines subject a particle to compression and shear simultaneously, because compressive force is applied at an angle that automatically leads to shear forces.

It is known that the product size distributions resulting from size reduction of a coal can be affected by the method of stress application. However, to define the breakage properties of a coal by HGI or WI results in an entirely empirical arbitrary number that gives no information about the size distribution of coal particles produced by size reduction. Size distributions produced by the Hardgrove test do not vary uniformly with HGI,[73] although the variance (spread) of the size distribution generally increases with increasing HGI.

A more detailed analysis of coal breakage properties can be achieved by using the concepts of primary progeny fragment distribution, specific rates of breakage, and by-pass values. Brown[74] claims that the primary B values can be explained from a random statistical flaw model. However, Austin and Klimpel[75] show that there are mathematical errors in this treatment that invalidate the result. Thus, while B values have a form that is consistent with concepts stemming from fracture theory, that is, Equation 3.7, their values must be measured for each coal type and each type of size reduction device. Higher values for γ and lower values for Φ of Equation 3.7 indicate that product size distributions will contain a greater proportion of fine particles; B values are relatively insensitive to β. Similarly, attempts have been made to deduce the relationship between specific rates of breakage and particle size from simple geometrical concepts.[76,77] These attempts have been successful only in that the consensus result,

$$S_i \propto x_i^\alpha \tag{3.20}$$

where α is an adjustable (positive) parameter, is one that is observed in practice. Higher values for specific rates of breakage indicate that the coal is more easily broken; lower values of α mean that fines will break more rapidly to give product size distributions containing a greater proportion of fine particles. Finally, by-pass values for coal have, to date, been reported only for double-roll crushers. However, the evidence is that the primary by-pass values a_i can be expressed as

$$\text{Fraction by-passing} = 1 - a_i = \frac{1}{1 + \left(\dfrac{x_i/x_g}{d_{50}/x_g}\right)^{6.6}} \tag{3.21}$$

where d_{50} is an adjustable parameter and x_g is the gap setting in millimeters. Higher values for d_{50}/x_g means that the crusher passes more particles through without breakage. Secondary by-pass parameters have been shown to be related to a_i values.

Repeated application of the concepts of primary progeny fragment distributions, specific rates of breakage, and by-pass values has led to the measurement of breakage parameters of coals in double-roll crushers, ball mills, ball-and-race mills, and roller-race mills, all of which apply stress to particles in different ways. The purpose of this section is to demonstrate the observed patterns of change of these coal breakage properties as a function of coal type and the type of breakage action caused by these machines.

3.5.1 Double-Roll Crushers

Tables 3.2 and 3.3 give data for a number of coals tested by Austin et al.[58] in a laboratory-scale smooth double-roll crusher. The data in Tables 3.2 and 3.3 show that the weaker coals tend to give a higher natural proportion of fines upon crushing, whereas the stronger coals tend to give a lower proportion. This would indicate a correlation between breakage parameters and HGI, but the effect is obscured by the ash content. Figure 3.18 shows the values of d_{50}/x_g plotted versus percentage volatile matter for the coals. Also shown is the correlation of hydraulic shape factor with the percent volatile matter of coals, as

TABLE 3.2 Coals Tested in a Double-Roll Crusher

Coal	ASTM Class	HGI	H_2O	Ash	V.M.
			(as received basis wt. %)		
Shamokin (PA)	Anthracite	35	2.1	8.9	8.4
Illinois No. 6	Bituminous	46	8.9	18.0	33.1
Ohio No. 9	Bituminous	54	2.3	23.5	34.3
Western Kentucky No. 9	Bituminous	55	2.3	10.2	36.8
Belle Ayre South (Wyoming)	Sub-bituminous	58	24.5	5.8	33.9
Pittsburgh No. 8	Bituminous	65	3.2	21.3	28.4
Upper Freeport (PA)	Bituminous	72	3.6	10.4	31.5
Lower Freeport (PA)	Bituminous	88	0.9	10.2	27.4

TABLE 3.3 Parametric Description of Breakage Properties for Coals Tested in a Double-Roll Crusher

Coal	Φ	γ	β	d_{50}/x_g	λ
Shamokin	0.30	1.05	5.0	1.70	6.6
Illinois No. 6	0.36	0.81	3.0	1.66	6.6
Ohio No. 9	0.33	0.95	4.2	1.93	6.6
Western Kentucky No. 9	0.47	1.05	4.0	1.81	6.6
Belley Ayre South	0.49	1.17	4.0	1.70	6.6
Pittsburgh No. 8	0.32	0.81	3.6	1.60	6.6
Upper Freeport	0.39	0.96	4.0	1.56	6.6
Lower Freeport	0.50	1.05	4.5	1.54	6.6

determined by Austin et al.[78] using a liquid permeability technique. The value of hydraulic shape factor k is defined by $S_o = 6/kx$, where S_o is the hydraulic surface area per unit solid volume and x is particle size; $k = 1$ for a sphere or cube and is less than 1 for more flakelike particles. As expected, there is a general inverse correlation between d_{50}/x_g and k. This is representative of the higher fraction that by-passes for more flakelike particles.

Rogers and Shoji[57] demonstrated that primary B values and by-pass values obtained for laboratory-scale smooth double-roll crushers adequately described coal breakage in an industrial-scale double-roll crusher operated over a wide range of test conditions. It is concluded that breakage properties of coals in double-roll crushers are material dependent and that the mechanisms of breakage in small- and large-scale crushers are the same.

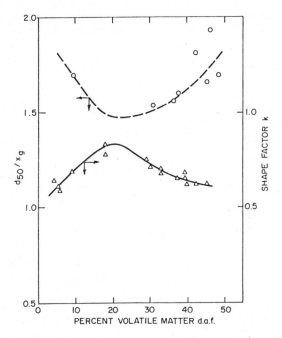

Figure 3.18   Variation of d_{50}/x_g and shape factor with volatile matter for coals.

3.5.2   Ball Mills

Figure 3.19 is a typical plot of specific rates of breakage versus particle size for coal grinding in a ball mill. The main features of this plot are the slope α, the size x_m at which S reaches a maximum value, and the sharpness of the bend-over in the S curve.

S values for ball mills can be fitted to the equation[79]

$$S_i = a(x_i/x_o)^\alpha Q_i, \qquad i = 1, n \qquad (3.22)$$

where a is an adjustable parameter, x_o is taken as 1 mm, and the Q_i are given by

$$Q_i = 1/[1 + (x_i/\mu)^\Lambda], \qquad i = 1, n \qquad (3.23)$$

The μ is an adjustable value and is the particle size at which Q_i has a value of 0.5. Λ is an index of how rapidly S_i values decrease with increasing size for size $> x_m$. The μ and x_m are related by

$$\mu = x_m[(\Lambda - \alpha)/\alpha]^{1/\Lambda} \qquad (3.24)$$

The fact that the S_i are a simple power function for smaller sizes has not been adequately explained theoretically, but it has been amply demonstrated by

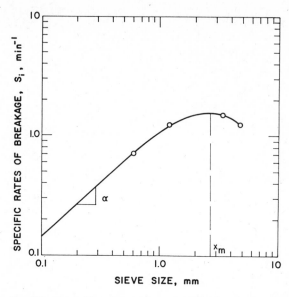

Figure 3.19 Specific rates of breakage versus particle size for a coal ground in a ball mill (S_i values plotted at top size of interval).

many experiments. For larger sizes, it is found that the rate of disappearance of material from a given top size interval is often not first order but appears to consist of a faster initial rate and a slower following rate. Some of the particles are too big and strong to be properly nipped and fractured by the balls, and the accumulation of fine material in the mill appears to cushion the breakage of these larger sizes. The Q_i are therefore introduced to allow for the slower mean rates of breakage of larger sizes with $Q_i = 1$ for small sizes and tending toward zero for larger sizes.

It has been found that the α of Equation 3.22 is constant for a given coal and independent of mill size or operating conditions. Thus α is a material characteristic that, along with Φ, γ, and β, provides an index of coal breakage properties. Austin et al.[80] has shown that for a given coal these characteristic parameters are the same for dry and wet (coal–water) ball milling, provided that the wet ball milling is performed on a slurry containing less than about 45 volume percent coal. The values of a, μ, and Λ are material dependent but also vary with mill size and operating conditions.[62] At otherwise identical conditions, a is higher for wet grinding than dry grinding of the same coal, indicating greater mill capacity for the former.

The coals listed in Tables 3.2 and 3.3 that have HGI values of 35, 54, 55, 58, and 88, respectively, were ground wet and dry in a 195 mm diameter batch ball mill to determine their primary B values and specific rates of breakage. It was found that the characteristic parameters γ, Φ, and α varied with the HGI of the coal as seen in Figure 3.20, and β was approximately constant at 2.8 for all of the

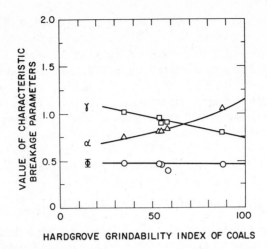

Figure 3.20 Variation of characteristic breakage parameters with Hardgrove Grindability Index for coals ground in a ball mill.

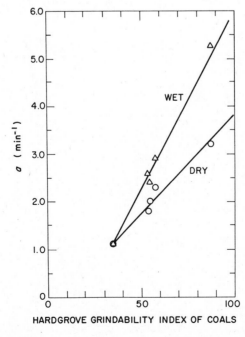

Figure 3.21 Variation of a of Equation 3.22 with Hardgrove Grindability Index for coals ground in a ball mill.

coals. The size at which S values are a maximum increased with increasing HGI, indicating that larger sizes of weaker coals are easier to grind than larger sizes of stronger coals. Λ was approximately constant at 3.0 for all of the coals. The value of a increased approximately linearly with HGI for both wet and dry grinding, as seen in Figure 3.21. Note in this figure that the difference in a values for wet versus dry grinding also increased with HGI. These results show that weaker coals give greater mill capacity and a larger proportion of fine particles than do stronger coals, and that the difference in mill capacity for wet versus dry grinding increases with increasing HGI. The test procedures used to generate these breakage property data are described in Austin and Bhatia.[81]

3.5.3 Ball-and-Race and Roller-Race Mills

Ball-and-race and roller-race mills mechanically are quite similar, and certain characteristics of coal breakage in these mills are similar. However, experiments have shown that the breakage actions caused by the ball-and-race versus roller-race configurations give different breakage parameters for a given coal and different variations of breakage parameters with coal type.

As with ball mills, coal grinding in ball-and-race and roller-race mills is first-order provided that particles are not too large to be acted on by the balls or rolls, and fines are not allowed to accumulate in the grinding zone and cause a slowing of breakage. Figure 3.22 is a typical plot of specific rates of breakage

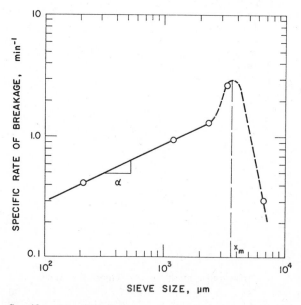

Figure 3.22 Specific rates of breakage versus particle size for a material ground in a roller-race mill.

versus particle size for first-order grinding in these mills. As seen, S values are a simple power function with size for small sizes, but deviate from the power function form for larger sizes of particles. It is convenient to express the S values in the form of Equation 3.22 where the a, x_o, α, and Q_i are defined as before. As Figure 3.22 indicates, the Q_i for these mills are 1 for small sizes, become greater than 1 for larger sizes, and decrease to below 1 for the largest sizes. Again, the fact that S_i are a simple power function with size has not been adequately explained theoretically, but it has been amply demonstrated by many experiments with these types of mills.

For a given machine type (ball-race versus roller-race), the α of Equation 3.22 is constant for a given coal and independent of the size or operating conditions of the machine. Thus, α is a material characteristic that, along with Φ, β, and γ, provides an index of coal breakage properties. The size x_m for which S values are a maximum and, thus, the correcting factors Q_i are coal independent but are different for ball-and-race versus roller-race mills. For each of the two machine types, x_m varies approximately linearly with the size of the balls or rolls, and for

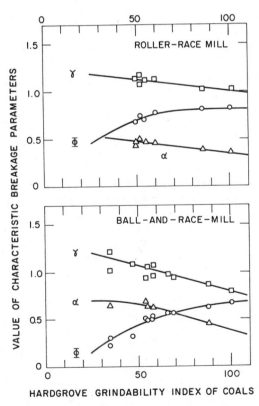

Figure 3.23 Variation of characteristic breakage parameters with Hardgrove Grindability Index for coals ground in roller-race and ball-and-race mills.

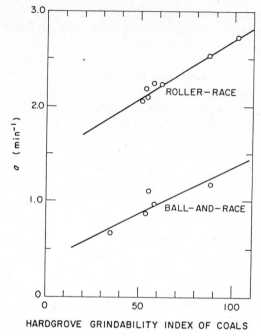

Figure 3.24 Variation of a of Equation 3.22 with Hardgrove Grindability Index for coals ground in laboratory scale, ball-and-race and roller-race mills.

TABLE 3.4 Empirical Correcting Factors Q_i for Coals Ground in Laboratory-Scale, Batch Ball-and-Race and Roller-Race Mills

Upper Size of Size Interval (mm)	Q_i Laboratory Ball-and-Race Mill	Q_i Laboratory Roller-Race Mill
1.19	1.00	1.00
1.68	1.03	1.02
2.38	1.11	1.23
3.36	1.44	1.89
4.76	1.86	2.27
6.73	1.18	1.53
9.51	0.26	0.69
13.2	0.004	0.31
19.0	—	0.13
27.2	—	0.05

a given coal and machine type the value of a depends on the machine size and operating conditions and the moisture content of the coal at the test conditions.

S and B values for coals have been measured for ball-and-race and roller-race milling.[69,72] This was done using laboratory-scale, batch versions of these two machine types. The laboratory-scale ball-and-race mill was a standard Hardgrove testing machine and the laboratory-scale roller-race mill is described by Austin et al.[72] The Φ, γ, α, and a varied consistently with HGI, as seen in Figures 3.23 and 3.24. β was 5 for all coals in both machines. The variation of the Q_i with size for each machine is given in Table 3.4. These data show that coal grinding in either type of machine will give higher mill capacity and a high proportion of fine particles for weaker versus stronger coals, but that the actual size distributions produced for a given coal will tend to be somewhat different.

3.5.4 Comments

Figure 3.25 compares B values for the size reduction of coals for each of these machine types. It is clear that weaker coals tend to give a larger proportion of fine particles than stronger coals for all of these different types of breakage actions. The differences in B values from one machine type to another implies that each causes size reduction to occur by different mechanisms, although B values do appear to be somewhat similar for stronger coals. It must therefore be concluded that measured breakage properties of coals will depend not only on the method by which stress is applied to particles but also on the actual design of a machine that employs a specific method.

It is important to note that although the Φ, γ, and α values for ball mills and vertical mills are directly comparable, the data given for a versus HGI are not indicative of capacity differences for industrial scale machines. The use of laboratory generated a values for ball mill and vertical mill scale-up and design is described by Austin et al.,[62,71] respectively.

3.6 APPLICATIONS OF COAL SIZE REDUCTION

The objectives of size reduction in industrial coal utilization are either to liberate coal from its impurities prior to separation (coal cleaning) or to meet product specifications for coal size distributions. In either case the degree of size reduction necessary is determined by the requirements of the processes downstream from the size reduction step. For some of these processes, the required feed size of coal can be rather loosely defined, whereas for others, proper size reduction of the coal feed stock is vital to successful process performance. As demonstrated in this section most large scale uses of coal, including coal cleaning, combustion, gasification, carbonization (coking), and pipeline transportation, fall into this latter category. For further discussions of these and other uses of coal, see Austin and Bhatia,[81] Luckie,[82] Babcock and Wilcox,[83] Leonard et al.,[84] and Anonymous.[85]

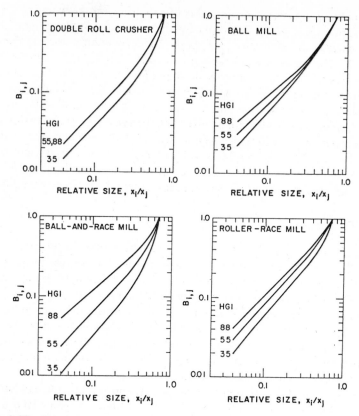

Figure 3.25 Primary progeny fragment distributions $B_{i,j}$ for coals having Hardgrove Grindability Index values of 35, 55, and 88.

3.6.1 Coal Cleaning

Coal mined today represents the deposition of phytogenic material 50 million to 350 million years ago. The resulting horizontal strata, referred to as seams, vary in thickness from several inches to several hundred feet. These seams are usually separated by varying thicknesses of sedimentary rocks such as shales, clays, sandstones, and sometimes even limestone. When combined with coal during mining, these materials represent impurities in the coal. Impurities can also be inherent in the coal (e.g., organic sulfur) or be in the form of fine dispersed particles locked in the coal matrix.

Modern *coal cleaning* is the removal of impurities from coal to produce a fuel or feedstock whose composition meets specifications required for its use. This is achieved by creating individual particles whose composition is predominantly coal or predominantly impurities (refuse) and separating them. The creation of particles prior to separation is termed *liberation*, and it is achieved through size reduction. Size reduction also creates particles that contain coal and refuse; these particles are referred to as *middlings*. Thus, separation, in practice, is the dividing of the particles created by size reduction into clean coal, refuse, and middlings.

Impurities in coal are numerous, but the largest number by far have specific gravities greater than coal. Therefore, the dominant method for separating the liberated coal is by gravity concentration, which relies on two physical property differences—size and specific gravity—to effect the separation of coal from impurities. Coal cleaning by gravity concentration is typically practiced for coal in the size range 200 mm by 0.5 mm, and a number of different unit operations are used. Each type of unit operation is designed to clean a specific size range of coal within the 200 mm by 0.5 mm interval; thus the degree of cleaning achievable with a particular unit operation is intimately related to the size of coal being cleaned. The cleaning of finer (−0.5 mm) coal can be achieved by froth flotation, oil agglomeration, or chemical cleaning methods. Froth flotation and oil agglomeration processes rely on differences in the surface properties of coal and impurities to affect separation. The intent of chemical cleaning is to dissolve impurities remaining in coal after physical cleaning. Each of these processes is most effective on a specific size range of particles smaller than 0.5 mm so that the degree of cleaning achievable with a particular process depends on the size of coal being cleaned.

Successful coal separation and recovery can be achieved only when the necessary degree of liberation has been attained. Thus, liberation is the backbone of any separation process. The interaction between the degree of liberation and the degree of size reduction necessary to obtain such liberation will, in most cases, determine which separation process can be used.

It is difficult to establish the capital and operational costs for coal cleaning, due to such things as market specifications, percentage of marketable coal in the raw coal, proportions of coarse, intermediate, and fine material in the raw coal, and separation method employed. It is known, however, that more impurities are liberated with finer coal size reduction, and, as particles become smaller, the cost of cleaning increases exponentially.[82] It is this trade-off that defines the size reduction for liberation objective—to achieve the maximum liberation at the coarsest possible product.

3.6.2 Coal Combustion

Coal combustion is the result of reactions between the combustible constituents of coal and oxygen, and the goal of size reduction is to produce coal particle size distributions that cause combustion to take place in as efficient and economical a manner as possible.

There are essentially two different coal burning schemes in current widespread use: *suspension firing* and *fuel-bed firing*. Both are sensitive to the degree to which the coal fuel is size reduced. For suspension firing, coal is ground, mixed with air, and blown into the furnace. For this type of firing the retention time of the coal fuel in the furnace is short, on the order of 1 second. Therefore, the fuel must have a high specific surface area (cm^2/g) for rapid ignition and must contain a minimum of coarse particles to avoid carbon loss through incomplete combustion. The actual retention time, and thus the required fuel size distribution, is influenced by the degree of turbulence in the combustor. Turbulence is a condition in which fuel and air whirl and eddy in irregular paths instead of flowing in streamlines; it can be mechanically or aerodynamically induced. Generally, more highly turbulent conditions are found in utility boilers than in rotary kilns; thus, fuels for kilns must generally be finer than those for utility boiler fuels. Coal type also has an influence on the degree of size reduction required for suspension firing. High-volatile coals ignite more readily than low-volatile coals. Since the ease of ignition dictates to some extent the fuel size distribution required, it follows that high-volatile coals need less fine pulverization than do low-volatile coals. Industrial experience shows that minimum carbon losses can be achieved for suspension firing of coal in boilers, kilns, and other furnaces if the fuel contains 1 weight percent or less of particles larger than 50 mesh (297 μm). The influence of furnace type and coal rank on the required amount of fine particles in the fuel is illustrated in Table 3.5.[83]

In fuel-bed firing, relatively coarse coal is pushed, dropped, or thrown onto a grate by a mechanical device known as a stoker to form a bed through which an upward flow of air passes. Fuel-bed firing gives a relatively long fuel retention time (several minutes); therefore, the size distribution of fuel is much coarser (about 95 weight percent <20 mm) than for suspension firing. Thus, fuel preparation costs are lower, but carbon losses are greater (4 to 8 percent for fuel-bed versus less than 0.4 percent for properly designed suspension firing). Fuel-bed combustion is therefore popular for industrial applications where the cost of relatively inefficient fuel usage is offset by lower fuel preparation costs. (For large-scale coal combustion, such as for steam generation in utility boiler plants, the converse situation generally holds true.) The most popular fuel-bed firing system is the spreader-stoker, which feeds coal into the furnace in a uniform

TABLE 5   Coal Size Distribution Data for Suspension Firing of Furnaces[a]

| | % through 200 U.S. Sieve[b] | | | | | |
| Type of Furnace | Fixed Carbon, (%) | | | Fixed Carbon Below (69%) | | |
	97.9–86	85.9–78	77.9–69	Btu/lb above 13,000	Btu/lb 12,900– 11,000	Btu/lb below 11,000
Water-cooled	80	75	70	70	65	60
Cement kiln	90	85	80	80	80	—
Metallurgical	(As determined by process, generally from 80 to 90%)					

[a]Extremely high ash content coals will require finer grinding than indicated.
[b]ASTM classification of coals is by rank.

spreading action above the combustion zone, allowing fine material to be burned while suspended in the upward flow of reactive gas. Larger, heavier particles fall to the grate and are burned in a thin bed. Coal size is probably the most important characteristic when firing with this stoker.[84] If the coal in the bed is too coarse, it will require too long a retention time and will result in unacceptable carbon losses. Closely sized coal is also not desirable, since it can tend to pile up on one portion of the grate. Too much fine coal may produce satisfactory burning conditions but result in a very high cinder carryover. Thus, the fuel must contain coarse particles as well as a sufficient amount of finer material (e.g., 20–30 weight percent <6 mm).

Coal size reduction is also important in some of the newer coal combustion techniques. One of these is *fluidized-bed* combustion. Fluidized-bed combustors support a bed of particles on a grate and the bed is kept in turbulent motion by bubbles created by air forced into the bed from below. Although the bed is ballasted (i.e., only about 1 percent of the particles are coal and the remainder is an inert material), all of the particles introduced as feed are quickly heated by the churning action of the bed. The retention time of fuel in this type of combustor is generally less than for fuel-bed firing; therefore, the fuel must be slightly finer (1 to 5 mm). A valuable feature of fluidized-bed combustion is that the inert bed material may consist of limestone. The calcium oxide from the calcination of the limestone absorbs the sulfur dioxide being released from the coal during combustion, thus reducing the direct cost of flue gas desulfurization.

Another relatively new development in coal combustion is the suspension firing of high density *coal–water* slurries. It is generally accepted that coal–water slurries represent a feasible option for the suspension firing of utility boilers, provided that the slurry be of high solids concentration (about 70 weight percent coal, or higher), pumpable at least over short distances, and stable (i.e., that the coal particles remain in suspension) at least for a short period of time. This can be achieved, in principle, by grinding to an appropriate particle size distribution and with chemical additives. The additives comprise about 1 percent of the slurry by weight and are used primarily to lower viscosity and promote stability. The coal particle size distribution must be suitable for suspension firing and, to achieve the high solids concentration, must contain a sufficient number of fine particles to fill the void spaces between larger particles. Figure 3.26 is a plot of coal size distributions for high density coal–water slurries suitable for utility boiler firing. Note that the coarser of these size distributions give weight percent passing values for the 50 and 200 mesh sizes similar to Table 3.5 for dry pulverized fuel. However, these are the only similarities—the dry pulverized fuel might also measure 40 to 45 weight percent <400 mesh (37 μm), whereas the coal–water slurries have 55 weight percent <400 mesh. This is an important distinction and is the key to why particle size distributions and their production are important in coal–slurry fuel preparation. It is noted that coal–slurries are currently being examined for various industrial applications (e.g., kilns, hot mix asphalt plants, and blast furnaces) and as diesel and turbine fuel, with the latter two requiring very fine coal size distributions (<30 μm).

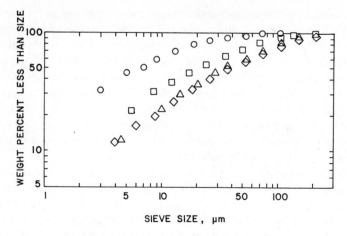

Coal HGI	Slurry Weight (Percent Solids)	Slurry Apparent Viscosity (cp)
◯ 50	72.4	960
☐ 46	70.6	1820
△ 50	68.5	1740
◇ 45	71.3	1940

Figure 3.26 Size distributions of coal–water fuels suitable for suspension firing.

3.6.3 Coal Gasification

Simply stated, *coal gasification* is the burning of coal but in a controlled deficiency of oxygen. The objective of coal gasification is to convert coal into an environmentally acceptable gaseous fuel or feedstock, and a variety of coal gasification processes exist. The first step involved in any coal gasification process involves size reducing the raw coal to meet the physical requirements of the gasifier. A common way of classifying types of gasifiers is according to the manner in which the coal and various reactive gases (i.e., steam and oxygen or air) are brought into contact. It is this feature of a gasification process that determines the degree to which the coal feedstock must be size reduced. In *fixed-bed* gasifiers, a bed of coarse (3 to 50 mm) coal is supported on a grate and the reactive gases are contacted as they sweep upward through the coal bed. These gasifiers are normally cylindrical in shape and fresh coal is added at the top through a lock hopper mechanism. Because fresh coal moves downward during the process, these gasifiers are also referred to as moving-bed gasifiers. Fixed-bed gasifiers have a low tolerance for fines, because fine coal can be carried out with the product gas stream. Thus, the feed must be precisely size reduced and screened to assure proper gasifier operation. In *fluidized-bed* gasifiers, reactive gases flow upward through the mass of coal particles at a sufficiently high velocity to keep them in suspension, but not so high as to carry solids out of the gasifier. The size of the coal feed is typically 10 to 100 mesh and, as with fixed-bed gasifiers, must be relatively free of fines. In *entrained-flow* gasifiers, coal particles entrained in a flow of reactive gases are introduced into the gasifier through one or more nozzles or burners, and the coal concentration is kept low to prevent particle–particle contact. This mode of feeding gives low coal retention times (less than 0.5 seconds); therefore, the coal feed is finely ground (40 to 70 weight percent smaller than 200 mesh) to obtain high conversion efficiency and avoid excessive gas flow requirements.

3.6.4   Coal Carbonization (Coking)

The destructive distillation or *carbonization* of certain coals occurs when the coals are heated in an atmosphere excluding oxygen, so as to expel their volatile content. This process decomposes the coal into gaseous and solid fractions, including a fused solid residue known as *coke*. A major use of coke is in the production of iron where the coke (termed *metallurgical coke*) acts as a source of carbon to provide the reducing agent in the conversion of iron oxides to elemental iron; acts as a fuel to generate high temperatures at which the iron oxide to iron reduction reaction is effectively accomplished; and functions as the primary solid constituent that supports the burden in blast furnaces. Thus, coke plays a fundamental role in blast furnace operation and as such must have certain physical and chemical properties: (1) strength, (2) an appropriate size to allow uniform gas flow through the furnace cross section, (3) uniform reactivity to reducing gases, (4) high carbon content, (5) low sulfur content to minimize iron contamination, and (6) low ash content. Unfortunately, there are very few coals that can be converted to a coke having these properties; therefore, it is generally necessary to blend two or more widely differing coals.

Coal size reduction plays an important role in the preparation of coal feed for metallurgical coke production. The strength of a coke is related to the size and strength of the parent coal or coals. (The feed to conversion processes are usually about 80 weight percent smaller than 3 mm with a slightly finer feed giving a stronger coke and a slightly coarser feed giving a weaker coke, for a given coal feedstock.) Thus, if the feed is to be a blend of a stronger coal and a weaker coal, achieving the desired coke strength will involve the controlled size reduction and blending of the two coals to give a feed with overall strength and size properties that will result in the desired coke strength. Additionally, virtually all coals suitable for coke production contain unacceptably high levels of sulfur and ash when mined. Thus, coking coals must be cleaned, a processing step that is intimately related to size reduction. Finally, higher quality cokes have higher bulk densities. For a given coal feed to the conversion process, the bulk density of the coke production can be altered by size reduction, with a finer coal feed giving a higher coke bulk density.

3.6.5 Coal Pipeline Transportation

The transport of coal in slurry form via pipelines has been established as being both technically and economically feasible in the United States and there are two operating domestic coal slurry pipelines. The longer of these (the Black Mesa pipline) transports a 50 weight percent coal-water slurry through a 0.46 m diameter pipeline over a distance of 455 km with the aid of four pumping stations. A description of this pipeline facility is given in Anonymous.[85]

Coal size reduction is of critical importance for coal pipeline transportation. A very coarse coal would require high slurry velocities in the pipeline to avoid deposition, but this would lead to unacceptable pumping costs and pipe wear. A very fine coal would require a more dilute slurry to avoid high pumping costs due to high slurry viscosity, but then the cost per ton of delivered coal would be unacceptably high. This would be caused by higher costs for size reduction, more water pumped per ton of coal, and higher slurry dewatering costs. The compromise between these extremes is a coal size distribution having a top size of 1.2 mm with a fines content of between 16 and 23 weight percent passing 44 μm. Experience has shown that this size distribution gives slurry stability in the sense that the fines provide sufficient slurry viscosity to hydraulically support the coarser coal and obtain relatively homogeneous flow.

3.7 COAL SIZE REDUCTION RESEARCH

It has been pointed out that there is very little fundamental understanding of the relationships between properties of coals and coal size reduction. Thus, there are areas where further investigation should prove profitable. Specific studies might include

- Defining the physical and chemical structures of coal as they relate to significant properties of size reduction,
- Defining the nature of the adhesion between organic and mineral matter and relating the nature of these bonds to size reduction,
- Improving the fundamental understanding of the mechanical properties of coal as they relate to size reduction,
- Determining shear and other moduli and their dependence on coal rank, material, and coal primary (chemical) structures; determining the applicability and utilization of various techniques to the determination of such moduli,
- Investigating the mechanism of fracture initiation and propagation in coal, and
- Developing more energy efficient size reduction processes.

There is also inadequate understanding and experimental information to construct *complete* process engineering design models. A major feature both in coal cleaning and in predicting the behavior of coal during combustion and conversion is the problem of describing liberation. Allowance has to be made for the effects of the heterogeneity of lumps or particles of coal, rock, pyrite, and all possible combinations thereof. This is at present treated primarily by a two-component approach, pure coal and rock, but it is known that coal–rock or coal–pyrite cannot be accurately treated as if complete liberation occurred on size reduction. Coal has to be described by a matrix of size and, for example, specific gravity or ash content. However, the knowledge of this matrix for a crushed coal does not automatically enable the calculation of the matrix for any other size distribution.

This is called the *liberation problem,* and its solution requires theoretical-experimental models to enable the prediction of the liberation matrix for any set of conditions from a knowledge of the distribution of, for example, pyrite measured under other conditions. One approach is the liberation model of Klimpel and Austin,[87] which basically consists of treating an original large mass of coal as having a random distribution of B component (e.g., pyrite) in an A matrix. Breaking is treated as the random placement of the volume size distribution over the volumes of B and A. The model is restricted because it allows for fracture of a maximum of only two B grains per fragment formed. This system uses a Monte Carlo simulation approach to predict the fraction of a mass of broken material of known size distribution that would be in each block of the size–pyrite content matrix, knowing the original grain size distribution of B in the AB matrix. The approach was experimentally validated by comminuting a synthetic coal made up of a known grain size distribution of pyrite hardened in an organic polymer, followed by float-and-sink analysis to construct the experimental matrix.

This Monte Carlo approach has been extended to three components by Bagga and Luckie,[88] using pyrite and silica in a matrix. However, it has two serious disadvantages: (1) It becomes numerically unstable and computationally unwieldy as the range of conditions is widened; and (2) for the same reasons, it can only be used to back-calculate the grain size distribution of B from an experimental matrix for very simple cases. Klimpel and Austin[87] also gave an analytical solution to the model with the further restriction that no more than one grain of B was contained per fragment of AB formed. This can only apply when size reduction has proceeded to the stage where significant liberation is occurring, and where the volume fraction of B in AB is small. Klimpel[89] has given the simple closed solutions of the equations for a single size of grain and for a grain size distribution following a Schumann distribution. The analytical form is particularly convenient for back-calculation of the original grain size distribution from experimental data, using a standard nonlinear regression routine. Klimpel claims[89] that it has been used routinely with success for ash in coals, providing the ash content of the total material is less than 10 percent by volume.

Although this model is restricted and requires further development, it appears to work well enough to enable the next step to be made. This would involve the treatment of size reduction as comminution and liberation of a suite of compositions ranging from pure *A* to pure *B*, each with its own breakage characteristics (*S* values and *B* values).

Research on the use of high-density coal–water slurries to replace fuel oil has opened new areas of research in the wet grinding of coal. As wet grinding to fine sizes is more efficient than dry grinding, it would appear economically attractive to produce the dense slurry in a single wet grinding operation. Wet ball mills are not normally operated at high slurry density because the grinding efficiency decreases due to high viscosity effects. The interest in high-density wet coal grinding in ball mills has led to renewed investigations of the rates of breakage as a function of slurry rheology,[64–66] mass transfer in the mills as a function of slurry density,[63] and correlation of slurry rheology with the size and size distribution of the powder.[90] Katzer et al.[91] have also examined the effect of chemical fluidity modifiers on slurry rheology and have patented[92] the use of anionic polymeric electrolytes as grinding aids for dense slurries in ball mills. The use of wet ball mills closed-circuited through high efficiency wet fine screens offers promise of an economic production of slurry.[93]

REFERENCES

1. F. C. Bond, *Br. Chem. Eng.* **6**, 378–391 (1965).

2. A. Nadai, *Theory of Flow and Fracture in Solids*, p. 89: McGraw-Hill, New York, 1950.

3. A. A. Griffith, *Philos. Trans. R. Soc. London, Ser. A* **221**, 163 (1920).

4. A. A. Griffith, *Proc. Int. Congr. Appl. Mech., 1st, 1924.*

5. R. P. von Rittinger, *Lehrbuch der Aufbereitungskunde.* Ernst V. Korn, Berlin, 1857.

6. I. Evans and C. D. Pomeroy, *Strength, Fracture and Workability of Coal.* Pergamon, Oxford, 1966.

7. P. F. Rad, *Inf. Circ—U.S. Bur. Mines* **8584** (1973).

8. J. D. McClung and M. R. Geer, in *Coal Preparation* (J. W. Leonard ed.). AIME, New York, 1979.

9. R. M. Hardgrove, *Trans. ASME* **14**, 37 (1932).

10. Anonymous, *Grindability of Coal by the Hardgrove-Machine Method*, ASTM Designation D409-71 (1983).

11. F. Agus and P. L. Waters, *Fuel* **51**, 38–43 (1972).

12. K. Jutte, *Brenst.-Waerme-Kraft* **6**, 92–93 (1954).

13. A. Ghosal et al., *J. Inst. Fuel* **31**, 50–55 (1958).

14. K. I. Savage, *Pulverizing Characteristics of Coal—Hardgrove Grindability Index*, Keystone Coal Ind. Manual, pp. 197–199. McGraw-Hill, New York, 1974.

15. A. Fitton, T. H. Hughes, and T. F. Hurtley, *J. Inst. Fuel* **30**, 54 (1957).

16. S. C. Sun and S. S. Hsieh, Ph.D. Thesis, Mineral Processing Section, Pennsylvania State University, University Park, 1976.

17. I. G. C. Dryden, *Fuel* **30**, 217 (1951).

18. M. Gomez and K. Hazen, *Rep. Invest.—U.S., Bur. Mines* **RI-7421** (1970).

19. C. Kroger and W. Ruland, *Brennst.-Chem.* **39**, 1 (1958).

20. D. W. Van Krevelen, *Coal.* Elsevier, Amsterdam, 1961.

21. P. B. Bonner and A. E. Abey, *Fuel* **54**, 165 (1975).

22. C. D. Pomeroy and P. Foote, *Colliery Eng.* (*London*) **37**, 146–154 (1960).

23. I. Evans and C. D. Pomeroy, *Mechanical Properties of Non-Metallic Brittle Materials.* Butterworth, London, 1958.

24. H. Bode, *Glueckauf* **69**, 296–297 (1933).

25. J. C. Macrae and A. R. Mitchell, *Fuel* **36**, 423–441 (1957).

26. H. Honda and Y. Sanada, *Fuel* **35**, 451 (1956); **36**, 403 (1957); **37**, 141 (1958).

27. D. English and F. J. Hiorns, *Trans.—Inst. Min. Metall., Sect. C* **75**, C87 (1966).

28. F. J. Hiorns, private communication with L. G. Austin.

29. L. G. Austin, *Powder Technol.* **15**, 1 (1971–1972).

30. P. T. Luckie and L. G. Austin, *Miner. Sci. Eng.* **4**, 24 (1972).

31. J. A. Herbst, G. A. Grandy, and T. S. Mika, *Trans.—Inst. Min. Metall., Sect. C* **80**, C193–C198 (19   ).

32. J. A. Herbst and D. W. Fuerstenau, *Trans.—Soc. Min. Eng. AIME* **254**, 343 (1976).

33. R. P. Gardner and T. Sukanjanjtee, *Powder Technol.* **7**, 169 (1973).

34. L. G. Austin, P. T. Luckie, and D. Wightman, *Int. J. Miner. Process.* **2**, 127 (1975).

35. A. J. Lynch, *Mineral Crushing and Grinding Circuits*, pp. 27–43, Elsevier, Amsterdam, 1977.

36. W. E. Horst and E. J. Freeh, *Trans.—Soc. Min. Eng. AIME* **252**, 160 (1972).

37. K. Schonert, *DECHEMA-Monogr.*, pp. 361–387 (1972).

38. D. F. Kelsall, K. J. Reid, and C. J. Restarick, *Powder Technol.* **1**, 291 (1967–1968).

39. A. L. Mular and J. A. Herbst, in *Mineral Processing Plant Design* (A. L. Mular and R. B. Bhappu, eds.), 2nd ed., pp. 306–338. SME-AIME, New York, 1978.

40. R. R. Klimpel and L. G. Austin, *DECHEMA-Monogr.*, pp. 449–473 (1971).

41. L. G. Austin, P. T. Luckie, and K. Shoji, *Powder Technol.* **33**, 127 (1982).

42. L. G. Austin, D. R. Van Orden, and J. W. Perez, *Int. J. Miner. Process.* **6**, 321 (1980).

43. R. S. C. Rogers, *Powder Technol.* **32**, 125 (1982).

44. R. P. Gardner, K. Verghese, and R. S. C. Rogers, *Min. Eng.* (*London*) **239**, 81 (1980).

45. J. A. Herbst and J. L. Sepulveda, in *Proceedings of the Powder and Bulk Solids Handling Conference.* Ind. Sci. Conf. Manage., Chicago, IL, 1978.

46. V. K. Gupta, D. Hodouin, and M. D. Everall, *Powder Technol.* **32**, 233 (1982).

47. L. G. Austin and K. A. Weller, *Prepr., Int. Miner. Process. Congr., 14th* I-8.1 (1982).

48. K. Shoji and L. G. Austin, *Powder Technol.* **10**, 29 (1974).

49. L. G. Austin, N. P. Weymont, K. A. Presbey, and M. Hoover, in *Proceedings of the 14th APCOM Symposium* (V. Ramani, ed.), p. 207, AIME, New York, 1977.

50. V. K. Jindal and L. G. Austin, *Powder Technol.* **14**, 35 (1976).

51. R. S. C. Rogers, *Powder Technol.* **36**, 137–143 (1983).

52. W. J. Whiten, *Proc. Appl. Comput. Methods Miner. Ind., Int. Symp., 10th*, pp. 317–323 (1973).

53. W. J. Whiten and M. E. White, *Prepr., Int. Miner. Process. Congr., 12th*, Vol. 2, p. 148 (1979).

54. R. S. C. Rogers, *Powder Technol.* **35**, 131 (1983).

55. L. G. Austin and P. Bagga, *Powder Technol.* **28**, 83–90 (1981).

56. *Coal Preparation.* McNally Pittsburgh, Inc., Pittsburgh, PA, 19

57. R. S. C. Rogers and K. Shoji, *Powder Technol.* **35**, 123 (1983).

58. L. G. Austin, D. R. Van Orden, B. McWilliams, J. W. Perez, and K. Shoji, *Powder Technol.* **28**, 245 (1981).

59. R. S. C. Rogers, *113th AIME Annu. Meet.* Pap. 84-156 (1984).

60. A. M. Gaudin, *Principles of Mineral Dressing*, pp. 41–43. McGraw-Hill, New York, 1939.

61. R. L. Milliken, private communication, Kennedy Van Saun Corp., Danville, PA.

62. L. G. Austin, R. R. Klimpel, and P. T. Luckie, *Process Engineering of Size Reduction: Ball Milling.* SME-AIME, New York, 1984.

63. R. S. C. Rogers and L. G. Austin, *Part. Sci. Technol.* (in press).

64. R. R. Klimpel, *Min. Eng. (London)* Dec., pp. 1665–1668 (1982).

65. R. R. Klimpel, Jan., *Min. Eng. (London)*, pp. 21–26 (1983).

66. R. R. Klimpel, *Powder Technol.* **32**, 267–277 (1982).

67. L. G. Austin, D. Bell, and R. S. C. Rogers, *Part. Sci. Technol.* (in press).

68. L. G. Austin, P. T. Luckie, K. Shoji, and R. S. C. Rogers, *Powder Technol.* (in press).

69. L. G. Austin, J. Shah, J. Wang, E. Gallagher, and P. T. Luckie, *Powder Technol.* **29**, 263–275 (1981).

70. L. G. Austin, P. T. Luckie, and K. Shoji, *Powder Technol.* **33**, 113–125 (1987).

71. L. G. Austin, P. T. Luckie, and K. Shoji, *Powder Technol.* **33**, 127–134 (1982).

72. L. G. Austin, C. Cannon, and O. Knobloch, in preparation.

73. T. G. Calcott, *J. Inst. Fuel* **29**, 207 (1956).

74. R. L. Brown, *Br. Coal Util. Res. Assoc., Mon. Bull.* **26**(9), Rev. 215 (1962).

75. L. G. Austin and R. R. Klimpel, *Trans. ASME* 219 (1968).

76. A. M. Gaudin and T. P. Meloy, *Trans.—Soc. Min. Eng. AIME* (1962).

77. H. E. Rose, DECHEMA-*Monogr.* **69**, Part L (1972).

78. L. G. Austin, R. P. Gardner, and P. L. Walker, Jr., *Fuel* **42**, 319–323 (1963).

79. L. G. Austin, K. Shoji, V. Bhatia, V. Sindhal, K. Savage, and R. R. Klimpel, *Ind. Eng. Chem. Process Des. Dev.* **15**, 187 (1976).

80. L. G. Austin, P. Bagga, and M. Celik, *Powder Technol.* **28**, 235–241 (1981).

81. L. G. Austin and V. V. Bhatia, *Powder Technol.* **5** 261–266 (1972).

82. P. T. Luckie, in *Pennsylvania Coal: Resources, Technology, and Utilization* (S. K. Majumdar and E. W. Miller, eds.). Pennsylvania Academy of Sciences, 1983.

83. Babcock and Wilcox, *Steam, Its Generation and Use*, 39th ed. 1978.

84. J. W. Leonard, W. F. Lawrence, and W. A. McCurdy, in *Coal Preparation* (J. W. Leonard, ed.). AIME, New York, 1979.

85. Anonymous, *Coal Min. Process.* Feb., pp. 37–54 (1971).

86. J. G. Singer (ed.), *Combustion*, Combustion Engineering, 1981.

87. R. R. Klimpel and L. G. Austin, *Powder Technol.* **34**, 121–130 (1983).

88. P. Bagga and P. T. Luckie, in *Use of Computers in the Coal Industry*, (Y. J. Yang and R. L. Sanford, eds.), pp. 247–250. AIME, New York, 1983.

89. R. R. Klimpel, *Powder Technol.* (in press).

90. L. G. Austin, C. Tangsathitkulchai, and R. R. Klimpel, in preparation.

91. M. Katzer, R. R. Klimpel, and J. Sewell, *Min. Eng.* (*Littleton, Colo.*) **33**, 1471–1476 (1981).

92. U.S. Patents 4,126,276; 4,126,277; 4,126,278; 4,136,830; 4,162,044; 4,162,045; and 4,274,599.

93. R. S. C. Rogers and K. A. Brame, *Powder Technol.* (in Press).

Commentary

Coal size reduction obviously increases the external area to volume ratio for a given mass and can lead to an increase in the water exchange rate between the coal and its surroundings. It is pertinent at this point to establish a somewhat detailed exposition of the physical structure of coal with consideration of the pore structure. This leads naturally to gas adsorption as a surface process and to drying and swelling. Particularly important is the water–coal interaction since the coal in its excursion from the mine through manipulations such as coal size reduction and beneficiation to its end use is exposed to diverse water vapor concentrations and liquid water. Clearly, the coal size reduction process produces a product whose characteristics are markedly affected by the environment in which the reduction occurs. Water (as well as oxygen) plays an important role. The following chapter on adsorption and pore structure and coal–water interactions will serve as an instructive framework for these concepts.

4

ADSORPTION AND PORE STRUCTURE AND COAL–WATER INTERACTIONS

O. P. Mahajan

Amoco Oil Research Department
Amoco Research Center

4.1 INTRODUCTION

Coal is a complex heterogeneous material with a rather extensive pore structure. Pores in coal vary in size from large cracks of micron dimensions to apertures that are even closed to helium at room temperature. Several important physical and chemical properties of coal are influenced by its pore structure. According to Grimes, "Details of the pore structure influence the behavior of coal more directly than does virtually any other property."[1] Since it is principally through the internal pore structure that coal comes into contact with the reacting environment, it is not surprising that porosity has a great influence on coal's behavior during mining, preparation, and utilization. As discussed recently by Mahajan and Walker,[2-4] coal porosity plays a major role in diffusion of methane from coal seams, coal beneficiation, gasification, liquefaction, transport of coal in water slurries, and the production of metallurgical coke, activated carbon, and carbon molecular sieves from coal.

This chapter has two important goals. The first is to define and understand the role of pore structure in the aging of coal, the changes in porosity due to aging, and the effects of these changes on coal properties and utilization. The second is to understand the role of water, inherently present in coal, on its aging characteristics and reactivity. Coal in its pristine state usually exists in a water-saturated environment. It has recently been observed that removal of water, irrespective of the mode of its removal, has a deleterious effect on coal liquefaction yield.[5]

Very little information is available on coal aging and its effect on reactivity. Two important reasons for this lack are that (1) it is only recently that researchers have become acutely aware of the adverse effects of aging of coals on their properties and utilization, and (2) because of the great practical implications of coal aging on utilization, the work done in industrial organizations on these aspects is mostly proprietary and not likely to be generally available. The major emphasis here is on the fundamentals of coal porosity and coal–water interactions and their implications for coal aging. This is followed by a discussion of some aspects of the removal of water and oxidative weathering on coal properties.

Pores in coal are distinguished according to the IUPAC classification,[7] namely, macropores (pores that have diameters > 500 Å), mesopores (20–500 Å in diameter), micropores (<8–20 Å), and submicropores (<8 Å). Until the above classification was introduced in 1972, pores with diameter <20 Å were considered as micropores. Since some of the work reported here was published prior to 1972, no distinction has been made in the text between submicropores and micropores.

For characterizing coal pore structure, the pore volume, surface area, and pore size distribution are needed. Displacement of gases and liquids by coals is used to measure densities and pore volumes, while physical adsorption of gases leads to the measure of surface area, pore volume, and area distribution. In addition, useful information on the nature of microporosity and adsorption characteristics of coal can be obtained from a study of diffusion of gases. These measurements are made on samples that must be dried initially. This can give rise to misleading information since drying coals of certain rank results in irreversible changes in their characteristics. A better understanding of the role of pore structure in coal utilization and conversion processes will come from quantitative characterization of porosity under dynanic conditions at temperatures and pressures typically existing in commercial processes. Realistically, we are still far from achieving these goals. Nevertheless, the use of several promising techniques such as small angle x-ray and neutron scattering, electron spin resonance spectroscopy, nuclear magnetic resonance spectroscopy, electron microscopy, heat capacity measurements, and x-ray computerized tomography hold exciting possibilities in the investigation of not only the inherent coal pore structure but also the interaction of molecular species with pore surfaces.

4.2 ADSORPTION AND PORE STRUCTURE

Adsorption is generally an exothermic process. Provided equilibrium has been established, the extent of adsorption at a given relative vapor pressure should decrease with increase in temperature. In microporous solids, such as coals and commercial zeolite molecular sieves containing pores commensurate in size with the adsorbate molecules, the magnitude of adsorption often increases with increase in temperature. This behavior is exemplified in Figure 4.1 for the

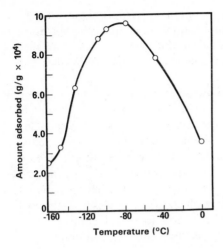

Figure 4.1 Variation in N_2 uptake on coal with temperature for an equilibrium time of 1 hour.[4]

adsorption of N_2 on a well-evacuated Welsh coal at temperatures between $-160°C$ and $0°C$ for an arbitrary equilibrium time of 1 hour for each data point.[8] Maximum adsorption occurs at $-80°C$. On 4 Å zeolite, which has an aperture opening of about 3.5 Å, maximum adsorption of N_2 also occurs at $-80°C$. Adsorption behavior depicted in Figure 4.1 is characteristic of a system that is not at equilibrium, but indicates a process that is activated in the reaction kinetic sense and occurs more rapidly at higher temperatures.

The adsorption behavior in Figure 4.1 is understandable from the following discussion. Coals have an aperture-cavity type pore system;[2,3] entrance to the pore system is governed by aperture size, and extent of adsorption is determined by the cavity size. When the size of an adsorbate molecule approaches that of the pore aperture through which it passes, physical interaction energy between the adsorbate and aperture becomes important. This energy represents the sum of dispersive and repulsive interactions.[9] When the aperture width becomes sufficiently small compared to the size of the adsorbate molecule, repulsive interaction becomes dominant. The adsorbate molecule then requires an activation energy to diffuse through the aperture. The rate of entry of the adsorbate molecule into the pore system therefore has a positive temperature coefficient.

With increase in temperature, two opposing factors will operate simultaneously. Since adsorption is generally exothermic, thermodynamic considerations would dictate a decrease in the extent of adsorption with increase in temperature. On the other hand, because of activated diffusion, adsorbate uptake would tend to increase. Adsorbate molecules in a given system will possess enough kinetic energy to permit attainment of adsorption equilibrium within the period of its measurements at a sufficiently high temperature. Above this temperature, adsorbate uptake should show the expected decrease with increase of temperature. The decrease commences above $-80°C$ in the system of Figure 4.1.

Diffusion of gases in coals provides useful information on coal porosity and adsorption behavior and has important implications for coal weathering. Some aspects of this phenomenon are discussed in this chapter. More detailed discussion may be found in reviews by Walker et al.[10] and by Walker and Mahajan.[11]

Flow of gases in coal particles is studied by using the unsteady state diffusion of gases either into or out of the coal particles.[12] For small particles, diffusion in the larger pores is relatively rapid. Hence, the dominant effect over an extended time period is unsteady state diffusion in the micropores. In one experimental procedure, the coal particles are charged with the given gas at the desired temperature and pressure. The pressure outside the particles is rapidly reduced to a constant value at the beginning of the diffusion experiment. This is achieved by increasing the volume of the collector into which the molecules from the coal micropores diffuse. Fick's second law of diffusion for short contact times, assuming noninteracting spherical particles, has the following solution:[12]

$$\frac{V_t - V_o}{V_e - V_o} = 6 \left(\frac{Dt}{\pi r_o^2} \right)^{1/2} \tag{4.1}$$

where D is the diffusion constant, and V_o, V_t, and V_e are the volumes of the gas collected at time $t = o$, $t = t$, and $t = \infty$. At $t = \infty$, the pressure inside and outside the coal particles is equal. The diffusion parameter, $D^{1/2}/r_o$, is calculated from the slope of the straight line of the $(V_t - V_o)/(V_e - V_o)$ versus $t^{1/2}$ plot. D for coals cannot be determined with certainty because r_0 is always less than the particle radius, and its value is unknown in all cases.[11]

Equation 4.1 is applicable for uniform particles of fixed pore size. Coals satisfy neither of these requirements. No matter how narrow the cut between sieve sizes, a ground material always has a particle size distribution. Further, coals do not have pores of single size but exhibit a trimodal distribution with macropores, mesopores, and micropores.[2,3] In unsteady state diffusion experiments, equilibrium is reached rapidly in macropores and mesopores. Thus, one would be measuring the approach to equilibrium in micropores as a function of time. Even in this case an average effect is measured because there is a distribution in the size of micropores.[2,3]

Oxidative weathering of coal produces subtle, as yet not completely understood, changes in coal structure. Among these is the formation of certain carbon–oxygen functional groups. Their formation must first involve physical adsorption of oxygen at the external surface of the coal particles followed by transport into the micropores (Section 4.2.3). The latter process is relatively slow and could be the rate-determining step in the overall oxidative weathering phenomenon. Information on the rates and activation energies of diffusion of oxygen into coal should provide a better understanding of the weathering phenomenon. Unfortunately such information is currently unavailable.

A qualitative indication of the role of oxygen diffusion in coal weathering can be obtained from diffusion of other gases. The work of Nandi and Walker[13] on the diffusion of N_2 and CO_2 out of coals is illustrative. They characterized the diffusion of the two gases in the temperature range 25–140°C from six coals varying in their carbon contents from 82.9 to 94.0 percent, expressed on a dry, moisture-free (DMF) basis. The activation energy for diffusion of N_2 was higher than that of CO_2 for each coal. The pre-exponential factor was greater for N_2 than that for CO_2 above about 30°C. The diffusion behavior is shown in Figure 4.2 for bituminous coal with 86 percent carbon content. It is noteworthy that only a 0.35 Å difference in the kinetic diameter ofan N_2 molecule (3.65 Å) and a CO_2 molecule (3.3 Å) causes over a threefold increase in the activation energy of diffusion. The kinetic diameter of an O_2 molecule (3.4 Å) is intermediate between that of N_2 and CO_2 molecules. One would expect that the diffusion behavior of O_2 would be intermediate between that of N_2 and CO_2.

4.2.1   Diffusion of Gases

The fact that diffusion of gases in coal pore structure is an activated process suggests that coals, prior to their utilization, should be stored at as low a temperature as possible to minimize the rate of diffusion of oxygen into the internal pore structure and hence the severity of weathering. Since a decrease in particle size increases the diffusion parameter,[11] the smaller the coal particle size the greater will be its rate of oxidation during weathering. This was shown by Mahajan et al.[14] in studies of the rate of oxidation of 40 × 70 and 200 × 250 mesh fractions of two coals, PSOC-337 and PSOC-127, containing 84.9 and 89.6 percent carbon content (dry, ash-free [DAF] basis) at temperatures up to 250°C. In this range, the coals lost little or no volatile matter. Results, listed in Table 4.1, show that at a given oxidation temperature an increase in particle size from 200 × 250 mesh to 40 × 70 mesh substantially increases (in some cases by an

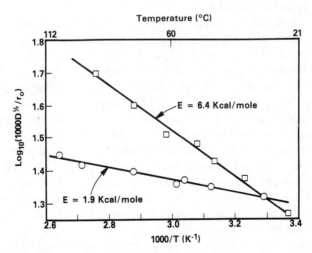

Figure 4.2   Arrhenius plots of diffusion for N_2 and CO_2.[13] ○, CO_2; □, N_2.

TABLE 4.1 Effect of Coal Particle Size on Rate of Oxidation

Oxidation Temperature (°C)	Weight Increase During Oxidation (%, DMF)	Oxidation Time (Minutes)	
		200×250 Mesh	40×70 Mesh
PSOC-337			
135	0.75	47	745
180	1.4	43	305
PSOC-127			
180	0.66	120	202
220	1.9	90	384
250	3.0	37	394
250	4.5	65	945

Source: O. P. Mahajan, M. Komatsu, and P. L. Walker, Jr., *Fuel* **59**, 3 (1980).

order of magnitude or more) the time required to achieve the same level of oxidation. It was also observed that coal particle size affected not only the rate but also the maximum level of oxidation. For example, a 7 percent weight increase was achieved during oxidation of 200×250 mesh fraction of PSOC-337 coal at 220°C. It was not possible to exceed 3.5 weight percent increase during oxidation of a 40×70 mesh fraction at temperatures up to 250°C. It follows that particle size has a marked effect on oxidative or weathering of coals. To minimize the oxidative weathering process coals should be stored as large lumps rather than small particles and comminuted to the desired particle size immediately prior to utilization.

There is no direct experimental evidence to compare coals regarding resistance to diffusion of oxygen during their oxidative weathering. There are data for diffusion of argon from coals of different rank, however. Kinetic diameters of O_2 and Ar are 3.46 and 3.40 Å, respectively. Because of the very small difference in the sizes of the two molecules, the argon diffusion results may be considered closely similar to those for O_2. Nandi and Walker[15] studied the diffusion of argon from 10 coals varying in carbon content (DAF) from 72 to 94 percent in the temperature range 25–100°C. The diffusion in each case was found to be temperature dependant. A volcano-shaped curve with a peak at about 86 percent carbon represents the activation energies for diffusion as a function of carbon (see Figure 4.3). The maximum indicates that the average size of the micropores is a minimum at 86 percent carbon and suggests that the internal microporosity in coals of this carbon content will be most inaccessible to oxygen during oxidative weathering.

In carbonaceous solids, pores are formed by the approach of basal planes of different crystallites. The size of the pores is determined by the crystallite size

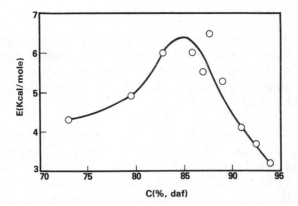

Figure 4.3 Relation between activation energy for diffusion of Ar in coals and their carbon content.[15]

and orientation. Walker et al.[10] have made theoretical calculations on the diffusion of rare gases between the basal planes of graphite. Using the Lennard–Jones 6–12 potential for dispersive and repulsive energies, they concluded that activated diffusion commences when the distance of separation between basal planes is less than the kinetic diameter of the diffusing species plus 1.6 Å. If it is assumed that the slit-shaped apertures in coal are formed from planar building blocks (or pseudographitic basal planes), one can estimate the distance of approach of these blocks at which activated diffusion would commence. Activated diffusion of O_2 (3.46 Å diameter) will commence when the planes approach each other more closely than about 5 Å. For N_2 (3.65 Å) and CO_2 (3.3 Å), activated diffusion will start at about 5.25 and 4.9 Å, respectively. It is noted that relatively small decreases in slit thickness cause very large increases in the activation energy for diffusion since the repulsive potential varies inversely with the twelfth power of distance. These considerations have important implications in adsorption properties and behavior of coal. Apertures in coal, being of molecular dimensions, exhibit molecular sieving effects and show major differences in capacity for adsorbing molecules differing only slightly in their sizes. Anderson et al.[16] observed that most coals at 0°C adsorb significantly more butane (kinetic diameter = 4.3 Å) than isobutane (kinetic diameter = 5.3 Å). These results have little practical importance because rates of adsorption require up to 24 hours for attainment of equilibrium.

One important implication of coal microporosity and diffusion phenomenon on coal behavior and utilization is that chemisorption of even small amounts of oxygen during oxidative weathering can either block some of the micropores or reduce their size. Under given process conditions, less internal surface is accessible to reactants, and lower reaction rates result. This is discussed in Section 4.4.

4.2.2 Densities

The relationship between porosity and density suggests that some emphasis be given to coal density and its determination. Three different densities can be considered for coal: true density, particle density, and apparent density. These densities are discussed briefly here. More detailed information is available in Mahajan and Walker.[2,3]

True Density. True density of a solid is defined as the weight of a unit volume of the pore-free solid. Its determination requires a noninteracting fluid that completely fills the pores. Such a fluid does not exist so that the true density of coals cannot be measured directly. Helium is the smallest fluid species and should give the closest approximation to the true density. Coals contain some closed porosity (pores with openings less than 4.2 Å) inaccessible to helium.[17] Obviously, the helium density of a coal will be less than its true density.

Helium densities of coals have been reported by several workers.[18-21] The variation of helium densities of American coals as a function of carbon content is shown in Figure 4.4. The density decreases with increase in carbon content, initially passes through a broad minimum in the 82–86 percent carbon content range, and rises sharply at about 90 percent carbon. Similar trends have been observed for British[18] and Japanese[19] coals. However, the minimum occurs at somewhat higher carbon contents than for the American coals. These results suggest that geological history has a marked bearing on coal structure.

Neavel et al.,[22a] using multivariant analysis techniques, found that the helium density (σ_{He}) of 66 coals is related to elemental composition* by the relationship

$$\rho_{He} = 0.023(C) + 0.0292(O) - 0.0261(H) = 0.0225(S_{org}) - 0.765 \quad (4.2a)$$

This accounted for 94 percent of the variance of the helium density. Parkash[22b] has reported that Equation (4.2a) accounts for only 63 percent of the variance of the densities of Canadian subbituminous coals (C content = 70.8–75.8 percent weight; reflectance = 0.24–0.48 percent). His multivariate analysis of the data on 32 low-rank coals showed that the helium density is related to elemental composition, expressed on a DAF basis by:

$$\rho_{He} = 3.5742 - (0.0197C + 0.0192O) - 0.0691H \quad (4.2b)$$

Low sulfur content in the coals tested had almost no effect on the relation in Equation (4.2b). This relation accounted for 83 percent of the variance of the samples tested.

Particle Density. Particle density is defined as the weight of a unit volume of the solid including pores and cracks. It is usually determined by mercury

---

*Expressed on a dry, mineral-matter-free (DMMF) basis.

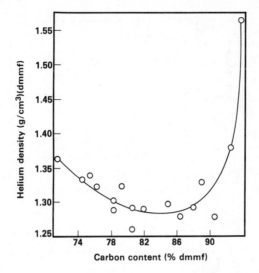

Figure 4.4   Variation of helium density of coals with carbon content.[21]

displacement. The pressure required to force mercury into a cylindrical pore of radius r is given by the Washburn equation.[23]

$$p = -(2\gamma \cos \theta)/r \tag{4.3}$$

where γ is the surface tension of mercury and θ is the contact angle between mercury and pore wall. For a wide variety of solids θ is assumed to be 140° and γ is 480 dyn/cm. Equation 4.3 then simplifies to

$$r = 106/p \tag{4.4}$$

with p in pounds per square inch and r in microns. Equation 4.4 indicates that in an outgassed coal immersed in mercury at atmospheric pressure the mercury cannot enter pores smaller than 7 μm in diameter.

Mahajan and Walker[24] measured mercury densities of 40 × 70 mesh fractions of 32 coals of different rank. The densities are plotted against carbon and volatile matter contents in Figures 4.5 and 4.6, respectively. There is a considerable data scatter in both figures, but some general trends are discernible. The densities of coals with carbon contents < 80 percent fall on a straight line (Figure 4.5), the slope and correlation coefficient are 0.0256 and 0.828, respectively. A sharp transition occurs at about 80 percent carbon content. For higher-rank coals, the mercury densities are represented by a straight line of considerably less slope (0.0063) and correlation coefficient (0.636). Figure 4.6 shows that particle density decreases sharply with increase in volatile matter up to about 20 percent. It then remains essentially constant in the 20–40 percent volatile matter range and decreases sharply thereafter.

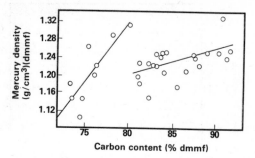

Figure 4.5 Variation of mercury density of coals with carbon content.[24]

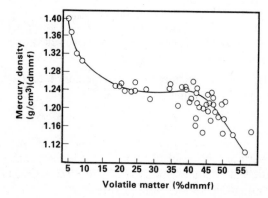

Figure 4.6 Variation of mercury density of coals with volatile matter content.[24]

Apparent Densities in Liquids. Apparent densities are normally deter-mined pycnometrically. These densities show drifts with time. This behavior is characteristic of microporous adsorbents. Since diffusion in micropores is a slow process, air present in micropores is very slowly displaced by the liquid. Apparent density of coal depends on its particle size, pore volume distribution, molecular size of the fluid, degree of interaction of the fluid with the solid, and time allowed for penetration of the fluid. Some solvents induce swelling in coals. Densities of coals in such solvents exceed their helium densities.[25] The total open pore volume of coals accessible to helium is calculated from their mercury and helium densities, whereas open pore volume accessible to a fluid other than helium is calculated from the mercury density and apparent density in the given fluid.[2,3]

4.2.3   Surface Area and Pore Volume Distribution

Adsorption of Gases.   Brunauer et al.[26] developed the BET equation for determining surface areas of solids from adsorption isotherms of N_2 measured at $-196°C$. Their approach has been used extensively for a wide variety of porous and nonporous solids. It is now generally agreed that adsorption of N_2 at $-196°C$ does not measure the total surface area of coals because of the activated diffusion phenomenon. That is, at $-196°C$, N_2 molecules do not have enough kinetic energy to penetrate readily into the coal micropores. This is supported by the observation that more Ar is adsorbed on a coal if the gas is first introduced at room temperature and the sample is subsequently cooled to $-196°C$.[27] Korta et al.[28] compared the adsorption of N_2 on a coal already cooled to $-196°C$ and on the same coal when it was exposed to N_2 during cooling to $-196°C$. A higher N_2 uptake and hence a higher surface area was obtained in the latter case. It has been suggested that N_2 adsorption at $-196°C$ measures the external surface area[29] and macropore area[30] and not the total surface area. Diffusional problems associated with the use of N_2 adsorption at $-196°C$ indicated the advantage of using adsorption of other gases at as high a temperature as possible to measure surface area. The various gases and conditions have been reviewed by Mahajan[2] and Mahajan and Walker.[3]

Adsorption of CO_2 at $-78, 0$, and $25°C$ has been used extensively for surface area measurements of coals since 1962.[29,31-36] These surface areas greatly exceed those determined from N_2 adsorption at $-196°C$ for at least two reasons. First, the minimum molecular dimension of CO_2 is smaller than that of N_2. Second, the kinetic energy of CO_2 at these adsorption temperatures far exceeds that of N_2 at $-196°C$ so that the rate of diffusion of CO_2 into the coal micropores will be significantly higher than that for N_2. It has been reported[33] that the time of passage of a CO_2 molecule through a capillary pore of 10 μm length and 5 Å diameter at $-78°C$ is less than that for N_2 by a factor of over 10^5. In studies of diffusion of N_2 and CO_2, Nandi and Walker[13] observed that the activation energy for diffusion of N_2 was higher than that for CO_2 below $30°$. The authors extrapolated the Arrhenius plots to -78 and $-196°C$. At these low temperatures the diffusion parameter for CO_2 was significantly greater than that for N_2. They calculated that for an arbitrary equilibrium time of 30 minutes for each adsorption point, a very large fraction of micropore volume was accessible to CO_2 at $-78°C$, and at $25°C$ essentially all was accessible. Only a negligible fraction of the micropore volume was accessible to N_2 at $-196°C$ in 30 minutes; any reasonable increase in diffusion time of N_2 at $-196°C$ was calculated to produce only an insignificant increase in the volume filled. It is thus clear why surface areas calculated from adsorption of N_2 at $-196°C$ are so low.

Walker and Kini[35] determined surface areas of several coals from adsorption of N_2 at $-196°C$, Kr at $-78°C$, Xe at $0°C$, and CO_2 at -78 and $25°C$. They concluded that adsorption of Xe at $0°C$, and of CO_2 at $-78°C$ should usually measure the total surface area whereas CO_2 adsorption at 25°C should essentially measure it. Marsh and Siemieniewska[36] used the Polanyi–Dubinin (P–D) equation to calculate surface area of coals from CO_2 adsorption at 0 and 20°C.

$$\log V = \log V_o - \frac{0.434BT^2}{\beta^2} \log \frac{p_s^2}{p} \qquad (4.5)$$

where

V = amount adsorbed at equilibrium pressure p

v_o = micropore volume

p_s = saturation vapor pressure of the adsorbate at adsorption temperature T

β = affinity coefficient of the adsorbate relative to N_2 or benzene

B = a constant that is a qualitative measure of the average micropore size

The micropore capacity, V_o, is obtained from the log V vs $\log(p_s^2/p)$ plot. The micropore volume can be converted to micropore surface area by multiplying with the adsorbate molecule area. The useful feature of the P–D equation is that it permits the calculation of surface area from adsorption data obtained below 1 atm in a conventional volumetric apparatus. Excellent agreement has been reported between surface areas calculated by the BET and P–D equations from CO_2 adsorption data obtained at 25°C for a number of coals and chars.[37] This agreement is quite significant because, in sharp contrast to the P–D surface areas, the BET CO_2 surface area is determined from adsorption data measured at about 200–300 psia.[35] The excellent agreement between P–D and BET surface areas indicates that a large percentage of the surface area in coals exists in micropores.

Higher-rank coals have higher CO_2 areas than the lower-rank coals[36]; a shallow minimum in area was observed at about 89 percent carbon. It was suggested that because surface areas of coals of different rank are not far from $200 \, m^2/g$, the observed variation in surface area with rank is caused by differences in accessibility rather than a difference in true surface area. The CO_2 surface areas (25°C) of a wide variety of coals have been determined.[24] The variation of these areas with carbon content is shown in Figure 4.7 where the CO_2 surface areas of 18 coals[20] are also included. The areas of different coals vary between 110 and 425 m^2/g. The areas of coals of similar carbon content fall within a band rather than on a line; coals with the same carbon content differ in their CO_2 areas by as much as 275 m^2/g. N_2 surface areas (Figure 4.7), were determined by Gan et al.[20] and Nandi and Walker[38] and are much lower than those corresponding to CO_2. Coals with N_2 areas > 10 m^2/g generally fall in the 76–84 percent carbon content range. Surface areas of most coals on both sides of this range are < 1 m^2/g; the anthracites are an exception and have areas of 5–8 m^2/g. CO_2 surface areas may be influenced by: (1) interaction of the quadrupole moment of CO_2 molecule with the oxygen functional groups

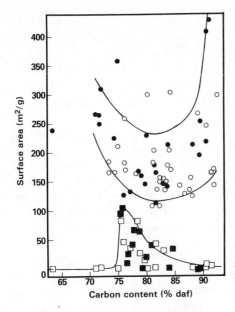

Figure 4.7 Variation of CO_2 and N_2 areas of coals with carbon content. CO_2 areas: ○, from Ref. 24; ●, from Ref. 20. N_2 areas: □, from Ref. 20; ■, from Ref. 38.

present on the coal surface, (2) solubilization or extraction of the low-boiling hydrocarbons adsorbed tightly within micropores, and (3) swelling.[2]

The concept of surface area of coals measured by gas adsorption has been criticized for two important reasons. First, surface area is calculated by multiplying the monolayer volume (as determined by the BET equation) or micropore volume (as determined by the P–D equation) with the cross-sectional area of an adsorbed molecule. One cannot assign a definite cross-sectional area for molecules adsorbed in coal micropores. Since pores in coal are slit-shaped, molecular areas in such pores are likely to be two to three times greater than those found on a flat surface.[35] Second, Dubinin[39] argues that, "Since there is an adsorption force field in the entire volume of micropores, adsorption of vapor in micropores leads to their volume filling. Therefore, the concept of layer-by-layer filling and of the micropore surface lose their physical significance." Spencer and Bond[40] emphasized that the surface area of coals has no physical significance and recommended that only monolayer volume or total volume of sorbate uptake should be reported.

Effect of Different Variables on Surface Area Measured by Gas Adsorption

Drying. As-mined coals may be saturated with water. It is essential to remove the water prior to measuring surface area by gas adsorption without changing the coal structure. The situation is quite complex with lower-rank coals because they undergo certain irreversible structural changes upon drying. It has been recommended that, prior to measuring surface area, the coal should be dried overnight in vacuum at 110°C.[3] CO_2 surface areas (25°C) of a lignite following outgassing at 90, 110, 130, and 150°C were found to be 206, 212, 198, and 184 m^2/g, respectively.[24] In contrast, a HV bituminous coal whether outgassed at 110 or 150°C gave essentially the same surface area. In spite of the uncertainties associated with the effect of water removal on the porosity of low-rank coals, outgassing overnight at 110°C is the most practical approach to drying.

Particle Size. The rate of adsorption of gases on coal is usually diffusion-controlled and can be enhanced by decreasing particle size. Coals generally show an increase in adsorption capacity, and hence, surface area with decreasing particle size. The magnitude of the increase for a given adsorbate depends also on the coal rank, but some of the increase can be attributed to other factors. Grinding can produce macrocracks or fissures, and grinding of caking coals in air results in the closure of some pores due to overheating and plastic flow.[41]

Swelling. Swelling of coals during adsorption is a well-known phenomenon with important implications in coal structure, reactivity, and utilization. The total sorbate uptake on a coal includes contributions from physical adsorption on pore walls, pore filling, and swelling. Sorbate uptake due to swelling, if included in the adsorption data used for calculating surface area of coals, will result in unrealistically high surface areas. Coal is now increasingly being viewed as having a crosslinked macromolecular structure.[42,43] The degree of solvent-induced swelling is used, among other approaches, to study the macromolecular structure and nature of the crosslinks. Some workers considered total sorbate uptake on coal as a measure of its swelling[44,45] but ignoring the contribution of physical adsorption and pore filling overestimates the degree of swelling.

The study of solvent-induced swelling is also important in understanding the effect of aging on coal behavior and utilization. It is well known that oxidative weathering of coking coals has an adverse effect on their plastic properties. This has been attributed to the introduction of crosslinks.[46,47] Nelson et al.[25] used a combination of gravimetric and pycnometric techniques to determine solvent-induced swelling of 8 coals varying in carbon content (DMMF) from 71.5 to 89.5 percent; the solvents used were methanol, benzene, and tetralin. The results obtained at 25°C are plotted in Figure 4.8. The contribution of swelling to total sorbate uptake varies from about 10 to 80 percent depending on the solvent used. Methanol-induced swelling remains approximately constant up to 78 percent carbon and then decreases sharply. Benzene-induced swelling increases with increasing carbon to about 77 percent carbon and then decreases to a minimum at about 82 percent carbon. Beyond that it rises sharply to give a second maximum at about 86 percent. At even higher carbon, a sharp decrease is manifest. Tetralin induces swelling only in coals with carbon contents in the 75–82 percent range with a maximum at about 78 percent.

Coal fluidity shows a maximum at about 86 percent carbon.[44] Larsen[48] suggested that a maximum in fluidity should coincide with a minimum in

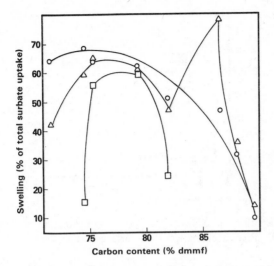

Figure 4.8   Contribution of swelling to total sorbate uptake on coals as a function of their carbon content.[25] ◯, methanol; △, benzene; ☐, tetralin.

crosslink density. A minimum in crosslink density should correspond to a maximum in swelling, and a maximum in solvent-induced swelling should be expected at about 86 percent. This is not observed in Figure 4.8. It will be shown that fluidity and swelling need not necessarily provide the same information about crosslink density.

Since swelling contributes substantially to total sorbate uptake (cf., Figure 4.8), surface areas of coals measured by sorption of organic vapors can be in error. It has been reported[49] that the exposure of a coal of 87 percent carbon content to saturated pyridine vapor increases the subsequent CO_2 surface area threefold and pore volume twofold. Methanol-induced swelling decreases linearly with decreasing oxygen content of the coal.[25] This casts doubt on the reliability of surface areas and pore size distributions determined from adsorption of methanol or any other polar molecule. The interaction of organic oxygen content of coal with polar water molecules plays an important role in moisture-holding capacity of coals and will be discussed in Section 4.3.4.

One of the objections to the use of CO_2 for measuring total surface area is its ability to induce swelling. The results of Reucroft and Patel[50] have important bearing on swelling and surface area. They determined the surface area of a Kentucky bituminous coal from adsorption isotherms of a number of vapors having a solubility parameter δ in the $6-12$ cal$^{0.5}$ cm$^{-1.5}$ range. Surface areas, calculated by both the BET and P–D equations, are plotted as a function of solubility parameter of the adsorbate in Figure 4.9. These results show that surface area is strongly dependent upon the sorbate used and increases sharply with increasing δ up to a maximum at $\delta = 10$ cal$^{0.5}$ cm$^{-1.5}$. Further increase in

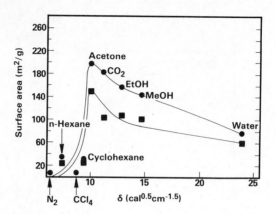

Figure 4.9 Variation of coal surface area as a function of sorbate solubility parameter.[50] ●, D–P surface area; ■, BET surface area.

δ reduces the surface area. The Xe surface area predicted from its δ (9.5 $cal^{0.5}$ $cm^{-1.5}$) is approximately equal to the CO_2 surface area. This agreement is consistent with the observation of Walker and Kini[35] that coals have similar CO_2 and Xe surface areas.

Reucroft and Patel[50] calculated a swelling parameter Q defined as swollen volume/unswollen volume for each sorbate from equilibrium vapor uptake at 0.90 relative vapor pressure. The volume of the swollen coal sample was considered to consist of the coal volume and the total sorbate uptake, with no allowance for physical adsorption and pore filling. Based on the adsorption behavior of n-hexane, cyclohexane, and carbon tetrachloride, the authors inferred that the Kentucky coal has a very low uptake due to pore filling. The variation of the swelling parameter as a function of the sorbate solubility parameter is shown in Figure 4.10. The maximum in Q is observed at $\delta \approx 11$ $cal^{0.5}$ $cm^{-1.5}$. This maximum may be taken as approximately equal to δ of the coal.[50] The maxima in Q (Figure 4.10) and surface area (Figure 4.9) occur at about the same δ. This fact in conjunction with similar trends in Figures 4.9 and 4.10 indicates a similarity between surface area and swelling parameter. The authors conclude that surface area of coals is influenced by solvent-induced swelling, and the true surface area of coals is measured only with solvent vapors having low or high δ values compared to that of the coal itself. In this view the BET N_2 surface areas are closer to the true surface area of coals, the CO_2 areas being influenced by swelling. In a later publication, Reucroft and Patel[51] adduced direct confirmation of the swelling effect of CO_2 on coal by dilatometric studies. Volume increases up to 1–3 percent were obtained on exposure to CO_2 at 1 atm; whereas negligible swelling effects were observed upon exposure to N_2, helium, and xenon. It was estimated that the magnitude of the CO_2-induced swelling could account for up to 14.5 percent of the CO_2 BET area. Surprisingly, Stacy and Jones[52] observed no measurable swelling effect upon exposure of a coal of 69.1 percent carbon content to 1 atm CO_2.

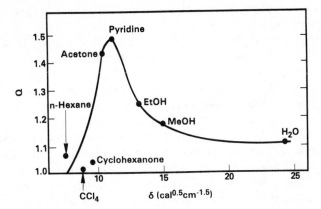

Figure 4.10   Variation of swelling parameter as a function of sorbate solubility parameter.[50]

Reucroft and Sethuraman[53] made dilatometric studies on coal samples exposed to CO_2 at 5, 10, and 15 atm. The sample dimension in each case increased upon exposure to CO_2, and the effect increased with increase in pressure and with decrease in carbon content of coal. It was estimated that CO_2-induced swelling may account for up to 50 percent of the CO_2 surface areas for lignites and subbituminous coals, and up to about 20 percent of the surface area for bituminous coals.

Walker et al.[54] made a comprehensive study of the interaction of gases and vapors with different coals and macerals with a view to characterizing their porosity. They also observed that coals expand to measurable extents upon exposure to CO_2 at room temperature. Their dilatometric studies show that expansion in CO_2 increases with increase in CO_2 pressure. At 1 atm CO_2 pressure, most coals studied expanded between 0.05 and 0.10 percent and showed no trend with rank. At a pressure of about 50 atm, coals expanded between 1.6 and 3.8 percent and there was a general trend toward increase in expansion with decrease in coal rank. Expansion due to CO_2 was found to be irreversible when an elevated pressure was reduced to atmospheric pressure. This irreversibility was also observed for a microporous Saran carbon (prepared by the carbonization of Saran, a copolymer of PVDC and PVC) that showed only limited imbibition of CO_2. The authors concluded that uptake of molecules in pores of molecular dimensions and/or their imbibition leads to irreversible changes in the solid structure. X-ray diffraction,[55] small angle neutron scattering,[56,57] and electron spin resonance spectroscopy[58,59] are now being used to evaluate structural changes resulting from coal swelling.

Expansion of coals upon exposure to CO_2 helps to explain why pore volumes accessible to CO_2 exceed those accessible to helium for some coals. Walker et al.[54] attribute greater CO_2 uptake to its access to pores closed to helium as well as to its imbibition into the coal solid structure. They point out the problem of apportioning the total CO_2 uptake into its various components, namely adsorption on to walls of pores open to helium, adsorption onto walls of pores closed to helium, and imbibition.

Walker et al. suggest that large CO_2 surface area of coals compared to N_2 areas cannot be attributed primarily to CO_2 imbibition because the ordering of dilatometric expansion values for coals varying in rank from anthracite to lignite do not parallel the ordering of the CO_2 surface area values (Table 4.2). Significantly, the anthracite sample (PSOC–85) shows the least expansion in 1 atm CO_2, yet it has the highest CO_2 surface area estimated from adsorption data obtained below 1 atm. On the basis of these results, Walker et al. endorse the views of others (Section 2.3.1.) that reporting of surface area of coals should be discontinued. They suggest that since surface areas are incorrect to varying degrees, they have only limited utility and recommend that densities of coals measured in different fluids and amounts of fluids taken up at particular relative pressures or concentrations are of more practical and fundamental importance.

It is appropriate at this stage to discuss the work of Larsen and Wernett[60] on the implications of coal swelling on surface area. They measured surface area of an Illinois No. 6 coal from CO_2, ethane, and cyclopropane adsorption data. The cyclopropane and ethane surface areas were similar, but both were substantially lower than the CO_2 surface area. Larsen and Wernett concluded that the surface area results measured by the three adsorbate molecules, which have similar cross-sectional areas, cannot be rationalized using a constricted pore model.

TABLE 4.2 CO_2 Surface Areas and Expansion of Coals in 1 atm CO_2[54]

PSOC Sample No.	Rank	CO_2 Surface Area (m^2/g)	Dilatometric Expansion, %
85	Anthracite	451	0.028
1197	LV bit.	223	0.105
1192	MV bit.	211	0.095
1201	MV bit.	248	0.043
1141	HVA bit.	216	0.116
1166	HVA bit.	183	0.139
223	HVB bit.	241	0.093
212	HVC bit.	262	0.094
248	Sbb-A	225	0.069
242	Sbb-B	234	0.048
246	Lignite	225	0.099

Source: P. L. Walker, Jr., A. Davis, S. K. Verma, J. Rivera-Ultrilla and M. R. Khan, "Interaction of Gases, Vapors and Liquids with Coals and Macerals at Ambient Temperatures—Their Use to Characterize Porosity." Final Report—Part 9, May 1984, Pennsylvania State University, Prepared for DOE Under Contract No. DE-AC22-80 PC 30013.

Ethane (cylindrical) and cyclopropane (planar) have different shapes, and either slit or cylindrical pore openings should discriminate between the two adsorbate molecules. Furthermore, ethane and CO_2 have similar shapes (cylindrical) and diameters but give very different coal surface areas concluding that diffusion through constricted openings cannot be responsible for this size discrimination.

To explain their surface area results, Larsen and Wernett suggest that most pores in coals are closed, and to reach them an adsorbate must diffuse through solid coal, rather than through the pore network. They postulate that CO_2 gives approximately accurate values for the total surface area because it dissolves in coal during the swelling process and diffuses rapidly to the pore network. In contrast, hydrocarbon gases which are not significantly soluble in coals can reach only a much smaller portion of the coal's pores and thus give a much smaller surface area. Larsen[61] is extending these studies to coals of different rank.

Pore Volume Distribution.   To characterize pore structure of coal, the surface area, various densities and pore volume, pore size, and pore volume distributions must be known. These distributions can be determined in several ways. Some of the approaches used for this purpose, such as capillary condensation of N_2, mercury porosimetry, adsorption of CO_2 and methanol have been reviewed. Gan et al.[20] used helium and mercury densities of coals and N_2 (77 K) adsorption and mercury penetration data to estimate pore volume distributions in the following pore diameter ranges:

1. Total open pore volume V_T, as estimated from helium and mercury densities.
2. Pore volume V_1 contained in pores > 300 Å in diameter, as estimated from mercury porosimetry.
3. Pore volume V_2 contained in pores having diameters in the 12–300 Å range, as estimated from the N_2 adsorption isotherms.
4. Pore volume V_3 contained in pores < 12 Å in diameter, as estimated from $V_3 = V_T - (V_1 + V_2)$.

Gross open pore distributions in various coals are listed in Table 4.3. The proportion of V_3 is significant for all the coals; its value is a maximum for the PSOC-80 anthracite, and a minimum for the PSOC-89 lignite. These results indicate that all coals show molecular sieve properties. The results in Table 4.2 can be summarized as follows: (1) porosity in coals with carbon contents <75 percent resides primarily in macropores, (2) porosity in coals with carbon contents in the 75–84 percent range is predominantly due to micropores and transitional pores, and (3) porosity in coals varying in carbon contents from 85 to 91 percent is due to the presence of micropores.

TABLE 4.3 Gross Open Pore Distributions in Coals

Sample	Rank	C (%, DAF)	V_T (cm³/g)	V_1 (cm³/g)	V_2 (cm³/g)	V_3 (cm³/g)	V_3 (%)	V_2 (%)	V_1 (%)
PSOC-80	Anthracite	90.8	0.076	0.009	0.010	0.057	75.0	13.1	11.9
PSOC-127	LV bit.	89.5	0.052	0.014	0.000	0.038	73.0	nil	27.0
PSOC-135	MV bit.	88.3	0.042	0.016	0.000	0.026	61.9	nil	38.1
PSOC-4	HVA bit.	83.8	0.033	0.017	0.000	0.016	48.5	nil	51.5
PSOC-105A	HVB bit.	81.3	0.144	0.036	0.065	0.043	29.9	45.1	25.0
Rand	HVC bit.	79.9	0.083	0.017	0.027	0.039	47.0	32.5	20.5
PSOC-26	HVC bit.	77.2	0.158	0.031	0.061	0.066	41.8	38.6	19.6
POC-197	HVB bit.	76.5	0.105	0.022	0.013	0.070	66.7	12.4	20.9
PSOC-190	HVC bit.	75.5	0.232	0.040	0.122	0.070	30.2	52.6	17.2
PSOC-141	Lignite	71.7	0.114	0.088	0.004	0.022	19.3	3.5	77.2
PSOC-87	Lignite	71.2	0.105	0.062	0.000	0.043	40.9	nil	59.1
PSOC-89	Lignite	63.3	0.073	0.064	0.000	0.009	12.3	nil	87.7

Source: H. Gan, S. P. Nandi, and P. L. Walker, Jr., *Fuel* **51**, 272 (1972).

4.2.4   Small Angle X-ray Scattering

The use of adsorption to investigate pore structure has at least three drawbacks. First, the gas adsorption techniques require a predrying step to remove water. It is well known that low-rank coals undergo severe shrinkage upon drying. For these coals gas adsorption techniques may underestimate their porosity. Second, gas adsorption techniques do not provide information on closed porosity. Third, chemical interactions may occur between surface functional groups and probe molecules. These problems can be overcome by using nonintrusive physical probes such as small angle x-ray scattering (SAXS). It has been reported[62] that SAXS can be used to characterize micropore structure of low-rank coals whatever their moisture content. Since x-radiation sees the entire pore radius, the SAXS technique shows promise for determining pore size distribution for both open and closed pores. It also has the potential for application to the measurement of pore size distribution under temperature and pressure conditions in commercial processes.

In the SAXS approach, most of the incident x-rays are either absorbed in the sample or pass through it without being affected. However, a small fraction of the incident radiation is emitted in directions other than that of the incoming beam. X-ray scattering occurs when there are differences in electron density in different regions of the sample. Coals scatter x-rays because of inhomogeneities in their electron density. Coals contain three phases: organic matrix, mineral matter, and pores. Mineral matter and pores scatter x-rays independently of each other.[63] The form of the scattered intensity scattering angle curve depends primarily on the coal rank.[63] Analyses of the scattering curves provide evidence for the presence of three classes of pores: macro, meso, and micro. This classification is consistent with that of IUPAC discussed in Section 4.1.

Surface areas of coals calculated from the scattering curves are given in Table 4.4; the uncertainty is about $\pm 40\%$. N_2 ($-196°C$) and CO_2 ($25°C$) surface areas of the coals are also included for comparison. Scattering curves do not include any contributions from the micropores[63] since SAXS measures surface area of only meso- and macropores.

Mineral matter in coal contributes to scattering and hence surface area. Kalliat et al. ashed four coals at low temperature in an oxygen plasma to ascertain the magnitude of this contribution. This ashing procedure burns the organic content at a temperature of 150°C or less, leaving the mineral matter in an essentially unchanged form.[64] It was concluded from SAXS studies that the mineral matter contributes about 10 to 25 percent of the total surface. Since this change is well within the limits of accuracy of the SAXS method for determining surface areas, the contribution of mineral matter can be neglected. The BET N_2 ($-196°C$) surface area of mineral matter isolated by low temperature ashing of coals has been reported by O'Gorman[65] to be $\leqslant 10\,m^2/g$. The N_2 areas of some whole coals are $< 1\,m^2/g$ (Figure 4.7). It may be concluded that the finely divided mineral particles distributed in the coal organic matrix are not accessible to N_2 at $-196°C$.

TABLE 4.4   Surface Areas of Coals

PSOC Coal No.	Rank	Surface Area (m^2/g)		
		Adsorption		SAXS
		N_2	CO_2	
81	Anthracite	ND[a]	ND[a]	3.91
318	LV bit.	<1.0	186	2.54
127	LV bit.	<1.0	271	2.51
130	MV bit.	<1.0	249	2.51
135	MV bit.	<1.0	227	2.42
134	MV bit.	<1.0	208	2.39
95	HVA bit.	<1.0	220	5.66
105	HVB bit.	16.0	105	46.58
185	HVB bit.	49.0	ND[a]	8.72
188	HVB bit.	32.0	ND[a]	15.95
197	HVC bit.	11.0	ND[a]	6.18
22	HVC bit.	99.0	189	18.4
212	HVC bit.	3.7	254	9.16
181	Subbit. A	49.0	ND[a]	24.01
138	Subbit. C	2.5	254	7.25

[a]Not determined.
Source: M. Kalliat, C. Y. Kwak, and P. W. Schmidt, *ACS Symp. Ser.* **169**, 3–22 (1981).

The SAXS areas of all the coals in Table 4.4 are appreciably lower (in some cases by an order of magnitude or more) than the CO_2 areas. It was concluded that coals contain many pores with minimum dimensions less than about 20 Å that are accessible to CO_2 at room temperature. The SAXS areas are larger than the N_2 areas for nine coal samples; for the remainder of the coals this trend is reversed. A reasonable suggestion is that in the former case the coals contain closed pores inaccessible to N_2 at $-196°C$, whereas in the latter there are many pores penetrated by N_2 but not detected by x-rays. The contention of Kalliat et al. that SAXS cannot measure surface area of micropores is not supported by the work of others. Spitzer and Ulicky[66] have compared SAXS areas of two coals, St. Nicholas anthracite and Illinois No. 6 bituminous, with the CO_2 areas (25°C). The SAXS and CO_2 areas were 275 and 234 m^2/g for the anthracite and 179 and 160 m^2/g for the bituminous coal.

Small angle neutron scattering (SANS) can complement SAXS studies on coal pore structure. SANS should have an advantage in resolving macropores because neutrons have a longer wavelength than x-rays. Neutrons can see all elements, distinguish them, and show changes of magnitude and sign of scattering length through the use of isotopes. SANS can presumably be used to observe molecular species contained in pores. Coal porosity has been examined by SANS in the dry state,[67–69] in nonswelling deutrated solvents,[67,69] and solvent-swollen coals.[56,57]

It cannot be overemphasized that a better understanding of the role of porosity in coal conversion and utilization processes will come from quantitative characterization of pore structure under dynamic conditions of temperature and pressure associated with commercial processes. We are still far from achieving this enviable goal, but the recent preliminary work of Maylotte et al.[70] on the use of x-ray computed tomography to follow changes in the average pore structure of coals and reacting coal chars in situ is promising.

4.2.5 Electron Microscopy

Gas adsorption and SAXS techniques do not provide information about geometric characteristics of the pore system. Transmission electron microscopy (TEM) permits direct observation of the geometric characteristics of pores and correlation of pore distribution with specific macerals. Harris and Yust[71] and Lin et al.[72] used the TEM technique to study the porosity associated with the vitrinite maceral in an Illinois No. 6 coal. Pore dimensions were seen to range from 10 to 100 Å. Three-dimensional viewing in a stero pair revealed that porosity was irregularly shaped and present as a network of highly interconnecting pores. In the region of locally high porosity, the degree of interconnecting was relatively large. Pore volumes were largely isolated in the surrounding regions. Lin et al.[72] also used the SAXS technique to study the microporosity of vitrinite in the Illinois No. 6 coal. SAXS gave a trimodal size distribution with peaks at 30, 100, and 220 Å. These peaks were attributed to micropores, microminerals, and mesopores, respectively. Average mesopore size determined by SAXS was in excellent agreement with that by TEM.

4.3. COAL–WATER INTERACTIONS

Coal is virtually saturated with water in its natural state within a seam. Once the coal has been mined its moisture-holding capacity is generally considered as the moisture content of the coal in equilibrium with 96 percent relative humidity at 30°C. This capacity is taken to be the bed moisture for bituminous coals and is used in the classification (ASTM D388). What really constitutes moisture in coals has long been a dilemma to coal researchers. The most widely accepted definition is that it is the water released from coal heated to 105–110°C. Unfortunately, the standard method for determining moisture content does not distinguish between water present as H_2O, that released by the dehydration of some minerals present in the coal, or from decomposition of some surface oxygen complexes. Thermal decomposition of functional groups to yield water and CO_2 can commence at temperatures well below 100°C for low-rank coals.[73]

Allardice and Evans[73] measured the amounts of CO_2 and H_2O released upon drying an Australian brown coal at increasingly high temperatures. The coal was outgassed at 30°C and 10^{-3} mm Hg for 24 hours initially. Product gases were condensed in a liquid N_2 trap and analyzed for H_2O and CO_2.

Further outgassing was done at 60, 90, and 120°C. Most of the water in the coal was found to be removed at 30°C. Additional weight loss up to 60°C was almost completely that of water. Measurable losses of CO_2 occurred above 60°C and were quite large by 120°C. Dack et al.[74a] used electron spin resonance spectroscopy to study the changes in the nature and concentration of free radicals in brown coals as a result of vacuum drying at -15, 20, 50, 100 and 150°C. Free radical concentration increased during drying and was attributed to breakdown of organic functional groups. Gethner[74b] has also shown by in-situ Fourier transform infrared spectroscopy that coals undergo changes in functional group distribution during vacuum drying at 100°C. Allardice and Evans[73] concluded from their observations that at least two types of water are associated with coal at a given temperature, that is, (1) weakly associated water that can be removed at low temperatures, and (2) more strongly associated water removable only at higher temperature. The latter is considered to be associated with the thermal decomposition of some oxygen functional groups such as carboxyl. However the strongly bound water cannot originate from a simple breakdown of the carboxylate structure if hydrogen ions of the carboxyl groups have been replaced by cations. The release of water and CO_2 from cation-exchanged coals above 110°C has been attributed to complex decomposition reactions involving groups in addition to carboxyl.[75] Since coals higher in rank than subbituminous are essentially carboxyl free, the water released from them up to 110°C closely represents their moisture content.

4.3.1   Water Adsorption Isotherms

Water adsorption isotherms on coals are measured as a function of relative vapor pressure (RVP). Two approaches have been used to study moisture sorption isotherms. In the first, bed-moist coal is equilibrated with water vapor at successively decreasing RVP to zero pressure. In the second approach, a predried coal (usually oven-dried at 105–110°C) is equilibrated with water vapor at successively increasing RVP. The first approach was used by Allardice and Evans[76] to obtain the isotherm on a bed-moist brown coal (Figure 4.11). Following completion of the desorption isotherm to zero pressure, the sample was re-exposed to water vapor to determine the adsorption isotherm. There is a distinct difference between the desorption and readsorption branches. The second desorption branch is a scanning curve. It crosses the hysteresis region and closes the loop by joining the initial desorption curve at a pressure of 25–30 mm Hg. Below this pressure, the second and initial desorptions are identical. Mahajan and Walker[77] used the second approach for measuring water isotherms at 20°C on 6 coals—one anthracite and five bituminous. Analyses of the coals are given in Table 4.5, and the water adsorption isotherms are plotted in Figure 4.12. Bituminous coals adsorb more water at low RVP than does anthracite. Sorption on bituminous coals is characterized according to the classification of Brunauer et al.[78] by Type II isotherms and on anthracite by Type III. The former are typical of physical adsorption of condensable vapors

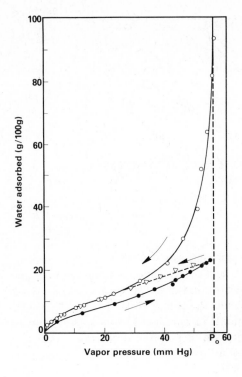

Figure 4.11 Water desorption–adsorption isotherms at 40°C on a brown coal.[61] ○, initial desorption; ●, readsorption; ▽, second desorption.

TABLE 4.5   Analyses and Surface Areas of Coals

Sample No.	Proximate Analysis (wt%, DMF)			Surface Area (m²/g, DMF)		
	Ash	Volatile Matter	Fixed Carbon	$N_2(77K)$	$CO_2(298K)$	H_2O
			Anthracite			
1	8.4	4.0	86.4	nil	224	33
			Bituminous Coals			
912	4.6	18.8	76.6	nil	146	17
956	6.3	33.7	60.0	nil	125	24
888	3.8	39.0	57.2	nil	104	38
885	5.4	37.6	57.0	11	132	42
6	7.6	45.4	—	22	139	82

Source: O. P. Mahajan and P. L. Walker, Jr., *Fuel* **50**, 308 (1971).

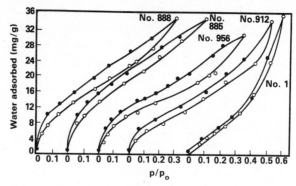

Figure 4.12 Water adsorption isotherms on coals at $20°C$.[77] Solid points denote desorption data.

on porous adsorbents and the latter characteristics of an adsorption process where pores are relatively large and the heat of adsorption is almost equal to the heat of vaporization. The sorption–desorption isotherms on various coals show hysteresis in each case (Figure 4.12). Anderson et al.[16] found that the hysteresis loops for coals usually do not close until RVP is reduced to zero. The results in Figure 4.12, however, show that the sorption–desorption isotherms do not meet even at zero RVP; a certain amount of adsorbed water could not be desorbed even on prolonged outgassing to constant weight at $20°C$. The irreversibly adsorbed water is presumably held tightly within the micropores or at the surface by a mechanism involving oxygen functional groups. After completion of the adsorption–desorption cycle, the sorption capacity of coals was fully restored after outgassing at $110°C$.

4.3.2 Effect of Inorganic Matter on Water Adsorption

Coals are associated with two broad classes of inorganic matter: discrete mineral matter usually present as particles larger than $1\ \mu m$, and minor and trace elements associated with the mineral or organic phase. Both classes can contribute to moisture sorption capacity. Some clays in coal can adsorb significant amounts of water. The contribution of inorganic matter to moisture sorption capacity is largely confined to low-rank coals such as subbituminous and lignitic. These coals contain significant amounts of carboxyl groups although the hydrogen ions of these groups have usually been exchanged for other cations such as Na^+, K^+, Ca^{2+} due to extended contact of embedded coals with water containing these cations. Replacement of surface H^+ ions of carboxyl groups of activated carbons and charcoals by alkali and alkaline earth cations significantly enhances their water adsorption capacity over the entire RVP range.[79,80] It might be expected that the presence of exchangeable cations on the surface of low-rank coals would also influence their moisture sorption characteristics. Naturally occurring low-rank coals adsorb more water than do the samples treated with HCl;[81] the acid treatment in addition to dissolving some discrete minerals (such as calcite), replaces surface cations by H^+. Schafer[82] made a detailed study of the effect of exchangeable cations on the moisture sorption capacity of 11 low-rank coals with carbon contents in the 64.1 to 79.1 percent range on a DMMF basis. The coals were acid-demineralized to remove organically bound cations and silicate minerals. The demineralized coals had carboxyl content from about 1 to 4 milli-eq/g, and phenolic groups from 2.5 to 4.0 milli-eq/g. The moisture sorption capacity of the acid-form coals at 0.52 RVP was related to the carboxylic and phenolic hydroxyl contents by the following relationship:

$$\% \text{ equilibrium water} = 2.181 \text{ [COOH]} + 0.599 \text{ [OH]} + 4.90 \qquad (4.6)$$

where [COOH] and [OH] are the carboxyl and phenolic hydroxyl contents in milli-eq/g of dry coal. The correlation coefficients for water adsorption with carboxyl and water with phenolic hydroxyl were 0.949 and 0.307, respectively. The mean correlation coefficient was 0.957. These results indicate that moisture sorption capacity of low-rank coals primarily depends on their carboxyl content.

Schafer exchanged H^+ ions of carboxyl groups of the acid-form coals with different cations. In each case the exchange increased the moisture sorption capacity. The moisture content of a given sample at 20°C and 0.52 RVP increased rectilinearly with the cation content of the coal. The type of exchangeable cation had a major influence on moisture sorption capacity. The equilibrium moisture content at 20°C and 0.52 RVP varied with the exchangeable cation in the following order:

$$Fe^{2+} > Mg^{2+} > Ca^{2+} > Cu^{2+} > Ba^{2+} > Al^{3+} \approx Fe^{3+} \approx Na^+ > K^+$$

This order corresponds roughly to that given by Stokes and Robinson[83] for the hydration number of cations, namely:

$$Mg^{2+} > Fe^{2+} > Ca^{2+} > Ba^{2+} > Na^+ > K^+$$

Schafer also observed that the cation-exchanged coals not only adsorbed more water but also retained it at $> 110°C$ more strongly than the acid-form coals.

4.3.3 Surface Area from Water Adsorption Isotherms

The water adsorption isotherms in Figure 4.12 show sensible BET plots even for the anthracite sample associated with a Type III isotherm. Isotherms of this type normally do not obey the BET equation. The water BET surface areas as well as the BET N_2 ($-196°C$) and CO_2 (25°C) areas of the coals are given in Table 4.5. The water areas are appreciably higher than those of the N_2 but are significantly lower than the CO_2. This cannot be explained by molecular sieve effects alone because the minimum dimension of a water molecule is smaller than that of N_2 or CO_2 molecules.

The results in Table 4.5 differ from those of Anderson et al.[16] who reported similar CO_2 and water surface areas. Mahajan and Walker[77] suggested that the disparity in the results is due to a difference in the mechanisms involved in the adsorption of CO_2 and water.

4.3.4 Mechanism of Water Adsorption

The nature of bonding of water to coals is still somewhat of an enigma to coal scientists. Various descriptive terms have been suggested for the bonding, including *free, bulk, surface, capillary, inherent, colloidal, hygroscopic, combined, bound, adsorbed,* and even *dissolved water*.[84] This confusion is understandable because there is a continuum in the release of water, ranging from evaporation of free water to the thermal decomposition of some functional groups and the release of water of hydration from certain minerals present.

The comparison of water and CO_2 surface areas (Table 4.5) indicates that different mechanisms govern the adsorption of CO_2 and H_2O. A pure carbon surface devoid of hetero-atom functionalities and inorganic impurities is essentially hydrophobic. Graphon, a graphitized carbon black, has such a surface with an area of $\sim 80 \, \text{m}^2/\text{g}$. It adsorbs little or no water up to about 0.60 RVP.[85] N_2 and other nonpolar molecules are adsorbed in proportion to the Graphon surface area. Dispersive forces, which largely determine the adsorption of nonpolar molecules such as CO_2 and N_2, play an insignificant role in water adsorption. The nature of a carbon surface rather than its magnitude largely determines its moisture sorption characteristics.[79] The moisture sorption isotherms on oxidized Graphon samples (Figure 4.13) support this view. The various Graphon samples referred to in Figure 4.13 were prepared by activating Graphon in oxygen at 600°C to different levels of burn-off.[86] These samples were subsequently exposed to oxygen at 300°C to chemisorb oxygen. The larger the level of burn-off, the larger was the subsequent capacity for oxygen chemisorption. The number of chemisorbed oxygen atoms was taken to be a measure of the active surface area (ASA) of the sample. Activation also increased the total BET N_2 surface area in addition to increasing the ASA. The increase in ASA upon activation greatly exceeded the increase in total surface area; the BET N_2 area increased by about 55 percent upon activation to 37.9 percent burn-off, whereas the ASA increased by a factor of about 20. Results (Fig. 4.13) show that moisture sorption capacity of activated Graphon samples increases with increasing chemisorbed oxygen content. It is of interest to note that when the sorption results are plotted as water molecules adsorbed per unit ASA as a function of RVP, the data points for all four isotherms in Figure 4.13 fall close to a common curve (see Figure 4.14). This clearly shows the insensitivity of moisture sorption capacity of carbons to surface area and the profound effect of oxygen complexes on water adsorption. The dependence of water sorption capacity of carbons, charcoals, and carbon blacks on their oxygen contents has also been ob-

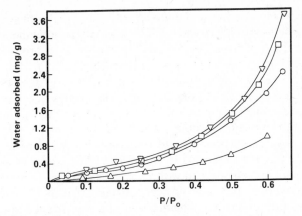

Figure 4.13 Water adsorption isotherms at 20°C on Graphon samples of different degrees of activation.[86] △, 9.1 percent; ○, 24.9 percent; □, 37.9 percent; ▽, 70.4 percent.

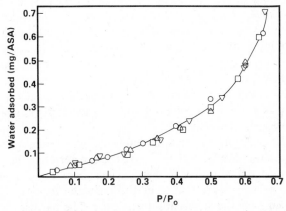

Figure 4.14 Water adsorption at 20°C on Graphon samples of different degrees of activation.[86] The amount adsorbed is normalized on a basis of active surface area covered by oxygen complexes. Degree of activation: △, 9.1 percent; ○, 24.9 percent; □, 37.9 percent; ▽, 70.4 percent.

served.[79,85–90] The porous carbonaceous solids, however, can potentially adsorb significantly more water per unit ASA than do the Graphon samples because of condensation of water vapor in the pore system.

Pierce et al.[85] proposed that initial sorption of water on carbons containing oxygen functional groups involves adsorption at the *primary* specific oxygen sites. Because of the strong intermolecular forces in water, water adsorbed on the primary sites acts as *secondary* sites for the adsorption of additional water. This leads to cluster or island formation. The clusters grow in size with increasing surface coverage and at some stage the clusters on neighboring adsorption sites merge to cover the surface with a monolayer of adsorbed water. With further increase in RVP, pores of increasing dimensions and ultimately the entire pore volume is filled. Such a mechanism has been proposed for water adsorption on carbons, charcoals, and carbon blacks[79,85–90] and should be applicable to coals as well. Coals, like other carbonaceous solids, contain varying amounts and types of oxygen functional groups at the edges of layer planes. Inherent mineral matter, as well as the functional groups, can also contribute to the moisture sorption capacity of coals.

That the oxygen functionalities in coal form hydrogen bonds with polar molecules like water is supported by the work of Mazumdar et al.[91] Phenolic OH groups in a vitrain associated with 13.5 percent oxygen were converted into the acetyl derivative. The moisture content (air-dried at 60 percent relative humidity) and heat of wetting in methanol of the raw vitrain and its acetylated derivative were measured. Both the moisture content and heat of wetting decreased sharply after acetylation. A decrease in these values proportional to the blockage of hydroxyl oxygen by acetyl was observed.

It has been suggested that the application of the BET equation to moisture sorption isotherms on coals gives a measure of specific adsorption sites, that is, carbon–oxygen surface complexes rather than the physical extent of surface. The concept of monolayer coverage and the surface area estimated from water sorption isotherms, particularly for microporous adsorbents, must be used with considerable reservation.[77] Dannenberg and Opie[87] are of the view that condensation of water in capillaries of porous carbon blacks probably starts before monolayer coverage is achieved and that it is not possible to estimate monolayer coverage accurately.

In spite of the generally accepted mechanism of water adsorption on carbons and coals discussed above, Kaji et al.[92] have tried to correlate the water-holding capacity of coals of different rank to their oxygen content and pore structure. In this study, coal particles were soaked in distilled water for 4 hours at $22 \pm 2°C$ and then washed until the supernatant was clear. The particles were removed, and wiped with filter paper to remove the water from their external surface. The water-holding capacity was measured as the loss in weight upon drying at 110°C to constant weight. It was ascertained that (1) the immersion of coal particles in water for 4 hours was enough for water sorption to reach equilibrium, and (2) the measured water-holding capacity was nearly equivalent to the moisture content at 100 percent relative humidity.

Pore size distribution and pore volume were measured using mercury intrusion porosimetry. Mercury intrusion was measured up to 2000 atm, corresponding to filling up of pores $\geqslant 37.5\,\text{Å}$. The pore surface area was calculated from the mercury intrusion data using the cylindrical pore model.

In their attempt to correlate water-holding capacity to coal oxygen content and porosity, Kaji et al. made two assumptions: first, that the organic oxygen content in coal is distributed uniformly and second, that the oxygen functional groups that influence water adsorption increase proportionately with the oxygen content. With these assumptions, the number of hydrophilic sites on coal surface was regarded as being proportional to the product of the oxygen content and pore surface area, $[0] \times S_g$, where $[0]$ is the weight fraction of oxygen in the coals, and S_g is the specific surface area. As seen in Figure 4.15, a fairly good linear relation is obtained between water-holding capacity and $[0] \times S_g$.

These results and conclusions referred must be viewed with skepticism for several reasons. First, physical adsorption of an adsorbate at close to saturation pressure is more a function of pore volume than surface area. In fact, adsorption at RVP ~ 1 is often used to calculate pore volume of an adsorbent (Gurvitsch rule). Second, Kaji et al. considered pore volume and surface area enclosed only in a part of the pore network, that is $\geqslant 37.5$ Å in size. As previously noted, most of the coal surface area is enclosed in micropores whereas most of the pore volume is enclosed in mesopores and macropores. If it is assumed, as Kaji et al. did, that oxygen is distributed uniformly within the coal structure, then most of the organic oxygen content should be contained on pores $\geqslant 37.5$ Å in size. In other words, surface area and oxygen content contained in pores $\leqslant 37.5$ Å are not included in the data in Figure 4.15. Third, the organic oxygen contents of coals calculated (by difference) from the dry-ash-free elemental analyses tend to be erroneous.[93,94] Thus, it appears that the relationship observed in Figure 4.15 is fortuitous. It cannot be denied, however, that coal porosity plays a major role in water adsorption, particularly at intermediate and high RVPs. Because of the complexity of the system it is difficult to relate water adsorption to coal surface area quantitatively.

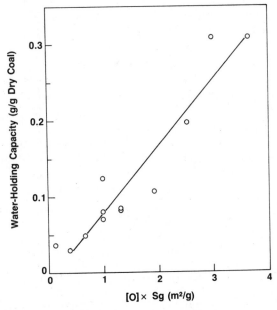

Figure 4.15 Water-holding capacity of coals in relation to their oxygen contents and surface areas.[92]

4.3.5 Hysteresis During Water Adsorption

Water adsorption–desorption isotherms on coals are characterized by hysteresis down to zero RVP, and several mechanisms have been proposed to explain this. The Kelvin equation is often invoked to explain hysteresis in porous adsorbents.

$$ln(p/p_o) = -(2V\gamma \cos\theta)/rRT \qquad (4.7)$$

where p_o is the saturation vapor pressure, γ the surface tension, p the equilibrium pressure, V the molar volume of the liquid adsorbate, and θ the contact angle between the liquid and pore wall. The Kelvin equation cannot explain hysteresis for adsorbents such as coals that contain pores of molecular dimensions. The equation implies meniscus formation and is based on the assumption that the size of the adsorbate molecule is negligible compared to the capillary radius. However, the concepts of a meniscus and surface tension lose their macroscopic sense in pores of molecular dimensions.[95] The cluster or domain model for water adsorption has been used to explain hysteresis in carbons.[79,85] In such a model desorption involves evaporation from the surface after the merger of clusters. This process persists until the continuous adsorbed water film breaks up. Thereafter adsorption would be reversible. This mechanism, unlike capillary condensation, can explain hysteresis for even nonporous adsorbents but does not account for it below monolayer coverage. The *independent domain* model proposed by Amberg et al.[96] attributes hysteresis to a large number of domains capable of existing in two physical states or microphases. The capillaries could define the boundary of such domains in coals. According to this model, hysteresis results from the conversion of one state to another.

Coals swell during adsorption and more surface is exposed for further adsorption. Swelling during water adsorption is ascribed to intermicellar penetration and is not reversed in desorption. Water molecules are desorbed in increasing order of bond strengths, and the collapse or shrinkage of capillary structure is delayed to very low coverages. Thus swelling–shrinkage could result in a hysteresis loop persisting well into the monolayer region. The proposed mechanisms for hysteresis are merely speculative and, as rightly pointed out by Allardice and Evans,[84] no test has been devised to check them rigorously. None of these mechanisms can explain hysteresis down to zero RVP. The irreversibly adsorbed water at zero RVP is either held tightly within the micropores or at the surface by a mechanism involving hydrogen bonding to certain functional groups.

4.3.6 Heat of Water Adsorption

Both desorption and adsorption isotherms have been used to calculate isosteric heats. Allardice and Evans[76] estimated isosteric heats of desorption from water isotherms on a brown coal. The isosteric heat is approximately equal to the heat

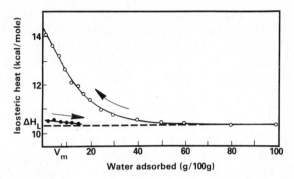

Figure 4.16 Isosteric heat of adsorption of water on brown coal as a function of moisture content.[76] O, desorption data; ●, readsorption data; dashed line denotes heat of vaporization of liquid water.

of vaporization down to ~ 60 g/100 g of dry coal (Figure 4.16). Water released in this range is believed to be present as free water between the coal particles. Isosteric heat increases from 10.4 to 11.9 k cal/mole between 60–15 g water/100 g of dry coal. This increase was attributed to the desorption of water from progressively smaller pores. Below 15 g water/100 g of dry coal the heat of desorption increases from 11.9 to 14.0 K cal/mole at the lowest surface coverage investigated, 2.5 g water/100 g dry coal. This increase was attributed to the desorption of strongly bound water in the multilayer and monolayer regions.

Allardice and Evans maintain that the net isosteric heat of desorption of ~ 4 K cal/mole (that is, the calculated isosteric heat minus the heat of vaporization of water) in the multilayer and monolayer regions is too low to be caused by strong chemical bonding of the coal surface to water. The heat is of the right order for hydrogen bonding such as that between water hydrogen and functional group oxygens. Allardice and Evans found the heats determined from the readsorption branches of the isotherms to be only slightly higher than the heat of vaporization of water even in the monolayer region, in sharp contrast to the desorption heats. Only a marginal dependence on surface coverage was shown. Mahajan and Walker[77] determined heats of adsorption of water in coals of different rank from adsorption isotherms at 0 and 20°C. Extent of adsorption at a given absolute pressure was larger at 0°C than at 20°C as expected. When the data for a given coal were plotted on a relative pressure basis, one smooth curve fitted the data at the two temperatures, implying that the isosteric heat of adsorption is equivalent to the heat of vaporization of water and consistent with the data of Allardice and Evans.

Heats of adsorption of water on microporous Saran carbon (prepared from the carbonization of Saran, a copolymer of polyvinylidene chloride and polyvinyl chloride) and Spheron (a carbon black) at low surface coverages are significantly greater than the heat of vaporization of water.[97–98] The two heats are almost equal at higher coverages. Dacey et al.[97] give the heat of adsorption of water on Saran carbon at a surface coverage of about 1 percent as 15 K cal/mole; at about 5 percent coverage, the value approached the heat of liquefaction. Higher heats of water adsorption on coals were not observed by Mahajan and Walker because the first adsorption point in their isotherms corresponded to much higher surface coverages, in some cases as high as 40–50 percent.

It appears then that water molecules initially adsorbed on the coal surface are hydrogen bonded to surface oxygen complexes and are more strongly bound than those adsorbed subsequently. Hydrogen bonds are stronger than those associated with physical adsorption. Hence, surface hydrogen-bonded water molecules require higher temperatures for their desorption. The occurrence of hysteresis down to zero RVP. (cf., Figure 4.12) can be explained on this basis. Puri et al.[99] state that the amount of water retained irreversibly at zero RVP by a series of sugar carbons containing different amounts of oxygen complexes is directly proportional to the CO_2-evolving complex. It is still not established whether the total oxygen content of a carbon or the fraction present as acidic oxygen complexes (such as phenolic and carboxylic) is the critical influence on its moisture sorption capacity.

An adsorption process can occur spontaneously only if it shows a decrease in free energy. The net heat of water adsorption is zero so an increase in entropy is required. Mahajan and Walker[77] found the entropy of adsorption to have a positive value over the entire pressure range, and its magnitude decreases with increasing adsorption.

4.3.7 Behavior of Water in Coal Pores

Heat Capacity Measurements. Low temperature heat capacity measurements have been used to understand the behavior of water present in the coal pore structure. Mraw and Naas-O'Rourke[100] and Mraw and Silbernagel[101] measured heat capacities of three bed-moist coals (Illinois bituminous, Wyodak subbituminous and Arkansas lignite) at low temperatures. Coal samples containing various amounts of water were prepared by exposing the bed-moist coals to an ambient laboratory atmosphere for different intervals of time. Dry samples were prepared by evacuating the bed-moist coals at room temperature in desiccators, first over $CaSO_4$ and then over P_2O_5. Heat capacity measurements on the subbituminous coal are illustrated in Figure 4.17. If water in the coal behaved as bulk water, a large sharp peak would have been observed at 273 K (melting point of ice) with a latent heat of fusion of ~ 80 cal/g water. Results (Figure 4.17) show that only a small fraction of the water in coal resembles bulk water. A phase transition is observed only in samples containing relatively large amounts of water, and the peaks in these samples represent only a fraction of the enthalpy change expected for bulk water. The apparent heat capacity of absorbed water was calculated by subtracting the heat capacity of the dry coal from that containing water. The assumption is that the heat

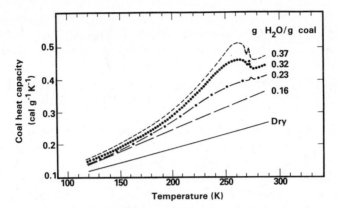

Figure 4.17   Total heat capacity for Wyodak coal as a function of water content.[100]

capacity of the coal matrix is unaffected by water. The apparent heat capacities for adsorbed water calculated for the samples in Figure 4.17 are shown as a function of temperature in Figure 4.18. Results for bulk water are also included. The heat capacity of bulk water is nearly independent of temperature above the melting point of ice. It shows a discontinuity of ~80 cal/g at 273 K and decreases linearly in the solid from 273 K to about 120 K. Near 120 K and at room temperature the heat capacity of water in the coal pores is similar to that of bulk water. The intermediate temperature behavior is different in two important respects. First, no latent heat of fusion is observed for pore water. Second, the temperature dependence of heat capacity is a function of the pore water content. At low water levels, for example, 16 percent, the heat capacity increases monotonically from the solidlike value near 125 K to liquidlike at room temperature. At higher water contents, such as 32 and 37 percent, a broad

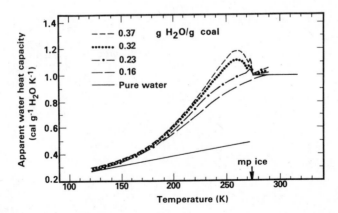

Figure 4.18   Apparent water heat capacity for Wyodak coal as a function of water content.[100]

heat capacity maximum is observed. The authors propose that there are two types of water, freezable and nonfreezable, the water fraction not exhibiting a heat capacity maximum being termed nonfreezable. The nonfreezable water is assumed to be absorbed in the smaller pores.

Distribution of freezable and nonfreezable water in the three coals is given in Table 4.6. Although the total water contents of the three coals are significantly different, the relative amounts of the freezable and nonfreezable water are comparable for all. The lignite differs from the other two in its capacity for additional water at >0.98 RVP. The surface areas of the three coals were calculated (Table 4.6) assuming the amount of nonfreezable water to be proportional to the coal surface area. The surface area of the bituminous coal is in excellent agreement with the CO_2 surface area ($128\, m^2/g$) as well as that measured by small angle x-ray scattering ($140\, m^2/g$). The surface areas of the subbituminous coal and lignite are appreciably higher than the total CO_2 areas expected from the general behavior in Figure 4.7. These results highlight once again an important question raised earlier, that is, what is the effect of removal of water from low-rank coals on their surface area? Clearly, there is need for further investigation.

Mraw and O'Rourke[102] compared the differential enthalpies of fusion of absorbed water in the three coals (listed in Table 4.6) containing different amounts of naturally occurring moisture as well as water reabsorbed on the dried coals. Separate batches of the three coals were dried at various temperatures between 24 and 100°C under an inert atmosphere and in vacuum; the dried coals were subsequently equilibrated with water vapor. The enthalpy of fusion of absorbed water was determined as the area between the peak in the heat capacity and a smooth base line drawn from about 220 to 280 K (cf., Figure 4.18). Results for the naturally occurring water are shown in Figure 4.19; the dashed curves were considered to be estimates of the true behavior. The slopes of the plots in Figure 4.19 represent the differential enthalpy of fusion of the

TABLE 4.6 Distribution of Water Types in Coals

	Water Content (wt% DMF)		
	Illinois Bituminous	Rawhide Subbituminous	Arkansas Lignite
Total	16	48	67
Nonfreezable	7	25	25
Freezable:			
<0.98 RVP	9	21	27
>0.98 RVP	0	~3	15
Effective H_2O surface area (m^2/g)	130	460	460

Source: S. C. Mraw and B. G. Silbernagel, *AIP Conf. Proc.* **70**, 332–343 (1981).

absorbed water at a given water content. Differential enthalpy of fusion of absorbed water in any system depends on the size of the cluster of water molecules participating in the fusion process. Mraw and O'Rourke propose that the apparent linearity of the curve for Illinois No. 6 coal in Figure 4.19 is due to the presence of pores of a relatively uniform size over a range of water contents. Curvatures in the plots for the lignite and subbituminous coal on the other hand imply a distribution of pore sizes; increasing slope of the curves reflects an increasing cluster size at progressively higher water contents.

Results of the enthalpy of fusion of reabsorbed water are compared to those of naturally occurring water (Figure 4.20). The higher-rank Illinois coal is apparently not affected by drying because the data for the reabsorbed water are similar to those for the naturally occurring moisture. This is not the case for the lignite and subbituminous coal. Even though the onset of the fusion peak for the lignite occurs at about the same water content for both the natural and reabsorbed water, the data points for the latter lie well above those of the natural water. Apparently, the drying process alters the original structure of lignite and relatively fewer small- and medium-sized water clusters are present. This phenomenon is analogous to what has been termed *pore collapse* in the gel-like structure of lignite.[103,104] The results for water reabsorbed on the subbituminous coal initially follow the natural water, implying that small-to-medium pores remain intact with drying, but total pore volume is altered.

Nuclear Magnetic Resonance Spectroscopy.   The heat capacity measurements help in understanding the long-term macroscopic average behavior of water in coal pore structure. A better understanding of the interaction of the resident molecules with pores and the transport of gases and liquids (as in liquefaction and gasification reactions) in the coal pore network

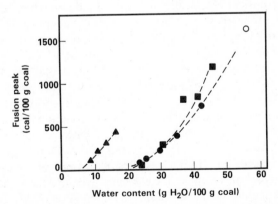

Figure 4.19   Enthalpy of fusion of naturally occurring absorbed water in relation to water content of the room-temperature air-dried coals.[102] ▲, Illinois coal; ■, Rawhide; ●, lignite. The open circle indicates a value for a lignite sample taken before equilibration at 98 percent relative humidity.

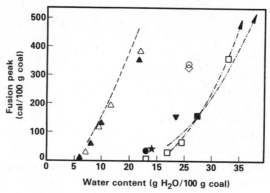

Figure 4.20   Enthalpy of fusion of reabsorbed water as a function of the amount of water reabsorbed on coals predried between 24 and 100°C.[102] Illinois coal dried at: △, 24°C; ▲, 100°C. Rawhide coal dried at: □, 24°C; ■, 100°C. Lignite dried at: ○, 24°C; ◇, 40°C; ▽, 60°C; ✳, 80°C; ●, 100°C. The curves are reproduced from Figure 4.18: ------ Illinois; -·-·-, Rawhide; -··-··-, lignite.

can be gained by coordinating heat capacity measurements with other suitable measurements that give information on short time molecular motions. Molecular motion in the $10^{-7}-10^{-8}$ second time range can be studied by nuclear magnetic resonance (NMR) spectroscopy.

Nuclear interactions are averaged by molecular motions occurring in the $10^{-5}-10^{-6}$ second range; this phenomenon is known as motional narrowing. Motions in the 10^{-7} to 10^{-8} second region produce fluctuations, resulting in energy exchange between the nuclei and their environments, the spin-lattice relaxation. Wide line NMR measurements of the shape and width of the NMR absorption and transient NMR measurements of the spin lattice relaxation time can be used to probe these motions. Mraw and Silbernagel[101] applied the technique to two types of coal samples, Wyodak coal with 31 and 18 percent water and Wyodak and Illinois No. 6 coals containing 17 and 12 percent D_2O. From heat capacity measurements, it was concluded that the Wyodak coal associated with 18 percent moisture contains the maximum amount of water in the nonfreezable form. Wyodak coal, with 31 percent water, contains a significant freezable component. The use of D_2O-containing samples has two important advantages. First, there is very little background signal from the organic material. Second, the NMR properties of the ^2D nucleus are distinctly different from those of the ^1H nucleus. Resonance properties of protons are governed by interaction of their nuclear spin with the local magnetic field, whereas those of deuterium are determined by interaction of the nuclear electric quadrupole moment with electric field inhomogeneities of the O–D bond in the molecule. This interaction is not only about an order of magnitude stronger than the magnetic interaction with the proton, but also intramolecular and highly local in nature. Proton line widths versus temperature plots for two Wyodak coal samples containing 0.18 and 0.30 g H_2O/g dry coal were virtually superimposable, indicating that there is no distinction between freezable and nonfreezable water at molecular distances on a $10\,\mu$ second time scale. In the 200–300 K temperature range, the relative intensity of the NMR signals from the two samples was constant. This shows that one component of the sample does not selectively freeze when the temperature is lowered. The activation energy for molecular diffusion was calculated to be 4.6 k cal/mole. Line broadening for the D_2O-containing samples occurred at higher temperatures than for the H_2O-containing samples. Broadening for the Wyodak coal occurred at higher temperature than for the Illinois coal, suggesting slower D_2O motion. The activation energy for molecular diffusion in Wyodak coal was calculated to be 3.4 k cal/mole.

The results of the transient NMR measurements of D_2O spin lattice relaxation imply an equivalent energy exchange process for all observable nuclei. The relaxation times show an exponential increase with temperature in the low temperature region corresponding to an activation energy of 4.0 K cal/mole. Transient NMR measurements of the H_2O spin relaxation for the two Wyodak coals showed a complex behavior. The evidence points to relaxation by coupling of the nuclei with paramagnetic impurities in the coal. Such effects do not occur in the D_2O observations because the quadrupole interaction is appreciably higher, and the impurity relaxation process is significantly weaker. It appears that the D_2O observations can be used to investigate molecular dynamics of resident molecules without any interference from the paramagnetic impurities.[101]

Important conclusions can be drawn from the heat capacity and NMR studies of Mraw and Silbernagel.[101] Whereas the heat capacity measurements show varying amounts of freezable and nonfreezable water in different coals, NMR studies show that the process of freezing does not necessarily involve discrete phase transitions but rather a continuous decrease in rate of molecular motion with decrease in temperature. Physical interactions of water with the surface of the pores and small pore size seem to be responsible for broadening of the transitions.

Lynch and Webster[105] studied the behavior of water associated with an air-dried brown coal using NMR pulse methods. The intensity and spin–spin relaxation of the magnetic resonance of water protons of specimens containing a range of water contents were measured during temperature cycles between 220 and 300 K. The ^1H NMR spin–spin relaxation decay of these samples could be resolved into a slowly decaying component attributed to water protons (present as mobile bound water + free water), and a faster decaying component attributable to coal protons and protons of any rigidly bound immobile water molecules; the slowly and faster decaying components approximate the freezable and nonfreezable water in the classification of Mraw and Silbernagel. Since the relaxation rates of the two parts were sufficiently different, it was possible to separate the water signal from the coal signal. The following significant conclusions can be drawn from the study.

1. The temperature dependence of the ^{1}H NMR signal intensity and spin–spin relaxation of water can distinguish between free and interacting or bound water; bound water can be present as immobile and/or mobile water.

2. There is little interaction between free and bound water.

3. Immobile water molecules do not contribute to the observed mobile water ^{1}H NMR signal. Their proportion increases with decreasing temperature and varies with water content. These water molecules do not exchange effectively with the rest of water during spin–spin relaxation.

4. The mobile bound water molecules constitute a single phase. This implies that they can effectively exchange with one another during the time of the spin–spin relaxation. It is unlikely that the bound mobile molecules exist in discrete phases with significantly different properties.

5. The immobile water molecules do not constitute a discrete phase but rather occupy extreme immobile states in a continuous distribution of states.

Lynch and Webster[106] later used ^{1}H NMR studies to measure changes in water-binding capacity of two bed-moist brown coal litho types subjected to heat treatment (HT) in the 0–300°C range. The bound water decreased appreciably with increase in HT temperature above 130°C for both litho types. This is consistent with the loss of oxygen functional groups upon HT.

4.4. EFFECTS OF WEATHERING ON COAL PROPERTIES AND BEHAVIOR

When bed-moist coal is exposed to air, it loses water until an equilibrium with the ambient atmosphere is established. Coal shrinks during desorption of water and swells on readsorption. The extent of shrinkage and swelling depends on coal rank. Bituminous coals shrink negligibly on drying, whereas low rank coals undergo severe shrinkage. The drying procedure can effect profound changes in pore structure and surface behavior.[103] Coals not only lose moisture but also chemisorb oxygen upon air exposure. The rate and extent of chemisorption depend on coal rank, particle size, maceral composition, surface area, pore size distribution, and possibly mineral matter. Oxidation results in the formation of surface oxygen functional groups such as carboxyl, hydroxyl, and carbonyl, and these complexes can also affect pore structure and surface behavior. It is reiterated that little specific information is available in the literature on the effects of weathering on coal properties by way of exposure of coal to the ambient atmosphere. A significant proportion of the work discussed here concerns weathering at relatively high temperatures. It may be assumed, though this is indeed speculative, that changes in structure and reactivity occurrring at high temperatures are similar in nature but different in magnitude than those at lower temperatures.

4.4.1 Porosity

Since low-rank coals undergo considerable shrinkage upon drying, gas adsorption techniques, which always necessitate a predrying step, may not measure their total surface area accurately. Small angle x-ray scattering (SAXS) is not affected by moisture present in coals (Section 4.2.4). Wegenfeld et al.[107] used the SAXS technique to evaluate changes in porosity and pore size distribution with drying, using two lithotypes of an Australian brown coal. They determined scattering curves from the dry and wet samples. Each sample gave a concave scattering curve attributed to the presence of a dilute system of micropores, polydisperse in size but approximately identical in shape. From these curves, radius of gyration, micropore volume, and surface area were calculated (Table 4.7). Micropore volumes of the dry samples determined from CO_2 (0°C) P–D plots are also included. SAXS and gas adsorption give comparable micropore volumes but significantly different surface areas for the dry lithotypes. In both methods the surface area is calculated from measured micropore volumes. The calculations are based on assumptions of uncertain validity. The area occupied by an adsorbed molecule must be used in the gas adsorption method, whereas with SAXS the geometrical form of the micropores is assumed for computing the surface area. Micropore volumes of both the dry

TABLE 4.7 Structural Parameters Determined by Gas Adsorption and SAXS Techniques

Lithotype	Micropore Volume (cm^3/g)		Surface Area (m^2/g)		Radius of Gyration (Å)
	Gas Adsorption	SAXS	Gas Adsorption	SAXS	
			I		
Wet	—	0.100	—	30–210	1580
Dry	0.079	0.095	298	90–1100	830
			II		
Wet	—	0.110	—	40–390	1360
Dry	0.058	0.060	216	70–830	720

Source: H. K. Wagenfeld, M. Setek, L. Kiss, and W. O. Stacy, *Proc. Int. Conf. Coal. Sci.*, 869–874 (1981).

and wet lithotype I are about the same, but the wet lithotype II has a significantly higher micropore volume than the dry sample. For both lithotypes the radius of gyration decreases sharply with drying and can be attributed to shrinkage in the microporous structure. Bale et al.[108], who also used SAXS to determine changes in pore structure of a North Dakota lignite upon vacuum drying at ambient temperature, reported an increase in total surface area by 57 percent from 13.7 to 21.6 percent. This result is in qualitative agreement, although of smaller magnitude, with the SAXS results on the dry and wet lithotypes of Victorian brown coals.[107,109]

Swann et al.[110] studied changes in surface area occurring during oxidative weathering of a "fresh" Australian brown coal. Oxidation of the bed-moist coal, $-6BS$ mesh, was not detectable until the surface layer was dry. The predried samples were exposed to pure oxygen for 45 days at 1 atm at 35 and 70°C. Blank samples for surface area measurements were prepared by outgassing at 35 and 70°C for seven days. Surface areas were determined from CO_2 isotherms (0°C), using the P–D equation (Equation 4.5).

Surface areas and values of D, the slope of the P–D plot, of various coal samples are given in Table 4.8. The results show that mild heat treatment of the unoxidized coal results in a slight decrease in surface area probably due to thermal decomposition of surface functional groups. The breakdown of these groups causes some collapse of micropores, reducing the surface area and increasing D, the latter being indicative of an increase in average micropore size. The results in Table 4.8 show that the effect of heat treatment alone is small compared to the effect of oxidation. Since most of the coal surface area is enclosed in micropores, adsorbed oxygen would tend to block the pores sufficiently to restrict access of CO_2 to the surface area contained within them.[110] The values of D in Table 4.8 show that oxidation tends to increase the mean diameter of those pores penetrated by CO_2. This is consistent with the

TABLE 4.8   Effect of Drying and Oxidation on Surface Area

Sample	Surface Area (m^2/g)	D
Raw Coal		
Degassed at 22°C	290	0.118
Degassed at 35°C for 7 days	262	0.105
Degassed at 70°C for 7 days	254	0.112
Oxidized Coal		
Oxidized at 35°C for 45 days	187	0.112
Oxidized at 70°C for 45 days	156	0.123

Source: P. D. Swann, D. J. Allardice, and D. G. Evans, *Fuel* **53**, 85 (1974).

adsorption of oxygen either closing some of the smaller pores or reducing their size sufficiently to exclude CO_2.

The effect of drying and oxidation of coals on pore structure has been investigated by Karsner and Perlmutter.[111] The coals examined varied in rank from anthracite to lignite. The separate effects of drying and oxidation on pore structure were determined by thermal drying in a nitrogen atmosphere at temperatures up to 300°C and by sequential drying plus oxidation. Simple vacuum drying at 25°C provided references. Pore volume distributions were measured by mercury porosimetry. Changes in pore volume distribution with drying and oxidation are given in Table 4.9. Comparison of the results show that PSOC-190 and PSOC-87 coals become more macroporous (pores > 300 Å) in vacuum drying and less microporous (pores between 35 and 300 Å) in thermal drying. (The authors used a different classification to distinguish pores of different sizes than the IUPAC classification discussed in Section 4.1). PSOC-197 and PSOC-127 coals on the other hand undergo a decrease in macroporosity upon thermal drying. The authors attributed the changes in pore volume resulting from loss of moisture to the formation of cracks, particle shrinkage, and particle breakage. The results shown in Table 4.9 have to be viewed with some skepticism. The drying conditions used in the study (300°C in N_2) are quite severe at least for low-rank coals. Lignites lose CO_2 (due to decarboxylation) at 100°C or below.[73] This fact was realized by the authors as well, and they monitored the release of CO_2 and CO during thermal drying. Coal weight losses due to release of the two gases in all cases were < 1 weight percent. Loss of even small amounts of CO_2 and CO can significantly affect the aperture size, and hence pore volume and surface area since coals have an aperture-cavity type porosity.

Before considering the combined effect of drying plus oxidation on pore volume distribution, it is emphasized that comparison of the data for the thermally dried samples with the dried plus oxidized samples may be misleading. In the former case, the maximum heat treatment temperature (HTT) during drying was 250°C, whereas in the latter it was mostly at 200°C. A 50°C increase in HTT can bring about a significant decrease in weight loss and change in the pore volume distribution particularly for lower-rank coals.[112]

Drying followed by oxidation increases the macroporosity of PSOC-87 and PSOC-190 coals relative to the vacuum-dried and thermally dried coals. Similar behavior is also indicated for the anthracite (PSOC-80). These trends would probably have been more pronounced if the coal samples during combined drying and oxidation were first dried at 250°C followed by oxidation at 200°C.

The macroporosity and microporosity changes were expected. Oxidation of coal at 200°C can lead to chemisorption of oxygen as well as gasification; gasification is always preceded by oxygen chemisorption.[113] The relative magnitude of the two reactions depends on the reaction temperature. Karsen and Perlmutter[111] reported a net weight loss of up to 5 weight percent during coal oxidation, indicating that gasification was more pronounced than oxygen chemisorption. It is well recognized that during the earlier stages of gasification

TABLE 4.9 Effect of Drying and Oxidation on Pore Volume Distribution

PSOC No.	Rank	Pore Volume (cm³/g)					
		> 300 Å			35–300 Å		
		Vacuum Dried	Thermally Dried[a]	Dried and Oxidized[b]	Vacuum Dried	Thermally Dried[a]	Dried and Oxidized[b]
87	Lignite	0.023	0.027	0.030	0.029	0.026	0.021
190	HVC bit.	0.025	0.028	0.029	0.089	0.067	0.061
197	HVB bit.	0.032	0.022	0.024	0.041	0.035	0.031
4	HVA bit.	0.012	—	0.012	0.023	—	0.024
127	LV bit.	0.014	0.010	0.014	0.026	0.026	0.024
80	Anthracite	0.009	—	0.011	0.019	—	0.014

[a]Dried at 250°C.
[b]PSOC-190 coal was dried and oxidized at 250°C; all other coals were dried and oxidized at 200°C.
Source: G. G. Karsner and D. D. Perlmutter, *Ind. Eng. Chem. Process Des. Dev.* **21**, 348 (1982).

of coals and chars there is an opening of closed pores and enlargement of smaller ones.[113] Dried and oxidized PSOC-4 and PSOC-127 coals show virtually no change in pore volume distribution from those of the vacuum-dried coals, consistent with their much lower reactivity to gasification. Mahajan and Walker[113] reported that in oxidizing atmospheres (CO_2, O_2, steam) gasification rates of chars prepared by heat treatment of low-rank PSOC-87 and PSOC-190 coals were appreciably higher than that of the char produced from the higher-rank PSOC-127 coal. Kaji et al.[114] have reported that pores with radii > 100 Å play an important role in the low temperature oxidation of coals varying in rank from anthracite to lignite. For an Australian bituminous coal (C content 76.9 percent, DAF) oxidized at 150°C, cumulative volume of pores $> \sim 100$ Å radius was larger than that of the original coal, but the trend was reversed for the smaller pores. As a result, surface area of the coal decreased by ~ 40 percent compared with the original coal. Similar results were obtained with the other coals as well. The results suggest that pores $\leqslant 100$ Å are blocked by the adsorption of oxygen and/or the reaction products at the early stages of the reaction. Kaji et al. concluded that pores < 100 Å do not contribute to the low temperature oxidation reaction indicating the lack of microdiffusional effects.

Oda et al.[115] observed the effect of air oxidation of several coals on their pore structure; the carbon contents (DAF) varied in the 77.8–89.9% range. The coals were oxidized with moist air at 150°C for 1, 5, and 10 hours and their methanol and hexane densities were determined. The methanol density drift was found to decrease with increase in the oxidation time, an effect very marked for the lower-rank coals. The hexane density drift increased after 1 hour of oxidation and decreased with further oxidation in each of the coals considered. Methanol and hexane densities of various samples versus carbon contents of coals are shown in Figure 4.21. Methanol density undergoes little or no change after 1 hour of oxidation but then increases with oxidation time. Hexane density increases significantly during the early stages of oxidation but decreases thereafter. There is a tendency for the hexane density to increase again after 5 hours of oxidation. Both the hexanol and methanol densities show a minimum at about 85 percent carbon content. The abnormal results for the coal with 87.6 percent carbon content were attributed to its relatively high fusinite content. The trends in methanol density and drift after 1 hour oxidation were explained by assuming that in the early stages the number of pores with sizes comparable to methanol do not change, but accessibility or wettability of the pore wall increases. The latter effect is due to the formation of polar oxygen functional groups on the surface. These groups readily form hydrogen bonds with methanol so that density drift decreases with increase in the level of oxidation. The buildup does not continue indefinitely, however. Beyond a certain level, the formation and decomposition of functional groups occur at comparable rates. The hexane density results are explained by the increase of the number of pores large enough to accept a straight chain hexane molecule. There is also an increase in the concentration of oxygen functional groups that increase the density drift by decreasing the accessibility of hexane to the pore network. The subsequent

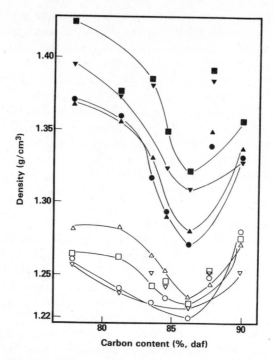

Figure 4.21 Changes in apparent densities of coals upon oxidation as a function of coal rank.[115] Methanol density: ●, parent coal; ▲, oxidized for 1 hour; ▽, oxidized for 5 hours; ■, oxidized for 10 hours. Hexane density: ○, parent coal; △, oxidized for 1 hour; ▽, oxidized for 5 hours; □, oxidized for 10 hours.

increase in density and decrease in the density drift result from the functional group decomposition as in the previous case. The CO_2 adsorption isotherms on various coal samples were also measured.[115] Micropore volumes, W_o, were calculated from the P–D plots, and even though W_o changed markedly with air oxidation, no general relationship was observed between W_o and the coal rank or the extent of oxidation.

4.4.2 Adsorption Characteristics

Paulson and Fuch[116] studied the effect of drying of a lignite on the exchange of Na^+ by Ca^{2+} ions. Samples with different moisture contents were prepared by air-drying the bed-moist lignite at 25°C. A reduction in the coal moisture content down to 34 percent had little or no effect on Na^+ exchange, but below 28 percent moisture, the exchange was reduced significantly. The authors suggest that loss of moisture from the lignite seals a portion of the capillary system, thereby reducing the ion exchange capacity.

Gorbaty[117] investigated the effect of drying of a Wyodak subbituminous coal on its adsorptive capacity for nickel sulfate. The as-received coal containing about 30 percent bed moisture was crushed to −20 mesh. Part of the sample was dried in vacuo at 105°C for 16 hours. Both the as-received and dried coals were soaked in 1 weight percent nickel sulfate, and the extent of adsorption was determined by atomic absorption. Adsorbed nickel on the as-received coal (expressed on a DMF basis) and the dried coal are plotted as a function of square root of time in Figure 4.22. The as-received coal adsorbs nickel much faster than does the dried coal particularly in the initial stages. The maximum nickel uptake on the dried coal is only 25 percent of that on the as-received sample. These results clearly show the deleterious effect of drying on the adsorption characteristics. Two hypotheses were suggested. First, during drying, the gel structure of the coal collapses. The pores within the structure collapse with a consequent loss of surface area. Second, when a coal is dried, the volume previously occupied by water is replaced by the ambient gas, and the wetting properties of the pore surfaces may be changed significantly. These changes tend to prevent or decrease wetting when the dried coal is again contacted with an aqueous solution. The two hypotheses were distinguished in the following manner. A slurry of 10 weight percent dried coal in 1 weight percent nickel sulfate solution was vacuum degassed for 1−2 minutes. The extent of adsorption on the vacuum-treated slurry sample (expressed on a DMF basis) as a function of square root of time is shown in Figure 4.23. Both the curves from Figure 4.22 are also reproduced in Figure 4.23. It is seen that the dried coal adsorbs significantly more nickel when the slurry is vacuum-treated. This observation substantiates the second hypothesis. Outgassing of the slurried coal does not completely restore the adsorption capacity to the same level as that of the as-received sample; some pores collapse irreversibly on drying. Gorbaty[117] suggested that if aqueous reagents are to be used in coal chemistry studies, as-received rather than dried coals should be used to maximize mass transfer.

Youssef and El-Wakil[118] monitored changes in adsorption properties of an

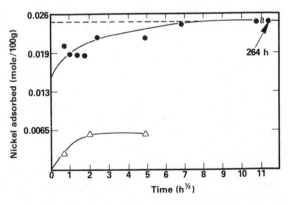

Figure 4.22 Adsorption of nickel sulfate by Wyodak coal slurries.[117] ●, as-received; △, dried.

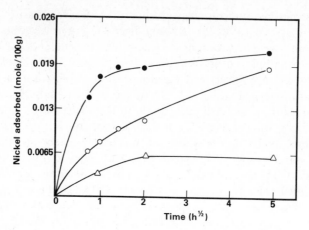

Figure 4.23 Adsorption of nickel sulfate by Wyodak coal slurries.[117] ●, as received; ○, dried, vacuum degassed slurry; △, dried.

Egyptian coal during storage. One aliquot of a sample (−0.6 to 0.2 mm) was stored in a glass bottle. No attempt was made to displace the air present by an inert gas. Another aliquot was stored in a sealed evacuated tube. Adsorption isotherms of water (30°C), methanol (35°C), and benzene (35°C) were measured on fresh samples and those that had been stored for 14 years. Adsorption on the sample stored in the evacuated tube showed little change from that on the fresh sample. These observations show that coal characteristics undergo a minimal change when air (oxygen) is excluded during storage. Storage in air, however, substantially increased water adsorption capacity over the entire RVP range. There was no noticeable effect on methanol and benzene adsorption.

That oxidative weathering of a coal increases its moisture sorption capacity is also supported by the work of Bhattacharaya.[119] He oxidized a coal with 80.7 percent carbon content (DMMF) with moist air at 30°C and measured equilibrium moisture sorption. The amounts of water adsorbed after 0, 30, 50, and 70 days of oxidation were 19.2, 20.8, 21.8, and 23.4 weight percent, respectively. This is in accord with the previous discussion (c.f., Section 4.3.4).

The methanol and benzene adsorption results of Youssef and El-Wakil are surprising in consideration of other work. Puri[79] showed that oxygen complexes on carbon surfaces promote sorption of both water and methanol. Isirikyan and Kiselev[120] reported that phenolic OH groups in carbons enhance adsorption of benzene. Puri et al.[121] observed that the presence of carbonyl groups (not a part of carboxylic and ester groups) on carbon surfaces favors benzene adsorption. This was attributed to the interaction of II electrons of the benzene ring with the partial positive charge on the carbonyl carbon. The results of Youssef and El-Wakil emphasize the complexity of the interpretation of sorption observations on coals. The formation of oxygen complexes can enhance sorption of water, methanol, and benzene. However there are two competing processes. First, oxygen chemisorption blocks some of the micropores and/or decreases the pore openings. The extent of pore exclusion is a function of the level of oxygen chemisorbed and the size of the adsorbate molecule. Second, some of the oxygen chemisorbed during oxidative weathering can introduce crosslinks in the coal structure,[46,47] which reduce swelling during sorption, and total sorbate uptake should decrease after oxidation.

Oxidative weathering in air of an Illinois No. 6 coal was studied by Liotta et al.[122] The compositional analyses of the weathered coal varied up to 56 days of exposure. No further changes in the analyses were observed up to 4 months. Weathering had no noticeable effect on the carbon, nitrogen, and organic sulfur contents of the coal, though the H/C ratio decreased and the O/C ratio increased with increasing time of exposure. Oxygen incorporated into the coal structure during weathering was shown to be present as ether crosslinks. To verify the crosslinking nature of the ether groups, tetrahydrofuran (THF) adsorption on the coal residues obtained after THF extraction was studied. For this study, the coal residues were exposed to saturated THF vapor for 48 hours (equilibrium condition) and the results are shown in Figure 4.24. Weathering is clearly detrimental to adsorption. The extent of adsorption decreases rectilinearly with increase in the severity of weathering, (i.e., the O/C ratio). A more highly crosslinked coal should adsorb less solvent since the structure cannot swell so much, but the observed decrease in THF adsorption could also be due to a decrease in surface area brought about by weathering.

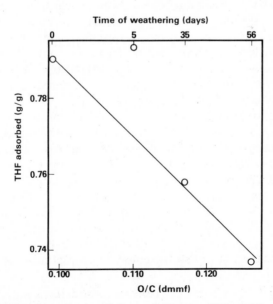

Figure 4.24   THF adsorption in relation to O/C ratio of Illinois No. 6 coal before and after oxidative weathering.[122]

The conclusions of Liotta et al. on the effect of weathering on swelling of coal are not supported by Larsen's results on a HVA Bruceton coal.[123] He oxidized the coal (-70 mesh) with air at 25 and 80°C up to a maximum of 190 days. Free swelling index (FSI) declined sharply over 39 days at 80°C; over a comparable time the coal exposed to air at 25°C exhibited no change in FSI. Weathering at 25 and 80°C reduced the pyridine extractability; the effect was more pronounced in the latter case. Larsen used the ratio of volumetric swelling of the coal before and after weathering as a measure of the changes in crosslink density. The swelling ratio was roughly constant for the sample weathered up to 190 days at 25°C, and the ratio decreased very slightly (indicating an increase in crosslink density) for the sample weathered up to 155 days at 80°C. The swelling ratio showed even less change with time when swelling was performed on the insoluble part of the pyridine-extracted coal compared to the same measurement on the whole coal. Larsen's results suggest that oxidation at the coal surface has a more significant influence on its plastic properties than on the crosslinking reactions that modify the macromolecular structure.

4.4.3 Surface Area Changes During Grinding of Coals in Presence of Water

It has been established that drying as well as oxidation effect some irreversible changes in coal properties; a decrease in particle size increases the magnitude of these changes. Grinding of coal particles in air not only causes surface oxidation but can also produce plastic flow due to heat generated during the grinding process.[41] There is an associated loss of surface area. Coals ground without predrying and in the absence of oxygen (for example under water) presumably would not show these deleterious effects. Two important aspects of grinding under water appropriate for discussion here are (1) rate of coal oxidation, and (2) surface area and porosity changes. To gain an insight into the role of water in the oxidation of coal, Mahajan et al.[14] studied the rate of oxidation of a bituminous coal at 150 and 180°C in 1 atm of dry air and in 1 atm of air containing 13 mm of water vapor. The coal was dried prior to its oxidation. At both temperatures oxidation was significantly slower in wet air: 242 minutes in dry air and 342 minutes in wet air were required to achieve a weight gain of 1.4 percent at 150°C. The corresponding times of oxidation at 180°C were 43 and 66 minutes. The presence of moisture in air affects not only the rate of oxidation of a carbon but also the nature of oxygen complexes formed. Mahajan et al.[124] found that oxidation of microporous carbons in wet air increases the concentration of acidic oxygen complexes that predominantly evolve CO_2 on degassing, whereas oxidation in dry air increases the concentration of CO-evolving complexes.

Water present during particle size reduction of coal has a marked deleterious effect on subsequent surface areas according to Lebedev et al.[125] Three coals—brown coal, hard coal, and anthracite—were ground with different amounts of water in a vibromill. Specific surface areas of the resultant samples were measured by low temperature Ar adsorption. Figure 4.25 shows the changes in surface area of two anthracite samples, one with a maximum amount of hygroscopic moisture (6.2 percent) and the other oven-dried at 150°C, as a function of grinding time. The largest relative increase in surface area for both the samples is observed after grinding for 5 minutes. There is only a slight difference in the surface areas of the two samples at this stage. A difference is clearly evident at longer grinding times. After 30–50 minutes, the surface area of the oven-dried anthracite is approximately twice that of the moist coal. Table 4.10 shows that an increase in water content during grinding significantly reduces the surface area of all three coals investigated. An increase in water content of the anthracite from 1.9 to 32.3 weight percent decreases the surface area from 312 to 102 m^2/g. Grinding of the oven-dried hard coal increases the surface area from 6.5 to 68.9 $m^{2/g.}$ When this coal is associated with the maximum hygroscopic moisture content of 4 weight percent, the ground sample has a surface area of only 31 m^2/g. For the brown coal with the maximum hygroscopic water content (23.1 weight percent), an increase in grinding time from 5 to 20 minutes increases the surface area by 65 percent. The corresponding increase for the oven-dried sample is less than 6 percent.

Anthracites have a BET N_2 ($-196°C$) area of $< 10 \, m^2/g$ and the CO_2 area is about 300–400 m^2/g.[20] It is significant that the Ar area of the oven-dried anthracite after grinding for 90 minutes is similar to the CO_2 surface area reported in the literature (cf. Figure 4.7). This suggests that a lack in reactivity of coals may possibly be overcome by a reduction in their particle size.

Measurement of meaningful properties in any fundamental coal study necessitates minimal changes in characteristics during grinding. To avoid the

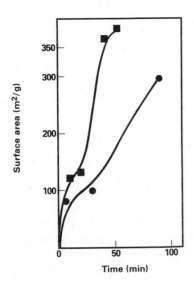

Figure 4.25 Changes in surface area of an anthracite associated with different amounts of water as a function of grinding time.[103] ■, dry coal; ●, coal containing 6.2 percent water.

TABLE 4.10   Effect of Water Content Associated with Coals During Grinding on Their Surface Area

Coal	Time of Grinding (minutes)	Water Content (wt%)	Surface Area (m²/g)
Anthracite	90	1.9	312
		6.2	298
		32.3	102
Hard coal	0	nil[a]	6.5
	5	nil[a]	68.9
		2.2	57.7
		4.0[b]	31.0
Brown coal	0	12.0	11.2
	5	nil[a]	87.0
		12.0	60.0
		23.1[b]	44.6
	20	nil[a]	92.0
		23.1[b]	73.5

[a]Dried at 105°C.
[b]Maximum hygroscopic moisture content.
Source: V. V. Lebedev, G. S. Golovina, and K. I. Cheredkova, *Solid Fuel Chem.* **13**, 35 (1979).

oxidative and thermal effects accompanying grinding in air, Solomon and Mains[126] ground an Illinois No. 6 coal under liquid N_2. The cryogenic procedure compared to the normal grinding procedure produced subtle changes in the coal characteristics. The normal grinding produced a particle size distribution averaging nearly 90 μm. Grinding in liquid N_2 produced particles closer to 38 μm. Nearly a third of the coal ground in liquid N_2 passed the 38 μm sieve compared with only 7.7 percent for the sample ground in air. The sample ground in air and liquid N_2 showed remarkably different sedimentation rates in a 1:1 toluene methanol mixture. Even though the cryogenically ground sample had a smaller particle size, it settled 100 times faster than the air-ground sample. Surface oxidation during grinding in air may produce polar surface groups that interact strongly with methanol, thus reducing the settling rate.

4.4.4   Char Properties

In coal conversion and utilization processes, coal usually undergoes heat treatment. It is useful to discuss some effects of oxidative weathering on pore structure and reactivity of chars produced from the oxidized coals. The changes in pore structure during heat treatment (HT) of oxidized coals are expected to vary in a complex manner with coal rank, particle size, amount and type of inorganic impurities, maceral composition, level of oxidation, heat treatment history, maximum HTT, soak time at maximum HTT, and atmosphere used during HT. Coals of a certain rank, the caking or coking coals, go through a plastic stage during their HT. In coal liquefaction development of plasticity is desirable[127] because it indicates breakage of crosslinks, reduction of molecular weight, and enhanced solubility in organic solvents. Development of plasticity is undesirable in coal combustion/gasification processes. Apart from interfering with the passage of gases through the reactor, development of plasticity results in the loss of surface area. It is desirable to maximize the surface area in coal/char gasification processes.[113] In coal combustion/gasification processes, caking coals are given a mild preoxidative treatment to decrease or destroy their plastic properties, and this treatment converts them from thermoplastic to thermosetting precursors.

Mahajan et al.[14] studied the effect of preoxidation of two highly caking coals, PSOC-337 and PSOC-127 with carbon contents (DAF) of 84.9 and 89.6 percent, on surface area and reactivity toward gasification of the chars produced from the oxidized coals. The coals (40×70 mesh) were preoxidized with air to different levels of weight gain at temperatures up to a maximum of $250°C$. The raw and oxidized coals were converted into chars in N_2 at $1000°C$ and subsequently gasified in 1 atm air at $470°C$. The CO_2 surface areas ($25°C$), and reactivities of the chars expressed as $t_{0.5}$, the time for 50 percent burn-off of the char sample, are given in Table 4.11. The preoxidation of caking coals leads to the production of chars of high surface areas; the greater the level of preoxidation, the greater the surface area.

Char reactivity in gasification increases sharply with preoxidation of the coal

TABLE 4.11   Effect of Preoxidation of Coals on Surface Areas and Reactivities of Chars Produced

Weight Increase During Preoxidation (%, DMF)	Char Surface Area (m^2/g, DMF)	Char Reactivity ($t_{0.5}$, minutes)
PSOC-337		
0.0	12	610
1.9	33	—
3.8	138	140
PSOC-127		
0.0	6	3720
1.7	12	343
4.2	45	170

Source: O. P. Mahajan, M. Komatsu, and P. L. Walker, Jr., *Fuel* **59**, 3 (1980).

precursors. The effect of preoxidation of the two coals on the reactivities of the chars produced is substantially different. A weight gain of 4 percent during preoxidation produces a subsequent fourfold decrease in $t_{0.5}$ for the PSOC-337 char but about a twenty-twofold decrease for the PSOC-127 char. The extent of the changes in reactivity cannot be correlated quantitatively with changes in surface areas of the chars produced as a result of preoxidation. Preoxidation of PSOC-337 coal to a 3.8 percent weight gain results in a twelvefold increase in surface area of the char produced. For PSOC-127 coal, a comparable weight gain during preoxidation results in less than an eightfold increase in char surface area. It may be concluded that surface area as measured by physical adsorption of CO_2 at 25°C is not a good predictor of char reactivity.

Oda et al.[115] made a detailed study of the effect of oxidation of coals on pore structure of the chars produced from them. The coals varied in carbon contents (DAF) from 77.8 to 89.9 percent. They were oxidized with moist air at 150°C for 1, 5, and 10 hours. The raw and oxidized coals were carbonized in a N_2 atmosphere at various temperatures between 300 and 1500°C. The methanol and hexane densities of the chars and their micropore volumes, W_o, were determined from the P–D plots of the CO_2 adsorption isotherms. The methanol density of chars produced from lower-rank coals passed through a maximum at about 700°C and decreased at higher HTT. Compared to the chars produced from the unoxidized coals, those from slightly oxidized coal precursors had lower methanol densities. Prolonged oxidation increased the methanol density. Slight oxidation did not produce a decrease in the methanol density of the chars produced from medium- and higher-rank coal. The hexane densities of the chars were always lower than the corresponding methanol densities. The difference was more marked for the chars produced from lower-rank coals at HTT $\approx$ 1000°C. The hexane density showed a maximum at about 1000°C. The effect of oxidation on hexane density was not significant for chars produced from lower-rank coals but was marked for chars from medium- and higher-rank coals. In parent coal chars, the volume of micropores accessible to methanol but inaccessible to hexane, that is $(V_{hexane} - V_{methanol})$ where V_{hexane} and $V_{methanol}$ represent the specific volumes of the two fluids displaced by coals, showed a minimum at $\sim$400°C and a maximum at $\sim$600–700°C. W_o was generally larger for lower-rank coal samples. It increased with HT at 400°C, passed through a maximum at higher HTT, and decreased abruptly thereafter. The variation of W_o increased with HTT and was dependent on the rank of the parent coal and its degree of oxidation. Air oxidation of the parent coal increased W_o of the char produced.

REFERENCES

1. W. R. Grimes, *Coal Sci.* **1**, 21–42 (1982).
2. O. P. Mahajan, in *Coal Structure* (R. A. Meyers, ed.), pp. 51–86, Academic Press, New York, 1982.
3. O. P. Mahajan and P. L. Walker, Jr., in *Analytical Methods for Coal and Coal*

Products (C. Karr, Jr., ed.), Vol. 1, pp. 125–162, Academic Press, New York, 1978.

4. P. L. Walker, Jr., *Philos. Trans. R. Soc. London, Ser. A* **300**, 65 (1981).

5. D. Spencer, *EPRI J.*, p. 31 (1982).

6. O. P. Mahajan and P. L. Walker, Jr., in *Chemistry of Coal Utilization, (Second Supplementary Volume,* (M. A. Elliot, ed.), Vol. 2, pp. 173–186, Wiley-Interscience), New York, 1981.

7. International Union of Pure and Applied Chemistry (IUPAC), *Manual of Symbols and Terminology for Physico Chemical Quantities and Units.* Butterworth, London, 1972.

8. F. A. P. Maggs, *Res. Corresp.* **6**, 13S (1953).

9. G. L. Kington and W. Laing, *Trans. Faraday Soc.* **51**, 287 (1955).

10. P. L. Walker, Jr., L. G. Austin, and S. P. Nandi, *Chem. Phys. Carbon* **2**, 257 (1966).

11. P. L. Walker, Jr. and O. P. Mahajan, in *Analytical Methods for Coal and Coal Products* (C. Karr, Jr., ed.), Vol. 1, pp. 163–188, Academic Press, New York, 1978.

12. R. M. Barrer and D. W. Brook, *Trans. Faraday Soc.* **49**, 1049 (1953).

13. S. P. Nandi and P. L. Walker, Jr., *Fuel* **43**, 385 (1964).

14. O. P. Mahajan, M. Komatsu, and P. L. Walker, Jr., *Fuel* **59**, 3 (1980).

15. S. P. Nandi and P. L. Walker, Jr., *Coal Science,* pp. 379–385, Am. Chem. Soc., Washington, DC, 1966.

16. R. B. Anderson, W. K. Hall, J. A. Leckey, and K. C. Stein, *J. Phys. Chem.* **60**, 1548 (1956).

17. W. V. Kotlensky and P. L. Walker, Jr., *Proc. Carbon Conf., 4th, 1959* p. 423 (1960).

18. R. E. Franklin, *Trans. Faraday Soc.* **45**, 274 (1949).

19. S. Fujii and H. Tsuboi, *Fuel* **46**, 361 (1967).

20. H. Gan, S. P. Nandi, and P. L. Walker, Jr., *Fuel* **51**, 272 (1972).

21. J. R. Nelson, Ph.D. Thesis, Pennsylvania State University, University Park, 1979.

22a. R. C. Neavel, E. J. Hippo, S. C. Smith, and R. N. Miller, *Prepr. Pap. J. Am. Chem. Soc., Div. Fuel Chem.* **25**(3), 246 (1980).

22b. S. Parkash, *Fuel*, **64**, 631 (1985).

23. E. W. Washburn, *Proc. Natl. Acad. Sci. U.S.A.* **7**, 115 (1921).

24. O. P. Mahajan and P. L. Walker, Jr., unpublished results, Pennsylvania State University, University Park, 1979.

25. J. R. Nelson, O. P. Mahajan, and P. L. Walker, Jr., *Fuel* **59**, 831 (1980).

26. S. Brunauer, P. H. Emmett, and E. Teller, *J. Am. Chem. Soc.* **60**, 309 (1938).

27. N. C. Ganguli, P. N. Mukherjee, and A. Lahiri, *Fuel* **40**, 525 (1961).

28. A. Korta, M. Lason, J. Kawceka, and J. Klosinska-Drwalova, *Arch. Gorn.* p. 229 (1959); through *Fuel Abstr. Curr. Titles* **2**, Abstr. No. 8 (1961).

29. R. B. Anderson, J. Bayer, and L. J. E. Hofer, *Fuel* **44**, 443 (1965).

30. P. L. Walker, Jr. and I. Geller, *Nature (London)* **178**, 1001 (1956).

31. H. Marsh and W. F. K. Wynne-Jones, *Carbon* **1**, 269 (1964).

32. J. J. Kipling, J. N. Sherwood, P. V. Shooter, N. R. Thompson, and R. N. Young *Carbon,* **4**, 5 (1964).

33. T. Thomas, Jr. and H. H. Damberger, *Circ.—Ill. State Geol. Surv.* **493** (1976).

34. K. A. Debelak and J. T. Schrodt, *Fuel* **58**, 732 (1979).

35. P. L. Walker, Jr. and K. A. Kini, *Fuel* **44**, 453 (1965).

36. H. Marsh and T. Siemieniewska, *Fuel* **44**, 355 (1965).

37. P. L. Walker, Jr. and R. L. Patel, *Fuel* **49**, 91 (1970).

38. S. P. Nandi and P. L. Walker, Jr., *Fuel* **50**, 345 (1971).

39. M. M. Dubinin, *Chem. Phys. Carbon* **2**, 51 (1966).

40. D. H. T. Spencer and R. L. Bond, *Adv. Chem. Ser.* **55**, 724 (1966).

41. P. L. Walker, Jr., O. Cariaso, and R. L. Patel, *Fuel*, **47**, 322 (1968).

42. L. M. Lucht and N. A. Peppas, *AIP Conf. Proc.* **70**, 28–48 (1981).

43. T. Green, J. Kovac, D. Brenner, and J. W. Larsen, in *Coal Structure* (R. A. Meyers, ed.), pp. 199–282, Academic Press, New York, 1982.

44. Y. Sanada and H. Honda, *Fuel* **45**, 295 (1966).

45. N. Y. Kirov, J. M. O'Shea, and G. D. Sergent *Fuel*, **46**, 415 (1967).

46. B. S. Ignasiak, A. J. Szladow, and D. S. Montgomery, *Fuel* **53**, 12 (1974).

47. H. M. Wachowska, B. N. Nandi, and D. S. Montgomery, *Fuel* **53**, 212 (1974).

48. J. W. Larsen, *AIP Conf. Proc.* **70**, 1–27 (1981).

49. R. G. Jenkins and S. C. Mitchell, *Fuel* **57**, 394 (1978).

50. P. J. Reucroft and K. B. Patel, *Fuel* **62**, 279 (1983).

51. P. J. Reucroft and H. Patel, *Fuel* **65**, 816 (1986).

52. W. O. Stacy and J. C. Jones, *Fuel* **65**, 1171 (1986).

53. P. J. Reucroft and A. R. Sethuraman, *Energy Fuels* **1**, 72 (1987).

54. P. L. Walker, Jr., A. Davis, S. K. Verma, J. Rivera-Ultrilla, and M. R. Khan, "Interaction of Gases, Vapors and Liquids With Coals and Macerals at Ambient Temperatures—Their Use to Characterize Porosity", Final Report—Part 9, May, 1984, Pennsylvania State University, prepared for DOE under Contract No. DE-AC22-80 PC 30013.

55. D. M. Bodily, "The Effect of Maceral Properties on the Comminution of Coal", Final Report, July 15, 1984–January 14, 1987, DOE/PC/70796-12.

56. R. E. Winan and P. Thiyagarajam, *Preprints ACS Fuel Chemistry Division*, New Orleans, **32**(4), 227 (1987).

57. R. E. Winan and P. Thiyagarajam, *Energy Fuels*, in press.

58. L. D. Kispert, "Catalyst Accessibility in High Volatile Bituminous Coal," Quarterly Report, Jan. 87, Contract FG 22-86 PC 90502, GPO Dep File No. DE 87004012.

59. S-K Wuu and L. D. Kispert, *Fuel* **64**, 1681 (1985).

60. J. W. Larsen and P. Wernett, *Energy Fuels* **2**, 719 (1988).

61. J. W. Larsen, personal communication.

62. H. K. Wagenfeld, M. Setek, L. Kiss, and W. O. Stacy, *Fuel* **62**, 480 (1983).

63. M. Kalliat, C. Y. Kwak, and P. W. Schmidt, *ACS Symp. Ser.* **169**, 3–22. (1981).

64. R. N. Miller, R. F. Yarzab, and P. H. Given *Fuel*, **58**, 4 (1979).

65. J. V. O'Gorman, Ph.D. Thesis, Pennsylvania State University, University Park, 1971.

66. Z. Spitzer and L. Ulicky, *Fuel* **55**, 21 (1976).

67. H. Kaiser and G. S. Gethner, *Proc. Int. Conf. Coal Sci., 1981* p. 300 (1981).

68. M. J. Tricker, A. Grint, G. J. Audley, S. M. Church, V. S. Rainey, and C. J. Wright, *Fuel* **62**, 1092 (1983).

69. J. S. Gethner, *J. Appl. Phys.* **59**, 1068 (1986).

70. D. H. Maylotte, E. J. Lamby, P. G. Kosky, and R. L. St. Peters, *Proc.—Int. Conf. Coal Sci., 1981* pp. 612–617 (1981).

71. L. A. Harris and C. S. Yust, *Adv. Chem. Ser.* **192**, 321 (1981).

72. J. S. Lin, R. W. Hendricks, L. A. Harris, and C. S. Yust, *J. Appl. Crystallogr.* **11**, 621 (1978).

73. D. J. Allardice and D. G. Evans, *Fuel* **50**, 201 (1971).

74a. S. W. Dack, M. D. Hobday, T. D. Smith, and J. R. Pilbrow, *Fuel* **63**, 39 (1984).

74b. J. S. Gethner, *Fuel* **64**, 1443 (1985).

75. H. N. S. Schafer, *Fuel* **58**, 667 (1979).

76. D. J. Allardice and D. G. Evans, *Fuel* **50**, 236 (1971).

77. O. P. Mahajan and P. L. Walker, Jr., *Fuel* **50**, 308 (1971).

78. S. Brunauer, L. S. Deming, W. E. Deming, and E. Teller, *J. Am. Chem. Soc.* **62**, 1723 (1940).

79. B. R. Puri, *Chem. Phys. Carbon* **6**, 191 (1970).

80. B. R. Puri and O. P. Mahajan, *Soil Sci.* **94**, 162 (1962).

81. I. Ubalidini and C. Sirinamed, *Atti Congr. Int. Chim., 10th* Vol. 3, p. 682 (1939); *Chem. Abstr.* **33**, 8621 (1939).

82. H. N. S. Schafer, *Fuel* **51**, 4 (1972).

83. R. H. Stokes and R. A. Robinson, *J. Am. Chem. Soc.* **70**, 1870 (1948).

84. D. G. Allardice and D. G. Evans, in *Analytical Methods for Coal and Coal Products* (C. Karr, Jr., ed.), Vol. 1, pp. 247–262, Academic Press, New York, 1978.

85. C. Pierce, R. N. Smith, J. W. Wiley, and H. Cordes, *J. Am. Chem. Soc.* **73**, 4551 (1951).

86. P. L. Walker, Jr. and J. Janov, *J. Colloid Interface Sci.* **28**, 449 (1968).

87. E. M. Dannenberg and W. H. Opie, Jr., *Rubber World* **137**, 847 (1958).

88. P. H. Emmett and R. B. Anderson, *J. Am. Chem. Soc.* **67**, 1492 (1945).

89. B. R. Puri, S. Singh, and O. P. Mahajan, *J. Indian Chem. Soc.* **42**, 427 (1965).

90. O. P. Mahajan, A. Youssef, and P. L. Walker, Jr., *Sep. Sci. Technol.* **17**, 1019 (1982).

91. B. K. Mazumdar, P. H. Bhangale, and A. Lahiri, *Fuel* **36**, 254 (1957).

92. R. Kaji, Y. Muranka, K. Otsuka, and Y. Hishinuma, *Fuel* **65**, 288 (1986).

93. P. H. Given and F. Yarzab in *Analytical Methods for Coal and Coal Products* (C. Karr, Jr., ed.), Vol. 2, pp. 3–41, Academic Press, 1978.

94. O. P. Mahajan, *Fuel* **64**, 973 (1985).

95. S. J. Gregg, *The Surface Chemistry of Solids*, Reinhold, New York, 1961.

96. G. H. Amberg, D. H. Everett, L. H. Ruiter, and F. W. Smith, in *The Solid-Gas Interface* (E. A. Flood, ed.), Vol. 2, pp. 3–16, Marcel Dekker, New York, 1967.

97. J. R. Dacey, J. C. Clunie, and D. G. Thomas, *Trans. Faraday Soc.* **54**, 250 (1958).

98. B. Millard, E. G. Caswell, E. F. Leger, and D. R. Mills, *J. Phys. Chem.* **59**, 976 (1955).

99. B. R. Puri, K. Murari, and D. D. Singh, *J. Phys. Chem.* **65**, 37 (1961).

100. S. C. Mraw and D. F. Naas-O'Rourke, *Science* **205**, 901 (1979).

101. S. C. Mraw and B. G. Silbernagel, *AIP Conf. Proc.* **70** 332–343 (1981).

102. S. C. Mraw and D. F. O'Rourke, *J. Colloid Interface Sci.* **89**, 268 (1982).

103. A. W. Gauger, *Chemistry of Coal Utilization* (H. H. Lowry, ed.), Vol. 1, Wiley, pp. 600–626, New York, 1945.

104. S. C. Deevi and E. M. Suuberg, *Fuel* **66**, 454 (1987).

105. L. J. Lynch and D. S. Webster, *Fuel* **58**, 429 (1979).

106. L. J. Lynch and D. S. Webster, *Fuel* **61**, 271 (1982).

107. H. K. Wegenfeld, M. Setek, L. Kiss, and W. O. Stacy, *Proc. Int. Conf. Coal Sci., 1981* pp. 869–874 (1981).

108. H. D. Bale, M. L. Carlson, and H. H. Schobert, *Fuel* **65**, 1185 (1986).

109. M. Setek, H. K. Wagenfeld, W. O. Stacy, and L. T. Kiss, *Fuel* **62**, 480 (1983).

110. P. D. Swann, D. J. Allardice, and D. G. Evans, *Fuel* **53**, 85 (1974).

111. G. G. Karsner and D. D. Perlmutter, *Ind. Eng. Chem. Process Des. Dev.* **21**, 348 (1982).

112. O. P. Mahajan, A. Tomita, and P. L. Walker, Jr., *Fuel* **55**, 63 (1976).

113. O. P. Mahajan and P. L. Walker, Jr., in *Analytical Methods for Coal and Coal Products* (C. Karr, Jr., ed.), Vol. 2, pp. 465–494. Academic Press, New York, 1978.

114. R. Kaji, Y. Hishinuma and Y. Nakamura, *Fuel* **64**, 297 (1985).

115. H. Oda, M. Tekeuchi and C. Yokokawa, *Fuel* **60**, 390 (1981).

116. L. E. Paulson and J. R. Fuch, *Prepr. Pap.—Am. Chem. Soc., Div. Fuel Chem.* **25**(1), 224 (1979).

117. M. L. Gorbaty, *Fuel* **57**, 796 (1978).

118. A. M. Youssef and A. M. El-Wakil, *Surf. Technol.* **10**, 303 (1980).

119. K. K. Bhattacharya, *Fuel* **50**, 367 (1971).

120. A. A. Isirikyan and A. V. Kiselev, *J. Phys. Chem.* **65**, 601 (1961).

121. B. R. Puri, B. C. Kaistha, Y. Vardhan, and O. P. Mahajan, *Carbon* **11**, 329 (1973).

122. R. Liotta, G. Brons, and J. Isaacs, *Proc. Int. Conf. Coal Sci., 1981* pp. 157–162 (1981).

123. J. W. Larsen, personal communication.

124. O. P. Mahajan, A. Youssef, and P. L. Walker, Jr., *Sep. Sci. Technol.* **13**, 487 (1978).

125. V. V. Lebedev, G. S. Golovina, and K. I. Cheredkova, *Solid Fuel Chem.* **13**, 35 (1979).

126. J. A. Solomon and G. J. Mains, *Fuel* **56**, 302 (1977).

127. R. C. Neavel, *Proc. Coal Agglom. Convers. Symp. 1975*, p. 120 (1976).

Commentary

Thus far we have been concerned with essentially the more physical aspects of coal, that is, sample selection, size reduction, pore structure, and so on. Now we turn to processes in which chemistry plays a more predominant role, these being oxidation, aging, and weathering, which are treated in the next three chapters. In this book aging or weathering of coal is considered as a natural atmospheric process. Accelerated aging involving higher temperatures or increased oxygen pressures is only of peripheral concern. In Chapter 5 the oxidation of coal is examined under atmospheric oxygen and ambient temperature conditions. Mechanisms of coal oxidation and detection methods for the extent of oxidation provide an appreciation of the aging and weathering processes on a molecular basis.

5

ATMOSPHERIC OXIDATION OF COAL

N. BERKOWITZ
Department of Mining, Metallurgical & Petroleum Engineering
University of Alberta

5.1 INTRODUCTION

When removed from its formative environment (the seam in which it lies) and exposed to air, coal progressively *weathers* (or *ages*) as it loses moisture and is attacked by atmospheric oxygen and as a result suffers pronounced quality deterioration.

Dehydration—from an initial bed-moisture content, which reflects the 100 percent relative humidity conditions in the seam, to air-dried status—induces shrinkage and fissuring that can seriously impair the mechanical integrity of the coal and, especially in low-rank coals, lead to extensive size degradation by *decrepitation*. *Oxidation*—which manifests itself in loss of carbon and hydrogen to offgas CO, CO_2, and H_2O—effectively devalues almost every other property that lends the coal a particular technical value as a fuel or feedstock.

There is a close connection between dehydration and oxidation. The loss of moisture facilitates oxidation by thinning a protective layer of water, which impedes access of oxygen to the coal surfaces and creates *additional* surface area for oxidation because it causes fissuring. Oxidation has been observed to promote spalling through causing surficial swelling of the coal material. How quickly and extensively these processes operate on any particular coal depend on the coal's rank and petrographic composition as well as the specific ambient conditions to which it is exposed. How serious the technical or economic consequences of quality deterioration prove to be depends in some measure on the purposes for which the coal is to be used.

This chapter discusses how oxidation proceeds, how it can be measured, and how it affects coals of different types. Since the mechanisms of oxidative degradation are still not fully understood and could vary in dependence on specific chemical features of different coals,[1] much of what can currently be said about these matters is necessarily tentative. In contrast to the substantial efforts centered on coal oxidation by *liquid* oxidants (such as nitric acid, alkaline potassium permanganate, and various peroxy acids), from which information about coal structure is being sought, atmospheric oxidation of coal at ambient air temperatures has elicited little interest from coal chemists in recent years, even though such studies might have been greatly aided by the more refined investigative tools that have come to hand since the 1950s. Earlier information is often contradictory or, where purporting to *quantify* oxidation effects, of questionable value. Current perceptions of weathering and weathering effects rest largely on a variety of qualitative observations and on broad inferences from laboratory studies.

5.2   OXYGEN AS A COAL COMPONENT

As an integral elemental constituent of fresh *humic* coals,* oxygen concentrations fall with increasing rank from ~25–30 weight percent (DMF) in lignites to 2–3 weight percent in low-volatile bituminous (LVB) coals (cf., Figure 5.1), and this progressive loss of oxygen is accompanied by a profound shift in the relative abundances of different O-*forms*.

In low-rank coals, over 80 percent of all oxygen is roughly equally partitioned between *carboxyl* (—COOH), phenolic *hydroxyl* (—OH), and *carbonyl* (=CO) functions, with the first two replacing hydrogen on C-atoms at the peripheries of (predominantly aromatic and/or hydroaromatic) core structures that collectively comprise the hypothetical coal macromolecule. In the very youngest coals, that is, in unconsolidated (soft) lignites, some oxygen may also occur in *methoxy* (—OCH$_3$) moieties. The remainder not encompassed by these forms, often termed *unreactive* because its disposition in the coal is very difficult to assess quantitatively by available means (cf., Section 5.4.1), is considered to lie in *ether*-linkages between components of the coal macromolecule, that is, in structures of the type

$$X\text{—}O\text{—}X, \; X\text{—}O\text{—}\hexagon, \quad \text{and} \quad \hexagon\text{—}O\text{—}\hexagon$$

Humic coals, which constitute the overwhelming bulk of all known coal occurrences, formed from plant debris that, at one time or another during its diagenetic (preburial) phase of coalification, was open to abiotic attack by atmospheric oxygen as well as to biochemical degradation. In contrast, so-called *sapropelic* coals, represented by cannels and bogheads, are products of putrefaction under wholly anaerobic (lacustrine and shallow marine) conditions.

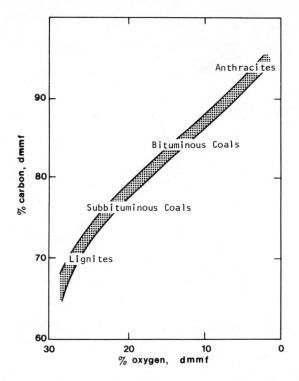

Figure 5.1   Variation of oxygen contents with coal rank.

where X is an alkyl moiety, and in O-*heterocycles*, aromatic entities in which an oxygen atom replaces a ring C-atom.

During metamorphic development to more mature coal types, most of the reactive oxygen is rapidly eliminated. —OCH$_3$ cannot be found in coals with more than ~ 70–72 weight percent carbon (DMF); —COOH is lost during approach to ~ 82 percent C; and above ~ 85 percent carbon, little if any —OH remains. In high-rank coals, in which it constitutes less than ~ 8 weight percent of the DMF coal material, oxygen exists therefore almost exclusively in =CO, —O—, and heterocyclic combinations.

These changes in total oxygen content and O-forms are illustrated in Figure 5.2 and are directly reflected in several industrially important coal properties. For example, due to loss of oxygen *per se* (and to the consequent relative enrichment of the coal substance in carbon and hydrogen), *calorific values* increase steadily with increasing rank. The progressive elimination of acidic O-forms causes the pronounced *hydrophilic* character of low-rank coals to give way to the *hydrophobicity* of bituminous coals. Because pyrolysis preferentially removes coal-O as H$_2$O, and for that purpose abstracts H that would otherwise be available for formation of volatile hydrocarbons, *tar* yields increase with

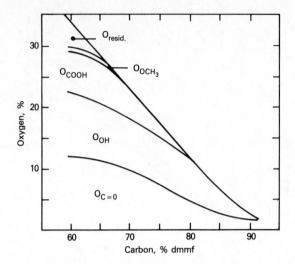

Figure 5.2   Distribution of oxygen in vitrinites of different rank (after Blom[72]).

increasing rank up to high-volatile bituminous (HVB) coals.† It has also been hypothesized that the appearance of *caking* propensities among bituminous coals can be associated with their relatively low oxygen contents and, more specifically, with their low concentrations of ether-oxygen, which links potentially mobile moieties to each other. Oxygen, if sufficiently abundant, precludes thermoplastic behavior.*

In view of such patterns, it occasions no surprise that atmospheric oxidation, which *introduces* oxygen into coal in various forms, should be found to reverse the directions in which progressive metamorphism shifts coal properties.

5.3   ATMOSPHERIC OXIDATION

5.3.1   Chemical Mechanisms

Even at ordinary temperatures virtually every oxygen molecule reaching a carbonaceous surface is *chemisorbed* and can not, as would be true of physically sorbed gas, be recovered as such.

†Significantly, tar plus aqueous liquor yields remain approximately constant up to hvAb rank.[2] Among higher-rank coals, tar yields fall again (presumably because of progressive aromatization of the coal substance and gradually falling H/C ratios), and anthracites furnish no tar at all.

*Evidence for this postulate is, however, far from convincing. Although caking properties do indeed generally find their strongest expression among medium- and low-volatile bituminous coals, there are countless exceptions to this rule; and even broad statistical correlations between caking properties and O-contents (or specific oxygen-forms) can be shown to possess no more than geographically limited regional validity.

If the substrate is a carbon, or a high-temperature coal char structurally characterized by fairly large aromatic or pseudographitic carbon platelets, such chemisorption is not significantly site-specific and can be represented by

$$
\begin{array}{ccccc}
 & O\!-\!O & & O\ \ O \\
 & :\quad : & & \|\ \ \| \\
-C\!=\!C\!- + O_2 \longrightarrow & -C\!=\!C\!- & \longrightarrow & -C\ \ C\!- \\
\ \ |\ \ \ | & \ \ |\ \ \ | & & \ \ |\ \ \ | \\
\end{array}
$$

or, more simply, by

$$2C + O_2 \longrightarrow 2C(O);\ C(O) \longrightarrow CO + C$$

where $\underline{C}$ denotes a free site, and $C(O)$ the resultant chemisorbed surface complex that eventually detaches and then exposes a new $\underline{C}$ below the initial site. Formation of surficial peroxides $=\!C\!-\!O\!-\!O\!-\!\underline{C}=$ is sterically hindered by the inability of the C-platelets to accommodate the accompanying lattice distortion. If moisture is present, hydroperoxides and carboxyl functions can be generated via concurrent

$$\underline{C} + H_2O \longrightarrow C(H)(OH)$$

$$C(H) + O_2 \longrightarrow\ \equiv\!C\!-\!O\!-\!OH$$

$$\text{and}\quad 2C(OH) + O_2 \longrightarrow 2\!-\!\underset{\displaystyle OH}{\overset{\displaystyle |}{C}}\!=\!O$$

Very similar processes proceed from oxygen chemisorption at coal surfaces exposed to air, except that the diversity of chemical configurations in coal and the smallness of aromatic units in coal macromolecules *allow* formation of peroxides (as well as of hydroperoxides) and facilitate at least two other sets of oxidation sequences.

Development of peroxides and hydroperoxides during atmospheric oxidation of coal was first reported by Yohe and Harman[3] and was subsequently confirmed by inter alios Jones and Townend[4,5] and Rafikov and Sibiryakova.[6] Such moieties form by

and

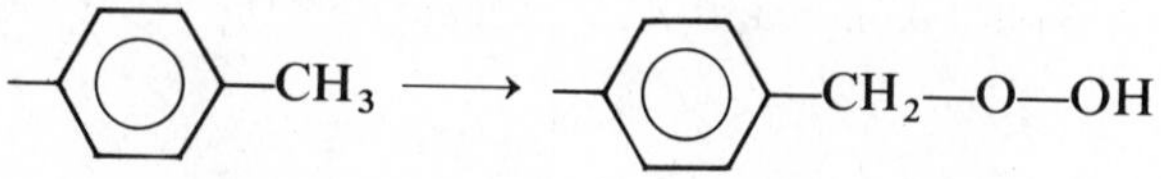

but are inherently unstable[3] and, when decomposing, can initiate a variety of free radical reactions via, for example,

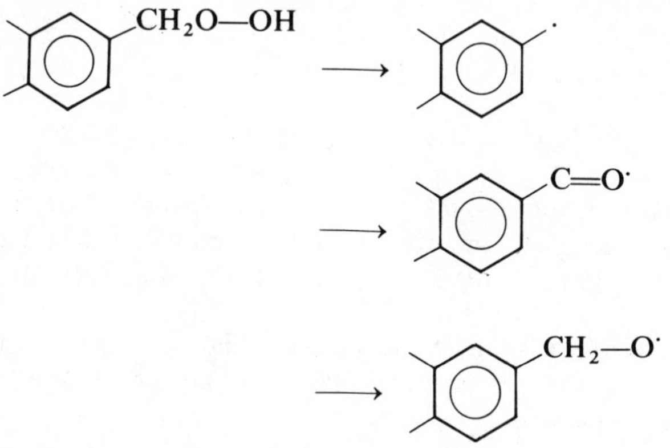

all of which furnish H_2O, CO, and/or CO_2 as by-products.

The —COOH and phenolic —OH functions are introduced into the coal by direct oxidative attack on peripheral alkyl substituents and saturated (—CH_2—) bridge structures.[7,8] Such reactions are illustrated by

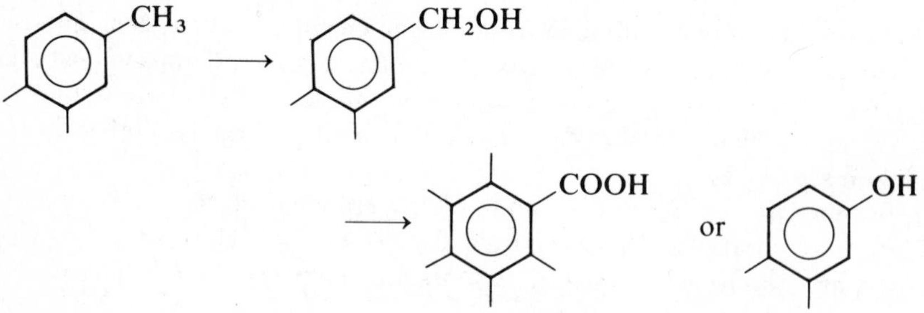

and

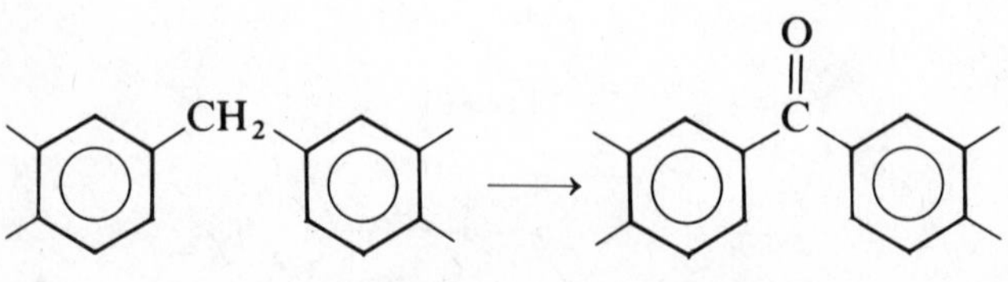

and can in certain circumstances, as suggested by Tronov,[9] degrade aromatic rings via a sequence of the form

This resembles the gas-phase oxidation of naphthalene to phthalic and maleic anhydrides. Some such mechanism, operating in conjunction with destruction of —CH$_2$— or similar bridges, probably also participates in oxidative disruption of skeletal structures that proceeds, although generally much more slowly, in parallel with formation of —OH, —COOH, and peroxides, and leads to generation of alkali-soluble *humic acids.*[10,*]

Like the chemical structure of coal, if indeed coal can be said to possess such, the structure of humic acids remains uncertain. When regenerated from their aqueous alkaline extract solutions by acidification, they form brown flocculates that dry to brittle, lustrous black solids with $C = 55{-}60$ percent, $H = 2{-}3$ percent. Since their infrared spectra as well as their x-ray diffraction patterns suggest that they are structurally not far removed from their precursor coal molecules, the wide range of (number-average) molecular weights reported for humic acids (600–10000) has been taken as evidence that they form fairly directly by random cleavage of coal macromolecules. However, humic acids are themselves subject to further oxidative degradation and, especially at high temperatures, undergo two kinds of simultaneous change. The first entails thermal decarboxylation and dehydroxylation, that is, loss of —COOH and —OH, with attendant loss of alkali solubility. The second involves progressive breakdown of humic acids via a series of lower molecular weight species, light brown, alcohol-soluble *hymatomelanic* acids and pale yellow, water-soluble *fulvic* acids, to simple benzene polycarboxylic acids. Schematically, these concurrent reactions can be represented by

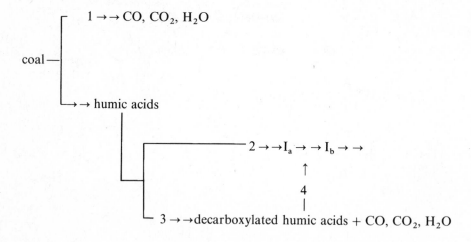

Here 1 denotes thermal decarboxylation and dehydroxylation of the oxidized alkali-*insoluble* coal material, 2 sequential secondary degradation of humic acids via the intermediates I_a, I_b, and 3 thermal decarboxylation and dehydro-

---

*By an almost 200-year old convention,[11,12] this term is applied to all dark brown or black alkali-soluble matter, regardless of whether it is produced by air-oxidation of coal or by the action of fungal oxidases on plant material (and, in the latter case, furnishing *soil* humic acids). How closely related, if at all, humic acids from different sources actually are, is, however, still quite speculative.

xylation of humic acids to coal-like alkali-insoluble solids that 4 can then oxidize to I_a, I_b, and so on.

Although details respecting these various oxidation processes are fragmentary, all three—transient formation of peroxides and hydroperoxides, direct development of —COOH, phenolic —OH and =CO functions, and molecular disruption to humic acids and their degradation products—have been clearly identified in numerous laboratory studies.[7,8,13-19] How strongly they express themselves *individually* depends on factors that influence their respective kinetics (cf., Section 5.3.3) and is, first and foremost, determined by time and temperature.

Peroxides and hydroperoxides, which can develop only if moisture is present,[4,5] are thus hardly observable at temperatures above $\sim 70-80°C$ because they will then decompose as fast as they form;[14-17] and within the limited time frame of a laboratory study, humic acids can only be generated to measurable concentration levels at temperatures above $100-150°C$ (cf., for example, Jensen et al.[10]). On the other hand, all three reactions, including secondary degradation of the humic acids, will manifest themselves in field situations, for example, when an outcrop, an abandoned mine face, or an inactive (dead) stockpile is open to unimpeded atmospheric oxidation. In such situations, time is not a limiting factor and oxidation then proceeds through three fairly distinct stages that clearly reflect the underlying oxidation chemistry. Farenden[19] has effectively simulated these in laboratory experiments in which they were expedited by elevated temperatures $(110-170°C)$. In the first stage, reacted oxygen, which leads to the formation of —COOH, —OH, =CO and peroxides is approximately equally partitioned between $\{O\}$ fixed in the coal and $\{O\}$ appearing in off-gas CO, CO_2, and H_2O. In the second, the development of humic acids is shown by the fairly rapidly increasing solubility of the coal in aqueous alkalis and, presumably due to more extensive decarboxylation and dehydroxylation, *more* $\{O\}$ appears in off-gas than is fixed in the coal. In the third, gradual breakdown of humic acids results in increasing yields of H_2O-soluble material.*

5.3.2 Oxidation of Associated Mineral Matter

Weathering to which mineral matter is prone, for example, conversion of primary alumino-silicates into clay minerals (kaolinite, montmorillonite, illite, etc.), or oxidation of igneous Fe minerals to hydrated Fe^{3+} oxides, occurred mostly in the (preburial) diagenetic phase of coal formation and during the early stages of metamorphic development when the immature coal mass was open to

---

*It is a matter of some interest that formation of humic acids in naturally weathering coal has been known for over 150 years. As early as 1819, Braconnot[20] noted that lignites, unlike bituminous coals, commonly furnished dark brown solutions when extracted with aqueous alkalis. It was later shown that weathered bituminous coals also did so, and that alkali-solubility could therefore not be used, as Braconnot had suggested, to discriminate between low- and high-rank coals.

percolating aerated waters. Atmospheric oxidation and concomitant hydrolysis* of mineral matter in *mined* coal (or coal outcrops) involves, therefore, little more than changing residual Fe^{2+}, Mn^{2+}, and S^{2-} (i.e., species that tend to predominate in deep-seated formations) to Fe^{3+}, Mn^{4+}, and S^{6-} in surficial locations. Although quantitatively of little significance, these reactions can hold serious environmental consequences. Of particular concern is the oxidation of pyrite (or marcasite†) that, when moisture is present, will generate H_2SO_4, Fe^{3+} ions, and limonite via

$$2FeS_2 + 7O_2 + 2H_2O \rightarrow 2Fe^{2+} + 4SO_4^{2-} + 4H^+$$

and

$$2Fe^{2+} + 3H_2O + 1/2O_2 \rightarrow Fe_2O_3 \cdot H_2O + 4H^+$$

as well as cause formation of variously hydrated ferrous sulfates via

$$2Fe^{2+} + SO_4^{2-} + xH_2O \rightarrow FeSO_4 \cdot xH_2O$$

These reactions are primarily responsible for producing acid liquid effluents from coal exposed to air. Oxidation of FeS_2 to ferrous sulfate is also a major factor in autogenous heating of coal (cf., Section 5.5). Analogous reactions involving arsenic, chromium, copper, mercury, molybdenum, selenium, tin, tungsten, and vanadium, common trace constituents of coal mineral matter, may also pose some ecological hazards. However, all these oxidation processes affect the *coal* only to the extent that progressive leaching of soluble mineral species enriches it in quartz and accessory minerals (most notably titanium oxides, tourmaline, and zircon). These tend to persist because of their chemical inertness and very low solubility in alkaline *or* acidic waters.

5.3.3   Kinetic Aspects of Atmospheric Oxidation

Like the rates of most other sorption-controlled processes, the rates at which coal will take up and react with oxygen at normal atmospheric temperatures can be represented by the Elovich equation

$$dx/dt = Ae^{-Bt}$$

where x is the amount of oxygen sorbed at time t, and A and B are constants.[21] The *values* that these constants assume depend entirely on the nature of the coal and on the conditions under which it oxidizes. They are therefore determined by a complex interplay of factors that include the rank and petrographic composition of the coal, its mean particle size, the extent (if any) of its prior oxidation, and of course the oxygen partial pressure and ambient temperature.

---

*Hydrolytic reactions are processes that produce or consume H^+ or OH^- and therefore directly affect the pH of any leachate.
†Pyrite and marcasite are dimorphs of FeS_2 that differ only in their crystal habits.

Because of the intrinsic heterogeneity of coal, and the inevitably very incomplete chemical descriptions of the samples used in laboratory studies, there are still insufficient *directly comparable* data for quantifying the impact of these variables. But extant literature does allow some broad semiquantitative inferences respecting them.

Generally, the oxidizability of coal, that is, *how fast* a coal responds to attack by oxygen, decreases rapidly with increasing rank.* In practical situations when exposure times may be lengthy but not infinitely long, substantial oxidative destruction of coal material, as distinct from often serious deterioration of coal properties, is therefore only observed when dealing with low-rank coals. One reason for this lies in the fact that the porosity of coal, and hence the surface area that it presents to oxygen, falls steeply from lignites to bituminous coals (cf., Figure 5.3).† Even more important are the *chemical* changes brought about in

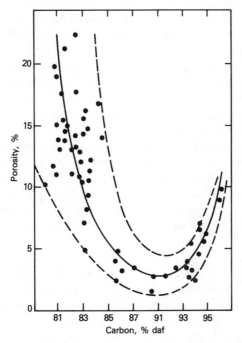

Figure 5.3   Variation of coal porosity with rank (after King and Wilkins, Proceedings, of Conference on "Ultra-Fine Structure of Coals and Cokes," *BCURA, London,* p. 46, (1944).

*One indication of this is the ease with which low-rank coals, especially lignites, will lose inherent, organically bound hydrogen as water when thermally dried in an atmosphere that contains small amounts of residual oxygen. In some cases, even traces of oxygen suffice to lower the (DAF) hydrogen content of the dried sample to as little as 2.5 or 3 percent.

†Since, by Hilt's rule, the transition from lignites to bituminous coals occurs under increasingly thick overburdens, falling porosities (and, consequently, bed moisture contents) can be ascribed to increasing compaction of the coal material.

coal by progressive metamorphism. Recent studies[22,23] suggest that a combination of factors, notably the different plant materials from which coal formed in successive geological periods, the diverse conditions under which the plant debris decomposed, and the different rates at which it subsequently coalified, make coal structure much more variable than has been and commonly still is supposed. Nevertheless, x-ray diffraction, ^{13}C-nmr, infrared and mass spectra, as well as observations on coal depolymerization and degradation by liquid oxidants all show that coal maturation is accompanied by gradual aromatization and homogenization through progressive elimination of non-aromatic and/or hydroaromatic moieties. It can therefore be concluded that an inverse variation of oxidizability with rank is also, and indeed *primarily*, due to steadily diminishing concentrations of chemical structures that oxygen can readily attack.

Direct support for such a conclusion is afforded by observations that the apparent activation energies of oxidation are rank-dependent[24] and that yields of water-soluble benzenoid acids from oxidative degradation of bituminous coal macerals with HNO_3 and/or alkaline $KMnO_4$ decrease with the atomic H/C ratio of the macerals in the order exinites > vitrinites > inertinites.[25-28] Because rank and petrographic compositions tend to be overriding determinants of oxidizability and oxidation rates (r), relationships between r and other rate-controlling factors are always subject to proportionality constants characteristic of the coal under test; excepting temperature, these other controls usually only gain practical importance in special situations.

Due to the fact that oxygen reaching the coal will, to some extent, also diffuse into the particle interior and there react at surfaces presented in pores and micro-fissures,[24] rates of oxygen uptake are generally directly proportional to the external (specific) surface area (S) of the coal, but there is still some disagreement about an appropriate $r–S$ relationship. Schmidt and Elder[29], Yamasaki,[30] and Carpenter and Giddings[31] have found good compliance with

$$r = kS^n$$

where n assumes values between 0.327 (Schmidt and Elder[29]) and 0.404 (Yamasaki[30]); this implies that a thousandfold increase in S would lead to an approximately tenfold increase in r. However, according to Scott and Jones[32] and Scott,[33] such formulation—with $n < 1.0$—may only hold for $> 850 \, \mu m$ particles (+20 mesh). There is in each case, as might perhaps be expected on theoretical grounds, a characteristic lower size limit* beyond which further comminution has *no* observable effect on r.

With respect to rate-control by particle size at different temperatures, a matter of considerable interest in prevention of autogenous heating (cf., Section 5.5), Merrill[34] has proposed a pseudo first order Arrhenius-type equation, namely,

---

*Having regard for the implications of Figure 5.3, it is probably more correct to speak of such lower size limits being characteristic of each of the different coal types or coal classes.

$$r = \frac{1}{S}\frac{n(O_2)}{dt} = Ae^{-E/RT} \cdot \{O_2\}$$

where $n(O_2)$ is the (mole) amount of oxygen sorbed, $\{O_2\}$ the oxygen concentration in the reaction system, and T the absolute temperature. Since the data from which this expression was developed were obtained at temperatures approaching the ignition point of the coal, that is, under conditions in which diffusion effects are negligible unless the particles are large,[35] that equation, too, may only possess limited validity.

Where the free flow of oxygen to the coal is impeded—as, for example, in tightly packed particle assemblages or when evaporating excess moisture partly shrouds the coal surface—or when, as in hot spots in a coal stockpile (cf., Section 5.5), oxygen is consumed faster than it can be replenished, r comes also to be controlled by the oxygen partial pressure $p_{(O)}$ in the immediate vicinity of the coal. Early laboratory observations of this effect[36] suggested that r varied directly with $p_{(O)}$ but later more reliable measurements that covered a wider range of different coals and initial oxygen concentrations between 1 and 20 percent,[29,37-39] show

$$r = kp_{(O)}^n$$

where n has values between ~ 0.5 and 0.66. However, regardless of the specific conditions under which a coal is attacked by oxygen, the more dominant influences on the observed oxidation rate are exerted by the extent of any prior oxidation and by the ambient temperature.

The effect of prior weathering has been systematically examined by, inter alios, Taylor,[40] Scott,[33] Elder et al.,[39] and Chalishazar and Spooner,[41] who consistently found aged coals to be distinctly less susceptible to oxidation than the corresponding fresh coals. Although it is often difficult to assess limited prior weathering (cf., Section 5.4), and some published data that purport to quantify it are therefore suspect, there is little doubt that the phenomenon reflects prior destruction of the most easily accessed and/or most sensitive reaction sites in the coal.

With respect to the effects of different reaction temperatures, which make r, like the rates of most other chemical reactions, increase exponentially with T, Carpenter and Giddings[31] have reported compliance with

$$d\ln r/T = E/RT^2$$

where E is the (apparent) activation energy of the oxidation reaction, R the gas constant (1.986 cal/°C), and T the absolute temperature. The fact that temperature coefficients of r from different sources often vary significantly is most likely due to different activation energies (which, as noted above, depend on coal rank). Schmidt and Elder[29] thus observed that a 10°C temperature rise caused oxidation rates of bituminous coal to increase by a factor of ~ 1.7. Thomas et al.[42] measured a similar coefficient for bituminous coal but recorded a much smaller value (1.17/10°C) for anthracite. Scott,[33] also working with an anthracite, found 1.23/10°C. If in fact attributable to unequal activation energies, these differences underscore a need for caution if temperature coefficients are used for estimating oxidation rates over *wide* temperature intervals in which the dominant oxidation reactions change. In kinetic studies of coal oxidation by air at 25–290°C, Oreshko[14–17] observed that E increased from some 3–4 kcal/mole at temperatures below 70–80°C to 6–8 kcal/mole between 80° and 160°C and then further rose to 12–17 kcal/mole in the 180–290°C interval.

5.3.4 Compositional Changes

In the initial stages of atmospheric oxidation, coals will generally show a slight weight *gain* due to (transient) net incorporation of oxygen. If weathering continues, this is soon superseded by pronounced progressive weight *losses* as some of the chemisorbed oxygen leaves the coal with carbon and hydrogen (in the form of carbon oxides and water). In an ongoing cyclical process this is replaced by newly sorbed oxygen that is subsequently similarly eliminated. These events shift the elemental compositions of coal to increasingly lower carbon and hydrogen contents and appear to establish some kind of pseudo-equilibrium when the (DAF) carbon and hydrogen contents have fallen to 60–65 percent and 2–3 percent.

Detailed laboratory measurements[43–45] show that these compositional changes follow distinctive oxidation *tracks* that closely parallel hydrocarbon tracks calculated by assuming that addition of one O-atom leads, on average, to the loss of one H-atom. An illustrative example for which this would be strictly true is an oxidation reaction such as

Figure 5.4 summarizes representative experimental data for coal (●) and shows how these relate to hypothetical hydrocarbon tracks (○). A notable feature here is that the coal tracks tend to converge slightly, that is, that high-rank coals tend

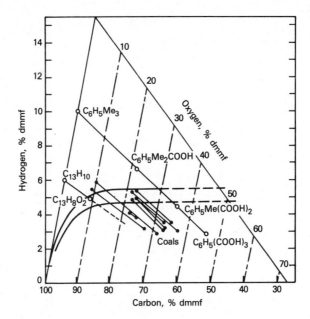

Figure 5.4 Changes in elemental composition during air-oxidation of coal (after Francis[45]).

to lose material with slightly higher C/H ratios.* Concurrently with and as a direct result of these shifts in elemental composition, the heat value of the coal declines steadily, and coal volatility changes[46]—with volatile matter contents of low-rank coals falling and those of more mature coals tending to increase somewhat (cf., Section 5.6).

Because the oxidation tracks imply predictability of the directions in which the elemental compositions will shift, it has been suggested[47] that the limits of probable composition of a fresh coal could be estimated from the carbon and hydrogen contents of a weathered sample, such as might be collected from an outcrop. A graphical procedure for this purpose, based on a coal "band" that can be delineated in a carbon versus hydrogen plot and accommodates the great majority of normal coals,[48] is shown in Figure 5.5. Starting from the composition of the weathered sample X, a line is drawn diagonally upward so that $a:b = a':b'$, and the (shaded) area defined by the intersection of this line with the coal band is assumed to define the range of most probable compositions of the corresponding unaltered coal. The oxidation tracks needed as reference for establishing $a:b = a':b'$ can be taken from Figure 5.4, or they can be calculated as in that diagram for any two hydrocarbons with suitable carbon and hydrogen contents. Other correlations, such as Seyler's 1938 coal chart,[48] can then be used for estimating the likely properties of the coal.

---

*This has also been observed by others (cf., for example, Howard[13] and may be due to the greater aromaticity of high-rank coals.

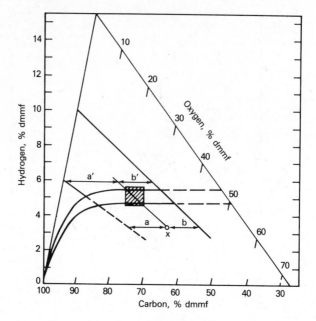

Figure 5.5 Estimation of fresh coal compositions from compositions of weathered (oxidized) coal (after Berkowitz[47]).

5.4 DETECTION OF OXIDATION

Total oxygen contents of coal are in principle directly and quite accurately measurable,[49-55] but in practice are routinely determined by difference, that is, as

$$\{O\}\ DMMF = 100 - (\{C\} + \{H\} + \{N\} + \{S_{org.}\})\ DMMF$$

and are subject to considerable errors (cf., for example, Abernethy and Gibson.[56] For reliable measurements of the (incremental) oxygen introduced into coal by weathering, recourse must therefore be made to other analytical methods, of which some can estimate specific O-forms. Although these make it possible to follow progressive atmospheric oxidation with considerable precision, difficulties are still encountered in the measurement of slight oxidation, which can only be positively identified by viewing the weathered sample against the corresponding fresh coal. Such comparisons are, in fact, mandatory in the case of low-rank coals that contain, in their *unaltered* states, high proportions of oxygen in forms that are also generated by weathering.

5.4.1 Spectroscopic and Wet Analytical Methods

Small amounts of added oxygen, commonly within the experimental error of a wet procedure, can be detected in the infrared spectra of weathered coal by the intensified absorption bands at 3300 cm^{-1} (due to —OH stretching) and 1700 cm^{-1} (carboxyl =CO stretching)* as well as in *weakened* bands at 2925 cm^{-1} (—CH$_3$, =CH$_2$ stretching) and 2940 or 2860 cm^{-1} (aliphatic CH stretching).† Reasonable, but scarcely better than semiquantitative, estimates of oxygen introduced by weathering can then be obtained from the relative intensities of the 3300 and/or 1700 cm^{-1} bands as they appear in the spectra of the weathered and fresh coal.

This method has been widely used in studies of air-oxidation of low-rank and bituminous coals,[7,57–60] and has more recently also been employed in an investigation of commercially stockpiled coal.[61] Great care must be taken to ensure that the test samples are thoroughly dried. Residual moisture will absorb at 3300 cm^{-1} and may totally mask contributions from —OH introduced by weathering.

More substantial oxidation can be equally well assessed by infrared spectroscopy but is also, and sometimes more precisely, encompassed by wet methods that employ specific chemical reactions for measuring *individual oxygen-forms*. Comprehensive reviews of these methods have been presented by, inter alios, Ihnatowicz,[62] van Krevelen,[63] Kröger et al.,[64,65] and Ignasiak et al.[66]

Most easily determined are concentrations of *carboxyl* functions. These are quantitatively measurable by ion-exchange when the pulverized coal is treated with excess aqueous calcium acetate or, preferably, barium acetate.[67] {—COOH} can then be calculated from the amount of acetic acid liberated by

$$2X \cdot COOH + (CH_3 \cdot COO)_2 Ba \rightarrow (X \cdot COO)_2 Ba + 2CH_3 \cdot COOH$$

and determined by titration. Alternatively, the treated coal is washed free of excess acetate, suspended in water, and acidified (e.g., with HCl). The —COOH concentrations are obtained from the Ba^{++} released by

$$(X \cdot COO)_2 Ba + 2HCl \rightarrow 2X \cdot COOH + BaCl_2$$

In either procedure it is imperative to prevent participation by phenolic —OH, and that requires either blocking —OH by prior methylation or conducting the exchange at pH = 8.2 at which —OH will not react.[67] If the coal contains substantial amounts of clay minerals whose acidic —OH is capable of ion exchange, it is also important to demineralize the test sample before use.

Total *hydroxyl* contents are most conveniently determined by acetylation, that is, by treating the test sample in pyridine with acetic anhydride at 95–100°C, and hydrolyzing the acetylated coal with barium hydroxide. {—OH} is

---

*The 1700 cm^{-1} band is usually seen as a shoulder on a very strong band at 1600 cm^{-1} that is characteristic of coal and certain coal-derived products and that has been variously assigned to aromatic C=C, =CO, and H-bonded —OH.

†The lowered intensities of these bands are due to partial oxidative destruction of —CH$_3$ and —CH$_2$— moieties (cf., Section 5.3.1).

then obtained by acidifying the hydrolysate with phosphoric acid, heating the mixture, and measuring the acetic acid that distills.[68] In a rather less tedious version of this procedure, in which ^{14}C-labeled acetic anhydride is used, the acetyl uptake, and hence {—OH}, is determined radiochemically[69] or the acetylated coal is combusted and the combustion gas collected for scintillation counting.[70] Clay minerals do not significantly affect the results of acetylation.[70]

An alternative method for total {—OH}, due to Friedman et al.[71] and also substantially unaffected by mineral matter, involves reacting the coal with hexamethyl disilazane in pyridine. This forms the trimethylsilyl ether derivatives, and {—OH} is calculated from the increased silicon contents of the sample.

However, neither of these procedures is appropriate if *alcoholic* as well as phenolic —OH is present or suspected, and if the analysis is required to discriminate between them. In that case, phenolic —OH must be separately determined by esterifying the coal with diazomethane in diethyl ether, saponifying the reaction product in order to nullify contributions from —COOH, and finally determining the concentration of the resultant —OCH$_3$ by a modified Zeisl method.[68] Alcoholic {—OH} is then given by {=OH}$_{total}$ − {—OH}$_{phenolic}$.

Measurements of *carbonyl* (=CO) functions, which may exist in ketonic and/or quinoid forms, that is, in structures such as

are generally less unequivocal because the reactivities of =CO depend in either case on their particular molecular environments. Satisfactory results have been claimed from reacting the coal with phenyl hydrazine hydrochloride[62] as well as from an oxime formation procedure.[68] Blom[72] and Kröger et al.[65] have also reported that quinoid =CO can be determined with reasonable accuracy by reduction with Ti^{3+} chloride, and according to Kröger et al.,[65] it is possible to measure ketonic (nonreducible) =CO by reaction with dinitrophenyl hydrazine.

Unresolved difficulties attach primarily to the measurement of {O} in *ethers*, which can also be generated in coal by atmospheric oxidation although to a much smaller extent than other O-forms. In infrared spectra, absorptions between 1300 and 1000 cm^{-1} have been assigned to diaryl, aryl-alkyl, and dialkyl ethers. Several investigators have sought to determine such structural entities directly. Bhaumik et al.[73] attempted this by cleaving ether linkages with HI at 135°C and then acetylating the phenols formed.

Takegami et al.,[74] Kröger et al.,[65] and Lazarov and Angelova[75] reduced ether-oxygen with Na in liquid ammonia. Wachowska and Pawlak[76] and Ignasiak and Gawlak[77] have estimated ether-O from the effects of reductive alkylation of coal in Na/tetrahydrofuran/naphthalene. Results reported from these efforts are very questionable. HI does not act on diaryl ethers (and probably not effectively on other types). Na in liquid NH_3 will cleave diaryl ethers but not operate on aliphatic —O—. Alkylation in Na/tetrahydrofuran/naphthalene, a coal solubilization technique developed by Sternberg et al.,[78,79] is accompanied by (at first unrecognized) side reactions that almost certainly quite vitiate any conclusions about ethers (cf., for example, Franz and Skiens[80] and Larsen and Urban[81]).*

5.4.2 Semiempirical and Other Methods

Partly in direct response to the needs of coal producers and consumers, and partly as the outcomes of less committed studies of coal properties, several largely empirical methods for determining the oxidation status of coal have been proposed. All furnish qualitative evidence of oxidation in a single test but show little of its extent unless the behavior of the fresh coal under the test conditions is known.

An example that illustrates this restriction particularly well is the measurement of peroxides and hydroperoxides (cf., Section 5.3.1). Procedures for determining the concentrations of these species use their oxidizing powers and are simple. Yohe and Harman[3] have shown that they will oxidize an aqueous solution of titanous chloride to T^{4+} chloride, which can be estimated by titration. Jones and Townend[4,5] obtained good results from shaking the pulverized coal with an aqueous solution of ferrous thiocyanate and recording the resultant ferric thiocyanate colorimetrically. Rafikov and Sibiryakova[6] made use of peroxide oxidation of hydroquinone to quinone, which was then measured iodochemically. However, since peroxides are *transient* species whose concentrations in an oxidizing coal tend over time to cycle through one or more maxima,[82] such measurements provide no indication of the severity of oxidation unless the *duration* of oxidation of the test specimen as well as the response of the fresh coal to exposure under similar conditions is known.

Pyrolytic procedures, several variants of which have been proposed from time to time, are based on Radspinner and Howard's[83] observation that a fairly

---

*It is probably no coincidence that a study of German lignites[65] failed to detect diaryl ethers, even though such moieties, given the high oxygen contents of the lignites and the fact that aromatic carbon would account for at least 50–55 percent of their total (DAF) carbon contents, might be expected to be fairly abundant in them.

constant proportion of oxygen entering the coal while it is exposed to air appears in —COOH groups or in moieties reacting similarly. Heating the coal *in vacuo* to 350°C (or in some versions to 550°C) consequently releases an amount of $CO + CO_2$ that is linearly related to {O} added by atmospheric oxidation.

The *permanganate number* (*n*) is defined and measured as the number of ml of 1N aq. $KMnO_4$ reduced by 0.5 g (DMMF) pyridine-extracted coal in 1 hour at 100°C.[84] Systematic measurements by Egorova[85] and by Khrisanfova[86] have confirmed a direct, though necessarily rank-dependent, relationship between *n* and the extent of oxidation of a coal.

Apparently due to the formation of —COOH and —OH groups at coal surfaces during atmospheric oxidation, and to the ability of such functions to hydrogen-bond molecules such as CH_3OH or H_2O, the *heats of wetting* of coal in these liquids have also been reported to increase proportionately with the degree of oxidation.[87] More recently it has been suggested that *vitrinite reflectance* (R_o), which in fresh coals is related to the (DMMF) volatile matter contents by

$$\%vm = 60.65 - (25.14R_o)$$

might likewise serve as means for detecting the quantifying oxidation.[88] In common with various other proposals for assessing the oxidation status of coal, such as measurements of ignition temperatures[89] and oxidation-induced luminescence,[90] neither of these possibilities appears to have been pursued. An intriguing device has been developed for rapid continuous monitoring of coal in industrial operations. This measures oxidation by the *static electric charge* that builds up on a metal plate when coal particles are made to impinge upon it[91] (see also Chapter 6).

As means for purely qualitative detection of oxidation, some interest has also been shown in the fact that adsorption of Safrino O (a red stain) makes oxidized coal appear green in reflected light.[92]

5.5   AUTOGENOUS HEATING

A particularly serious corollary of atmospheric oxidation lies in the tendency of certain coals to heat up and eventually spontaneously ignite. More properly termed *autogenous* heating, this phenomenon is a consequence of the exothermicity of oxidation reactions and, more specifically, of the large heat releases that accompany the oxidation of carbon to carbon dioxide (395 kJ/mole). If not effectively dissipated by radiation or convection, such heat releases will exponentially accelerate oxidation (and the associated further temperature increases). Autogenous heating can in principle be induced in *any* coal susceptible to oxidation, but unless unusual conditions intervene to favor this process, only *low-rank* coals with high moisture-carrying capacities are prone to it, and this recognition has contributed much toward a better understanding of its mechanisms.

Because oxidation at ambient air temperatures does not proceed fast enough or under conditions that foster accumulation of reaction heats, it has been postulated that autogenous heating is triggered by oxidation of pyrite in the coal (cf., Section 5.3.2). One such process that envisages oxidation of FeS_x to a pyrophoric Fe^{2+} sulfide as well as to Fe^{2+} and Fe^{3+} sulfates has been detailed by Mapstone.[93] Reactions of this type would eventually generate some 840 kJ/mole and thereby form pyrite/coal interfaces at which oxidation of the coal material could become self-accelerating. The fact that substantially S-free low rank coals are no less prone to autogenous heating than those containing massive amounts of (sometimes finely disseminated) pyrite makes this an unsatisfactory explanation except insofar as it points to a potential *contributory* cause. A more general trigger has been seen in the heat releases that accompany the wetting of partly dried-out coal.[94] Laboratory experiments in which H_2O-saturated *nitrogen* was passed through coals dried to below their normal air-dry moisture contents showed that consequent heats of wetting sufficed to raise coal temperatures by as much as 25–30°C and therefore to increase oxidation rates six- to eightfold.[94-99]

Qualitative evidence that this trigger operates in practical situations somes in part from observations on coal storage piles and can be traced back as far as the early 1900s. Erdtmann and Stoltzenberg[100] and Threllfall[101] reported that open piles most often tended to heat up when rained on after an extended period of dry sunny weather. Hoskin[102] and Francis[103] have drawn attention to the fact that when wet coal is placed on a dry stockpile, heating usually starts at the *interface* between wet and dry coal.

Since the heats generated by wetting are proportional to the wetted surface areas,* and therefore functions of coal porosities (cf., Figure 5.3), the proposition that autogenous heating is initiated by wetting of excessively dried coal explains why it is in general only encountered with hvCb and lower-rank coals—that is, porous coals with bed moisture contents greater than 10–12 percent—and why it becomes an increasingly serious potential hazard as moisture-carrying capacities rise beyond that limit.

Factors that significantly influence oxidation rates (cf., Section 5.3.3) can also sometimes, singly or in combination, promote autogenous heating in otherwise *safe* coals and heighten this hazard even in coals already judged to be *dangerous*. Aside from higher than normal ambient temperatures, these include

1. Excessive concentrations of coal *fines*, which increase the surface area open to wetting and oxidation;
2. *Packing densities*, which affect the accessibility of the coal surfaces, and
3. *Air pressures*, which govern air flow through or into the coal.

These variables can in certain circumstances engender autogenous heating in *mine environments* as well as in coal stockpiles exposed to the atmosphere.

---

*Because of the ability of —COOH and —OH to H-bond H_2O, relationships between heats of wetting and surface areas are affected by coal rank, with heats of wetting per unit surface increasing with the concentration of acid O-bearing functions.

Griffith et al.[104] Gill and Browning,[105] and Feng et al.[106] have itemized a number of conditions that would contribute to heating in *underground* mines. They have drawn particular attention to the role of coal fines* and high ventilation pressure differentials. Specific instances in which improvements to the ventilating system of an underground mine with *no* previous history of autogenous heating caused the coal to heat up have been reported by, inter alios, Riddell,[107] Vardy,[108] and Jeger.[109]

The circumstances that promote autogenous heating do *ipso facto* point to some practical steps for minimizing, if not actually eliminating, this hazard. Such steps would include measures to reduce size degradation and segregation as well as measures to minimize air flow through the coal. Where, as is common practice in the maintenance of coal stockpiles, watering programs are implemented in order to prevent or control heating, it is obviously also imperative to adhere to schedules that preclude the drying out of the coal between successive applications of water.

Some of these (and other) precautions against autogenous heating are more specifically addressed in Chapter 6 and are implicitly built into recommended procedures for constructing and maintaining coal stockpiles in open air.[110] They are also emphasized in the proceedings of numerous symposia on prevention of autogenous heating in active as well as abandoned mines.[111,112] Beyond broad recognition of the kinds of coal that pose a threat, and of the conditions under which they do so, little progress has been made toward defining the liability to autogenous heating *in particular circumstances*, especially in mine environments where preventive measures are generally more difficult to implement than in the case of stockpiles.

Guney and Hodges[113] and Prégermain[114] have reviewed laboratory tests, which included an ignitability test. Feng et al.[106] have stressed that the risk of autogenous heating in an underground mine is influenced by geological factors such as faults and weak or disturbed strata as well as by coal properties and mining practices. A *risk index* ought consequently to be thought of as a product of a *liability index* (I_a), which refers to the coal per se, and an *environmental index* (I_b). For practical purposes they suggest that a useful I_a (with values to > 10) can be obtained from the *relative ignition temperature*, that is, the lowest temperature at which self-heating of the coal begins to manifest itself in a standard ignitability test[114–117] by writing

$$I_a = \frac{\text{average heating rate between } 110° \text{ and } 220°C}{\text{relative ignition temperature}}$$

For I_b, which qualitatively incorporates the extent of fissuring of the coal and its enveloping strata, the relative amounts of coal fines generated by the mining operation and the ventilation pressure differential, they propose an intuitive

---

*Feng et al.[106] note that modern mining practices tend to produce substantially greater amounts of coal fines than were generated by older methods.

range of 1 (low) to 4 (high). A risk index, which can assume values between 1 and 40, then rates the risk of autogenous heating as *low* (< 10), *medium* (10–20) and *high* (> 20). Feng et al.[106] note, however, that considerable further field work is required before the utility of this formulation can be fully assessed.

In these circumstances, considerable interest attaches to a number of more recently developed mathematical models that purport to describe autogenous heating in coal stockpiles.[118–121]

Generally, for the sake of simplicity, assuming absence of radial heat and mass transfer, these one-dimensional models represent what are deemed to be essential parameters in the heating process (e.g., heats and rates of coal oxidation, oxygen concentrations, temperature, coal drying characteristics, and/or pressure regimes within the stockpile) by a set of three or four differential—or, in one instance,[119] by two coupled partial differential—equations; and to the extent that such models have actually been tested, they seem in some instances to predict behavior aspects with gratifying fidelity. Norden and Bainbridge[122] have thus claimed that measured distributions of free oxygen in a stockpile agree with predictions from the Norden[119] model. And a prediction from the Quan[118] model—namely, that in a one-dimensional condition, an air approach velocity of 1 cm min^{-1} in a subbituminous coal could produce temperatures on the order of 100°C at distances of 2 m from the air entry after 200 hours—has been experimentally (albeit so far only tentatively) confirmed by Stott[123] and Stott et al.[124]

5.6 EFFECTS OF OXIDATION ON COAL PROPERTIES

Except for impairment of mechanical properties, primarily a consequence of dehydration of the coal during weathering (cf., Section 5.1), property changes caused by atmospheric oxidation can be directly traced to the chemical alteration of the coal substance. Some such changes can be quantitatively assessed from the accompanying shifts in elemental composition. An illustrative example is the *calorific value* (*Q*), a function of the carbon and hydrogen contents of a coal, which therefore falls as the proportions of these coal constituents decline during exposure to the air. In practice, this form of deterioration is usually not a matter of great concern. Losses in *Q* from well-constructed stockpiles are generally modest and certainly not large enough to impair seriously the value of the coal as a fuel. In some situations, it is nevertheless useful to assess the *potential* for such losses over time, and this can be done by estimating rates of change, which depend on the specific storage conditions, from *initial* changes in percent H or percent C and recalling that hydrogen contents fall linearly with carbon contents (cf., Figure 5.4). Given the elemental analysis of the fresh coal, the expected value of *Q* after any (arbitrarily chosen) length of exposure can then be calculated from the Dulong equation

$$Q = 144.4\{C\} + 610.2\{H\} - 65.9\{O\} + 0.39\{O\}^2$$

where $\{C\}$, $\{H\}$, and $\{O\}$ are the DMMF carbon, hydrogen, and oxygen contents, respectively.

A similar calculation allows estimation of changes in the *true density* (D). This parameter is normally measured by helium-displacement[125] or can be closely approximated by much simpler pyknometric water-displacement[126] but can also be obtained with good accuracy from

$$1/D = 0.54 + 0.043H$$

where H is the DMMF hydrogen contents of the sample under test.[125,127-129]* This expression has been validated for fresh coals with (DMMF) carbon contents between 80.5 and 95 percent as well as for oxidized coals with 73–85.5 percent C (DMMF) and implies that true densities gradually increase during progressive oxidation.

It seems also possible to predict oxidative alteration of the *specific heat* (c_p) from elemental compositions by making use of

$$c_p(\text{cal/g}) = f \cdot R/2m$$

where f denotes the number of vibrational degrees of freedom of each atom, R is the gas constant (1.98 cal/°abs.), and m is the molecular weight. By assuming that each carbon and hydrogen atom contributes only one degree of freedom† and replacing m by a reduced molecular weight m_c, that is,

$$m_c = 12.01 + 1.008(\text{H/C}) + 16.00(\text{O/C}) + 14.008(\text{N/C}) + 32.06(\text{S/C})$$

where H/C, O/C, and so on represent the (DMMF) hydrogen:carbon, oxygen:carbon, and so on atomic ratios, very good agreement between calculated and measured values for fresh coals has been reported.[63]

However, considerable problems arise when seeking to assess the impacts of atmospheric oxidation on properties that determine how a coal responds to preparation, handling, and particular end-use processing (such as carbonization and coking). From an industrial standpoint, these aspects of coal behavior are of more immediate (and greater) importance than individual intrinsic physical coal properties. Because they are compounded of several interacting physical and chemical characteristics, they are, for practical purposes, assessed by empirical test methods that commonly allow results to be expressed in different and not always truly equivalent terms. The data in the technical literature therefore only illustrate the directions of change. In virtually all cases, more specific information still depends on specific testing.

---

*Mentser and Ergun[129] have used the equivalent form $1/D = 0.44 + 0.84x_H$, where x_H is the hydrogen atom-percentage.

†This assumption reflects the fact that the specific heat of graphite corresponds to the equipartition value of one vibrational degree of freedom.

The *mechanical strength* of coal, in effect a measure of its resistance to breakage, is usually obtained by standardized empirical procedures such as the drop shatter or tumbler tests detailed in ASTM D3038-72 and D3042-72. Both yield a nominal *size stability s* ($= 100y/x$) and *friability f* ($= 100 - s$) from the average screen sizes of the coal before (x) and after (y) the test. Parry et al.[130,131] have illustrated the destructive impact of air-exposure on these parameters by large-scale trials with a Colorado subbituminous coal.

The friability of this coal when stored in a ventilated bin increased from an initial 16.1 to 21.6 percent after 10 days, and to 38.4 percent after 20 days.[130] In a subsequent test in which another bulk sample of the same coal* was stored in a pit (through which air flow was much reduced), the friability rose only from 23.6 to 30.2 percent over an eight month period.[131]

In tests in which the size stability of Colorado subbituminous coal under various storage conditions was followed over 35 days,[130] the $-1/4$ in. (-6.35 mm) fraction increased from an initial 8 percent to very nearly 60 percent during storage in open air but remained substantially unchanged when the coal was stored in a closed bin and rose only to ~ 21 percent in a ventilated silo. Analogous results have been reported by Eckhardt and Yates[132] from tests in which the strength of a lignite during 160 days' storage under variously severe weathering conditions was observed. The relevant end-of-test data are summarized in Table 5.1.

No large-scale trials to assess the corresponding behavior of more mature bituminous coals, generally assumed to be much less prone to physical decay, can be cited. However, laboratory investigations[133,134] confirm that significant size degradation by decrepitation (or slacking) tends, like pronounced liability to autogenous heating (see Section 5.5), to be confined to low-rank coals with bed moisture contents greater than 10–12 percent (cf., Figure 5.6).

A more explicit impact of atmospheric oxidation, in the sense of being directly attributable to surficial O-bearing functions such as —COOH and/or —OH, manifests itself in coal *cleaning* operations, where prior oxidation is found to impair the flotability of the coal in froth flotation circuits.[135,136]† This arises from the fact that the O-functions make the surfaces of the coal more hydrophilic (cf., Section 5.2) and consequently inhibit attachment of the nonpolar parts of the frothing agent molecules, such as amyl or butyl alcohols, terpinols, or cresols‡, to the surfaces (cf., also Chapter 8).

How greatly flotability is affected depends on the degree of oxidation and is quantitatively not well documented. However, even very slightly oxidized coal, or a coal feed containing a significant proportion of oxidized coal, has been

---

*The significantly higher initial friability of this sample (23.6 percent) suggests that it was *similar* to, rather than the same as, the coal used in the earlier work. It was probably taken from a different mine working the same deposit.

†For a treatment of the fundamentals of froth flotation, reference should be made to Gaudin[137] or Klassen and Mokourov.[138]

‡Particularly favored in commercial operations is methyl isobutyl carbinol (MIBC), applied at rates of ~ 50 g/ton of coal feed.

TABLE 5.1   Effects of Storage Conditions on Mechanical Properties of a Lignite (160 days storage)

Storage Conditions	Size Degradation (%)	Size Stability (%)	Friability (%)	Moisture Loss (%)
Fresh coal, at start of tests	3.2	88.3	15.7	
Open air; full exposure to elements	81.1	a	64.9	n.a.
Open air; protected against sunshine	70.0	48.8	42.4	n.a.
Open air; protected against rain	28.2	59.9	36.5	21.0
Open air; protected against sunshine and rain	17.1	58.8	35.5	21.0
Stored in basement	9.6	69.2	32.6	18.7
Stored in sealed metal canisters	8.8	81.0	21.1	1.3
Stored under water	5.2	87.5	24.7	

aNot possible to obtain an appropriate sample after 84 days.
Source: C. J. Eckhardt and C. W. Yates, *Trans. ASME* **65**, 829 (1943).

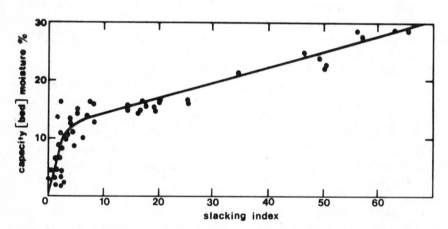

Figure 5.6 Variation of the slacking index with coal rank as measured by bed moisture contents (after Yancey et al.[133]).

repeatedly found to severely reduce the yield of clean coal even when its residence time in the flotation cells was extended. Although not specifically commented on in extant literature, such coal is likely to prove particularly troublesome in spherical agglomeration processes[139–141] that employ flotation principles, but go beyond conventional flotation by seeking to produce a consolidated material rather than cleaned *fines.*

Remedial measures have included attempts to neutralize surficial O-complexes with alkalis,[142] but greater (though still only partial) successes have been reported from the use of a surface conditioner (or collector)—commonly kerosene or a diesel oil,* used in a 10:1 ratio with a frothing agent—and *staged* addition of this material in order to ensure equitable distribution to all size fractions.[143,144] Some advantages may also accrue from *electrostatic* cleaning of oxidized coal fines.[92,145]

Another aspect of the more pronounced hydrophilic character of oxidized coal is an enhanced moisture-holding capacity expressed in an ability to accommodate more water than the corresponding fresh coal and *retain it more tenaciously*. In part, this arises from an oxidative widening of pore constrictions† whose small (5–8 nm) diameters in unaltered coals impede movement of water through a particle and in MVB and LVB coals actually exclude it from part of the pore volume. Equally important is that water molecules, like such other polar molecules as CH_3OH, can interact with —COOH and —OH to form H-bonded surface complexes and then require greater energy for their removal. This matter is not of concern in industrial drying operations but contributes to autogenous heating (cf., Section 5.5) through higher heats of adsorption (or correspondingly higher heats of wetting) that accompany such interactions. Even modest oxidation can suffice to raise heats of wetting by a factor of 2–3.[146]

Confirmation that the increased capacity for sorbing and retaining water is mostly due to interaction between H_2O molecules and —COOH and/or —OH functions can be seen in observations that blocking of these groups by acetylation reduces moisture sorption by, and heats of wetting of, low-rank coals.[147] The moisture-holding capacity can also be decreased by replacing H˙ on acid O-functions by cationic elements.[148]

The more deep-seated chemical alteration of coal by prolonged atmospheric oxidation, that is, progressive destruction of nonaromatic and/or hydroaromatic coal components (cf., Section 5.3.1), manifests itself in an altered *pyrolytic behavior*.

Without exception, though always depending quantitatively on the nature of the coal and on the particular conditions under which it oxidized, the yields of *tar* and, to a lesser extent, hydrocarbon gases fall while the amounts of CO, CO_2 and aqueous liquor increase. Concurrently with these changes in the product slate, pronounced changes in devolatilization *rates* are observed.[149] Substantial devolatilization begins at progressively lower temperatures with increasing oxidation, but maximum discharge rates, usually recorded in the first half of the active decomposition interval, that is, between ~400° and 450°C, are sharply reduced. These effects can be broadly associated with the thermal instability of oxygenated structures that generate oxides of carbon and H_2O, and with the

---

*The Dowell Division of Dow Chemical Corporation has more recently announced what it claims to be a singularly effective conditioner, M210, but has provided no details respecting its composition.
†Mechanistically this parallels the burning out of carbon particles by high-temperature oxidation, a process that activates the carbon by enlarging the surface area that an adsorbate can access. The dominant feature is presumably the removal of carbon as CO and CO_2.

lower yields of heavier (tarry) hydrocarbons whose diffusion-controlled discharge is a complex function of heating rates and coal particle size (cf., for example, Howard.)[150]

Extensive oxidation, which destroys the bulk of nonaromatic and hydroaromatic entities in coal, leads thus—much like dehydrogenation of coal[151-153]—to near-total loss of tar and lighter hydrocarbons. An extreme case presents itself in humic acids that furnish very little, if any, tars. They behave like unaltered anthracite in this respect.

Analogous changes are observed in what are collectively referred to as the *plastic properties* of caking coals. These are phenomenologically closely related to the composition and discharge rates of volatile matter,* and are measured empirically by the free swelling index (FSI), the Gieseler fluidity (ϕ), dilatation (*d*) and/or the plastic range, that is, the temperature interval between initial softening and final resolidification. The vast technical literature reporting on these aspects of coal behavior† shows that oxidation invariably causes rapid deterioration of all manifestations of plasticity. In other words, as oxidation proceeds, the free swelling index and fluidity, as well as the extent of initial softening and the subsequent dilatation *d* observed in dilatometric measurements, decrease;‡ the plastic range narrows progressively; and eventually, caking propensities, as reflected in these parameters, disappear completely, that is, the coal becomes noncaking.

Because of this gradual destruction of plastic properties, even a slightly oxidized caking coal will generally furnish a more extensively fissured coke. Since the oxidized particle surface precludes effective fusion along grain boundaries during carbonization,[24,157] the mechanical properties of such coke are also seriously impaired.

The ability to weaken or destroy plastic properties *deliberately* by oxidation has been utilized in coal gasification and combustion where caking coals would pose problems by blocking injection mechanisms operated at temperatures above ~ 300–$350°C$ and/or transiently depositing as (relatively unreactive) coke on reactor walls. The common practice is to pretreat the coal in a fluidized bed with air or air-enriched recycled flue gas at 300–400°C when there are no alternatives to caking coals. Incorporation of limited amounts of oxidized coal in a coke oven blend has also been explored as means for reducing otherwise

---

*For example, the plastic range is virtually coincident with the temperature range in which the coal actively decomposes; the FSI and dilatation (which measure the same phenomenon by different means) can be directly related to the amounts and discharge rates of volatile matter; and the fluidity may be governed by the chemical quality of tarry matter (or certain components thereof).

†Most of this literature dates from the 1930s and 1940s and has been extensively reviewed by Brewer,[154] Loison et al.,[155] and Habermehl et al.[156] More recent experimental studies have merely repeated the earlier work and confirmed the findings reported from it.

‡On occasion, the FSI of a highly swelling coal that also possesses a high *fluidity* will appear to *increase* somewhat after slight oxidation, but this is merely due to a lower fluidity and, consequently, to a lesser tendency of the swelling coke button to partly collapse upon itself during the test.

excessive fluidity of the blend and/or dangerous coking pressures that the blend would exert on oven walls.[158]

How fast caking propensites will deteriorate during exposure of a coal to the atmosphere is still quite unpredictable and seems to depend on an interplay of coal characteristics that can at present only be guessed at. Where one coal may deteriorate only modestly, another apparently very similar one will, under virtually identical storage conditions, lose almost *all* its plastic properties.* All that has so far been established is that moderately caking coals with FSIs in the range 3–5 tend to be much more prone to losing their caking properties completely than coals with higher FSIs.

In this connection, and as a matter of some practical importance per se, it should also be noted that oxidation has an adverse impact on the extractability of coal by organic solvents, especially those used for solvolytic extraction at temperatures between 200° and 400°C.[46,159] It likewise lowers the yield of so-called primary coal liquids (or soluble matter) obtainable from reaction of coal with an H-donor such as tetralin.[160] Both effects are almost certainly rooted in the chemical changes that simultaneously alter the pyrolytic behavior of coal.

5.7   SOME IMPLICATIONS OF ATMOSPHERIC OXIDATION

As noted in the introduction to this chapter and detailed in Sections 5.3.4 and 5.6, exposure of coal to the atmosphere will significantly alter its elemental composition and chemical structure over time, and thereby degrade its properties, or quality, to an extent that may render it unsuitable for the purposes for which it was originally intended. Also implicit in those discussions was that these changes are effectively *irreversible*. Although it is in principle possible to neutralize some effects of oxidation by chemical intervention—for example, caking properties can be restored and even improved upon by mild hydrogenation,[161–163] and unacceptable surface acidity can be countered by ion-exchange or-otherwise blocking O-bearing functions—such procedures are unlikely to prove practical outside the laboratory.

In such circumstances, emphasis must obviously always be placed on *avoidance* of oxidation (and associated hazards, such as autogenous heating), that is, on precautionary measures taken during mining and subsequent handling (or storage) of coal. Since the response to oxidation varies widely from one coal type to another, and even among coals of the *same* type or class, the cost-effectiveness of such measures must be site-specifically assessed in light of

---

*A major difficulty in the way of clarifying the reasons for such behavioral differences lies in the fact that the different aspects of coal plasticity, for example, FSIs (or dilatation) and fluidities, bear no discernible quantitative relationships to each other; that they respond differently to oxidation; and that their individual roles in determining plasticity and coke quality are still not fully understood.

the particular coal and the conditions under which it is being won, moved, and/or used.

Great care must also be taken in sampling coal for analytical purposes and in preparing or storing such samples. This is a matter of special importance in coal exploration programs where preliminary assessments of coal prospects are commonly based on coal samples taken from or via outcrops or abandoned earlier mine workings. Since oxygen can suffuse a coal body by traveling into it through fractures and bedding planes, as well as diffuse through its pore system, weathering effects can extend as much as 50 m into an undisturbed seam. Definitive sampling, difficult enough because of the intrinsic heterogeneity of coal, must therefore always be conducted at considerable distance from an exposed well-aged surface. With regard to sample preservation, it should also be borne in mind that antioxidants such as hydroquinones or amines, which have proven beneficial in other cases, appear to be quite ineffective in protecting coal.[164,165]

REFERENCES

1. R. Liotta, G. Brons, and J. Isaacs, *Fuel* **62**, 781 (1983).
2. W. A. Selvig and W. H. Ode, *Bull.—U.S. Bur. Mines* **571** (1957).
3. G. R. Yohe and C. A. Harman, *J. Am. Chem. Soc.* **63**, 555 (1941).
4. R. E. Jones and D. T. A. Townend, *Nature (London)* **155**, 424 (1945).
5. R. E. Jones and D. T. A. Townend, *Trans. Faraday Soc.* **42**, 297 (1946).
6. S. R. Rafikov and N. Y. Sibiryakova, *Izv. Akad. Nauk SSSR, Ser. Khim.* **9**, 13 (1956); *Chem. Abstr.* **50**, 8992 (1956).
7. W. N. Adams and G. J. Pitt, *Fuel* **34**, 383 (1955).
8. O. Grosskinsky and B. Jüttner, *Brennst.-Chem.* **39**, S7 (1958).
9. B. V. Tronov, *J. Appl. Chem. USSR (Engl. Transl.)* **13**, 1053 (1940); *Chem. Abstr.* **35**, 1966 (1941).
10. E. J. Jensen, N. Melnyk, J. C. Wood, and N. Berkowitz, *Adv. Chem. Ser.* **55**, 621 (1966).
11. C. Vauquelin, *Ann. Chim. (Paris)* **21**, 39 (1797).
12. A. Klaproth, *Gehlen's J.* **4**, 329 (1804).
13. H. C. Howard, *Trans. Am. Inst. Min. Metall. Eng.*, **177**, 523 (1947).
14. V. F. Oreshko, *Bull. Acad. Sci. USSR, Cl. Sci. Tech.* **7**, 52 (1943); *Chem. Abstr.* **39**, 2633 (1945).
15. V. F. Oreshko, *Izv. Akad. Nauk SSSR, Otd. Tekh. Nauk*, pp. 249, 748, 1642 (1949); *Chem. Abstr.* **44**, 2200, 2201, 2202 (1950).
16. V. F. Oreshko, *Dokl. Akad. Nauk SSSR* **70**, 445; **71**, 331; **74**, 327 (1950); *Chem. Abstr.* **44**, 10295 (1950); **45**, 2175, 5388 (1951).
17. V. F. Oreshko, *Izv. Akad. Nauk SSSR, Otd. Tekh. Nauk* p. 1031 (1951); *Chem. Abstr.* **46**, 10579 (1952).
18. L. D. Friedman and C. R. Kinney, *Ind. Eng. Chem.* **42**, 2525 (1950).

19. P. J. Farenden, cited by I.G.C. Dryden, in *Chemistry of Coal Utilization* (H. H. Lowry, ed.), Suppl. Vol. 1, p. 273. Wiley, New York, 1963.

20. H. Braconnot, *Ann. Chim. Phys.* **12**, 172 (1819).

21. T. Wood, *J. Appl. Chem.* **8**, 565 (1958); R. Kaji, Y. Hishinuma, and Y. Nakamura, *Fuel* **66**, 154 (1987).

22. I. Wender, L. A. Heredy, M. B. Neuworth, and I. G. C. Dryden, in *Chemistry of Coal Utilization, Second Supplementary Volume* (M. A. Elliott, ed.), pp. 425 et seq. Wiley (Interscience), New York, 1981.

23. N. Berkowitz, J. F. Fryer, B. S. Ignasiak, and A. J. Szladow, *Fuel* **53**, 141 (1974).

24. H. Sommers and W. Peters, *Chem. Ing.-Tech.* **26**, 441 (1954).

25. G. Schulz and H. C. Howard, *J. Am. Chem. Soc.* **68**, 994 (1946).

26. R. Q. Shotts, *Min. Eng. (London)* **187**, 889 (1950).

27. T. A. Kukharenko and Z. A. Ryzhova, *Khim. Tekhnol. Topl.* **4**, 20 (1956); *Chem. Abstr.* **50**, 12440 (1956).

28. D. Chandra, *Econ. Geol.* **53**, 102 (1958).

29. L. D. Schmidt and J. L. Elder, *Ind. Eng. Chem.* **32**, 249 (1940).

30. T. Yamasaki, *J. Min. Inst. Jpn.* **69**, 253 (1953).

31. D. L. Carpenter and D. G. Giddings, *Fuel* **43**, 247 (1964).

32. G. S. Scott and G. W. Jones, *Rep. Invest.—U.S. Bur. Mines* **RI-3546** (1941).

33. G. S. Scott, *Bull.—U.S. Bur. Mines* **455** (1944).

34. L. S. Merill, *Fuel* **52**, 61 (1973).

35. C. P. Fenimore and G. W. Jones, *J. Phys. Chem.* **71**, 593 (1967).

36. J. S. Haldane and F. G. Meacham, *Trans. Inst. Min. Eng.* **16**, 457 (1898).

37. T. F. Winmill, *Trans. Inst. Min. Eng.* **46**, 563 (1913).

38. T. F. Winmill, *Trans. Inst. Min. Eng.* **48**, 503, 514 (1914).

39. J. L. Elder, L. D. Schmidt, W. A. Steiner, and J. D. Davis, *U.S. Bur. Mines, Tech. Publ.* **681** (1945).

40. R. A. A. Taylor, *J. Inst. Fuel* **14**, 144 (1941).

41. B. H. Chalishazar and C. E. Spooner, *Fuel* **36**, 127 (1957).

42. W. M. Thomas, T. D. Jones, and J. I. Graham, *Proc. S. Wales Inst. Eng.* **49**, 201 (1933).

43. W. Francis and R. V. Wheeler, *J. Chem. Soc.* **127**, 112 (1925).

44. W. Francis and H. M. Morris, *Bull.—U.S., Bur. Mines* **340** (1931).

45. W. Francis; *Coal*, pp. 493–508. Edward Arnold, London, 1961.

46. A. I. Khrisanfova, *Tr. Inst. Goryuch. Iskop. Moscow* **2**, 179 (1950); *Chem. Abstr.* **49**, 4259 (1955).

47. N. Berkowitz, *Introduction to Coal Technology*, pp. 101–102. Academic Press, New York, 1979.

48. C. A. Seyler, *Proc. S. Wales Inst. Eng.* **53**, 254, 396 (1938).

49. M. Schütze, *Z. Anal. Chem.* **118**, 245 (1939).

50. J. Unterzaucher, *Ber. Dtsch. Chem. Ges. B* **73B**, 391 (1940).

51. W. Radmacher and A. Hoverath, *Brennst.-Chem.* **40**, 97 (1959).

52. A. Crawford, M. Glover, and J. H. Wood, *Mikrochim. Acta* **1**, 46 (1961).

53. M. S. Burns, R. Macara, and D. J. Swaine, *Fuel* **43**, 349 (1964).

54. B. S. Ignasiak, B. N. Nandi, and D. S. Montgomery, *Anal. Chem.* **41**, 1676 (1969).

55. ISO, *ISO R 1994*. Am. Natl. Stand. Inst., Washington, DC, 1970.

56. R. F. Abernethy and F. H. Gibson, *Rep. Invest.—U.S., Bur. Mines* **RI-6753** (1966).

57. G. Bergmann, G. Huck, J. Karweil, and H. Luther, *Brennst.-Chem.* **35**,f 175 (1954).

58. G. Bergmann, G. Huck, J. Karweil, and H. Luther, *Brennst.-Chem.* **39**, 20 (1958).

59. M. M. Roy, *Fuel* **36**, 250 (1957).

60. T. P. Maher, J. M. Harris, and G. R. Yohe, *Rep. Invest.—Ill. State Geol. Surv.* **212** (1959).

61. R. Bouwman and I. D. Frericks, *Fuel* **59**, 315 (1980).

62. A. Ihnatowicz, *Pr. Gl. Inst. Gorn.* **125** (1952).

63. D. W. Van Krevelen, *Coal*, pp. 160–176, 419. Elsevier, Amsterdam, 1961.

64. C. Kröger and G. Darsow, *Erdoel Kohle* **17**, 88 (1964).

65. C. Kröger, K. Fuhr, and G. Darsow, *Erdoel Kohle* **18**, 36, 701 (1965).

66. B. S. Ignasiak, T. M. Ignasiak, and N. Berkowitz, *Rev. Anal. Chem.* **2**, 278 (1975).

67. H. N. S. Schafer, *Fuel* **49**, 197 (1970).

68. L. Blom, L. Edelhausen, and D. W. Van Krevelen, *Fuel* **36**, 135 (1957).

69. L. W. Hill and P. H. Given, *Carbon* **7**, 649 (1969).

70. Z. Abdel-Baset, P. H. Given, and R. F. Yarzab, *Fuel* **57**, 95 (1978).

71. S. Friedman, M. L. Kaufman, W. A. Steiner, and I. Wender, *Fuel* **40**, 33 (1961).

72. L. Blom, Ph.D. Thesis, University of Delft, Netherlands, 1960.

73. J. N. Bhaumik, A. Lahiri, and P. N. Mukherjee, *Chem. Ind. (London)* p. 1998 (1960).

74. Y. Takegami, S. Kajiyama, and C. Yokokawa, *Fuel* **42**, 291 (1963).

75. L. Lazarov and G. Angelova, *Fuel* **47**, 333 (1968).

76. H. Wachowska and W. Pawlak, *Fuel* **56**, 422 (1977).

77. B. S. Ignasiak and M. Gawlak, *Fuel* **56**, 261 (1977).

78. H. W. Sternberg, C. L. Delle Donne, P. Pantages, E. C. Moroni, and R. E. Markby, *Fuel* **50**, 432 (1971).

79. H. W. Sternberg and C. L. Delle Donne, *Fuel* **53**, 172 (1974).

80. J. A. Franz and W. E. Skiens, *Fuel* **57**, 502 (1978).

81. J. W. Larsen and L. O. Urban, *J. Org. Chem.* **44**, 3219 (1979).

82. G. R. Yohe and M. H. Wilt, *J. Am. Chem. Soc.* **64**, 1809 (1942).

83. J. A. Radspinner and H. C. Howard, *Ind. Eng. Chem., Anal. Ed.* **15**, 566 (1943).

84. F. Heathcoat, *Fuel* **12**, 4 (1933).

85. O. I. Egorova, *Bull. Acad. Sci. URSS, Cl. Sci. Tech.* **7-8**, 107 (1942).

86. A. I. Khrisanfova, *Izv. Akad. Nauk SSSR, Otdl. Tekh. Nauk* p. 1116 (1949); *Chem. Abstr.* **49**, 7221 (1955).

87. B. A. Onusaitis and A. I. Khrisanfova, *Izv. Akad. Nauk SSSR, Otd. Tekh. Nauk* p. 895 (1947); *Chem. Abstr.* **43**, 3593 (1949).

88. L. G. Benedict and W. F. Berry, *Adv. Chem. Ser.* **55**, 577 (1966).

89. E. G. Lamba and I. L. Ettinger, *Izv. Akad. Nauk SSSR, Otd. Tekh. Nauk.* **4**, 110 (1955); *Chem. Abstr.* **49**, 15209 (1955); V. S. Veselovskii and E. A. Terpogosova, *ibid.* **12**, 140 (1954); *Chem. Abstr.* **50**, 11641 (1956).

90. F. S. Men'shchikov and Z. E. Rozmanova, *Zavod. Lab.* **21**, 1471 (1955); *Chem. Abstr.* **50**, 8993 (1956).

91. D. G. A. Thomas, *Br. J. Appl. Phys.* **4**, 55 (1953).

92. R. J. Gray, A. H. Rhoades, and D. T. King, *SME-AIME Prepr.* No. 74F63 (1974).

93. G. E. Mapstone, *Chem. Ind. (London)* p. 658 (1954).

94. N. Berkowitz and H. G. Schein, *Fuel* **30**, 94 (1951).

95. D. J. Hodges and F. B. Hinsley, *Trans. Inst. Min. Eng.* **123**, 211 (1963).

96. D. J. Hodges and B. Acherjee, *Trans. Inst. Min. Eng.* **126**, 121 (1966).

97. K. K. Bhattacharyya, D. J. Hodges, and F. B. Hinsley, *Min. Eng. (London)* **101**, 274 (1969).

98. M. Guney, *Bull. Can. Inst. Min. Metall.* **64** (No. 707), 138 (1971).

99. F. L. Shea and H. L. Hsu, *Ind. Eng. Chem. Prod. Res. Dev.* **11**, 184 (1972).

100. E. Erdtmann and H. Stoltzenberg, *Braunkohle (Duesseldorf)* **7**, 69 (1908).

101. R. Threllfall, *J. Soc. Chem. Ind., London* **28**, 759 (1909).

102. A. J. Hoskin, *Purdue Univ. Exp. Stn., Bull.* **30** (1928).

103. W. Francis, *Fuel* **17**, 363 (1938).

104. F. E. Griffith, M. O. Magnusen, and G. J. R. Toothman, *Bull.—U.S., Bur. Mines* **590** (1960).

105. F. S. Gill and E. J. Browning, *Colliery Guardian* **107**, 79; 134 (1971).

106. K. K. Feng, R. N. Chakravorty, and T. S. Cochrane, *Bull. Can. Inst. Min. Metall.* **66** (No. 738), 75 (1973).

107. M. Riddell, *Trans. Inst. Min. Eng.* p. 33 (1970).

108. S. Vardy, *Trans. Inst. Min. Eng.* p. 79 (1970).

109. C. Jeger, CERCHAR Publ. No. 2262. Charbonnages de France, 1972.

110. R. R. Allen and V. F. Parry, *Rep. Invest.—U.S., Bur. Mines* **RI-5034** (1954).

111. E. Bredenbuch, *Glueckauf* **90**, 373 (1954).

112. Institute of Mining Engineers, *Prevention of Spontaneous Combustion*. IME, London, 1970.

113. M. Guneg and D. J. Hodges, *Colliery Guardian* **105**, 173 (1969).

114. S. Prégermain, *Ind. Miner. (Paris)* September (1972).

115. R. V. Wheeler, *J. Chem. Soc.* **113**, 945 (1918).

116. H. F. Coward, *SMRE Res. Rep.* **142** (1957).

117. D. K. Nandy, D. D. Banerjee, and R. N. Chakravorty, *J. Mines, Met. Fuels* February (1972).

118. N. T. Quan, Ph.D. Thesis, University of Canterbury, New Zealand, 1971.

119. P. Norden, *Fuel* **58**, 456 (1979).

120. D. Schmal, J. H. Duyzer, and J. W. Van Heuven, *Fuel* **64**, 963 (1985).

121. K. Brooks and D. Glasser, *Fuel* **65**, 1035 (1986).

122. P. Norden and N. W. Bainbridge, *Fuel* **63**, 943 (1984).

123. J. B. Stott, *Proc. N.Z. Inst. Min. Eng.* pp. 1–17 (1974).

124. J. B. Stott, B. J. Harris, and P. L. Hansen, *Fuel* **66**, 1012 (1987).

125. R. E. Franklin, *Fuel* **27**, 46 (1948); *Trans. Faraday Soc.* **45**, 274 (1949).

126. J. A. Dulhunty and R. E. Penrose, *Fuel* **30**, 109 (1951).

127. J. W. Whitaker, *Fuel* **29**, 33 (1950).

128. P. G. Sevenster, *J. S. Afr. Chem. Inst.* **7**, 41 (1954).

129. M. Mentser and S. Ergun, *Fuel* **39**, 509 (1960).

130. V. F. Parry and J. B. Goodman, *Rep. Invest.—U.S., Bur. Mines* **RI-3587** (1941).

131. J. B. Goodman, V. F. Parry, and W. S. Landers, *Rep. Invest.—U.S. Bur. Mines* **RI-3915** (1946).

132. C. J. Eckhardt, Jr. and C. W. Yates, *Trans. ASME* **65**, 829 (1943).

133. H. F. Yancy, N. J. F. Johnson, and W. A. Selvig, *U.S. Bur. Mines, Tech. Pap.* **512** (1932).

134. B. Roga and P. Pampuch, *Pr. Gl. Inst. Gorn.* **189** (1956).

135. S. C. Sun, *Trans. AIME* **199**, 306 (1954).

136. J. B. Gayle, W. H. Eddy, and R. Q. Shotts, *Rep. Invest.—U.S. Bur. Mines* **RI-6620** (1965).

137. A. M. Gaudin, *Flotation*, 2nd ed. McGraw-Hill, New York, 1957.

138. V. I. Klassen and V. A. Mokourov, *Introduction to the Theory of Flotation* (transl. by J. Leja and G. W. Poling). Butterworth, London, 1963.

139. H. M. Smith and I. E. Puddington, *Can. J. Chem. Eng.* **39**, 94 (1960).

140. A. F. Sirianni, C. E. Capes, and I. E. Puddington, *Can. J. Chem. Eng.* **47**, 166 (1969).

141. C. E. Capes, A. E. Smith, and I. E. Puddington, *Bull. Can. Inst. Min. Metall.* **67** (No. 747), 115 (1974).

142. W. W. Wen and S. C. Sun, Paper presented at SME-AIME Meeting, Denver, CO, September (1976).

143. B. A. Firth, A. R. Swanson, and S. K. Nicol, *Proc. Australas. Inst. Min. Metall.* **267**, 49 (1978).

144. B. A. Firth, A. R. Swanson, and S. K. Nicol, *Int. J. Miner. Process.* **5**, 321 (1979).

145. R. M. Inculet, Paper presented at 82nd Annual Meeting CIM, Toronto, April (1980).

146. R. B. Anderson, W. K. Hall, J. A. Lecky, and K. C. Stein, *J. Chem. Phys.* **60**, 1548 (1956).

147. B. K. Mazumdar, D. H. Bhangale, and A. Lahiri, *Fuel* **36**, 254 (1957).

148. E. I. Ruskin and T. P. Gorskaya, *Khim. Tverd. Topl. (Moscow)* **8**(5), 28 (1974).

149. A. F. Boyer, *Ann. Mines Belg.* p. 908 (1956).

150. J. B. Howard, in *Chemistry of Coal Utilization, Second Supplementary Volume* (M. A. Eliott, ed.), pp. 688–701. Wiley (Interscience), New York, 1981.

151. B. K. Mazumdar, S. K. Chakrabartty, and A. Lahiri, *Fuel* **38**, 112 (1959).

152. L. Reggel, I. Wender, and R. Raymond, *Fuel* **47**, 373 (1968).

153. L. Reggel, I. Wender, and R. Raymond, *Fuel* **49**, 287 (1970).

154. R. E. Brewer, in *The Chemistry of Coal Utilization* (H. H. Lowry, ed.), Vol. 1, p. 160. Wiley, New York, 1945.

155. R. Loison, A. Peytavy, A. F. Boyer, and R. Grillot, in *The Chemistry of Coal Utilization* (H. H. Lowry, ed.) First Suppl. p. 150. Wiley, New York, 1963.

156. D. Habermehl, F. Orywal, and H. D. Beyer, in *Chemistry of Coal Utilization, Second Supplementary Volume* (M. A. Elliott, ed.), p. 317. Wiley (Interscience), New York, 1981.

157. E. P. Carman, M. R. McGeer, and H. L. Riley, *Inf. Circ.—U.S., Bur. Mines* **7794** (1957).

158. B. S. Ignasiak, D. Carson, P. Jadernik, and N. Berkowitz, *Bull. Can. Inst. Min. Metall.* **72** (No. 803), 154 (1979).

159. I. G. C. Dryden, in *The Chemistry of Coal Utilization* (H. H. Lowry, ed.), First Suppl. Vol., p. 243. Wiley, New York, 1963.

160. R. C. Neavel, *Fuel* **55**, 237 (1976).

161. C. H. Lander, F. S. Sinnatt, J. G. King, and A. Crawford, British Patent 301,720 (1928).

162. T. A. Kukharenko and A. I. Khrisanfova, *Izv. Akad. Nauk SSSR, Otd. Tekh. Nauk*, p. 863 (1947); *Chem. Abstr.* **43**, 4830 (1949).

163. M. Orchin, C. Golumbic, J. E. Anderson, and H. H. Storch, *Bull.*—U.S. Bur. Mines **505** (1951).

164. G. R. Yohe, R. H. Organist, and M. W. Lansford, *Circ.—Ill. State Geol. Surv.* **201** (1952).

165. G. R. Yohe, *Fuel* **44**, 135 (1965).

Commentary

Chapter 5 considered the atmospheric oxidation of coal from the point of view of chemical processes on the molecular level. The following two chapters treat the technological aspects of aging, the first on aging and weathering and the second on aging and beneficiation. Chapter 6 is a wide-ranging exposition of aging and weathering in which methods for the evaluation of the aging process, the deterioration of coking properties, and the weathering characteristics of coal in storage are presented in depth. Whether the interest of the reader is inclined toward the preservation of samples for research or the storage of coal for use in industrial processes, he or she is afforded an encyclopedic overview of a universal theme in the science and technology of coal, aging and weathering.

6

AGING AND WEATHERING

R. J. Gray
and
D. E. Lowenhaupt
Monroeville, PA
Conoco Coal Development Co.

6.1 INTRODUCTION

The extent of alteration of the physical and chemical properties of coal due to oxidation has been a perplexing problem for the coal producer and user as well as the coal scientist. Considerable study has been devoted to the effect of natural and induced oxidation on coal properties including the work of Gray, et al.,[1] Gray and Krupinski,[2] and Crelling, et al.[3] Most of this work compares the chemical and physical properties of fresh and oxidized coal. To a large extent, this work has been aimed at understanding the effect of coal oxidation on technological properties rather than the mechanism of oxidation. Coal oxidation studies have required the development of methods to detect oxidized coal.[4] Once oxidized coal can be identified, its effect on characterization properties and industrial processes such as carbonization and combustion can be determined. In general, oxidation of coal adversely affects the coking quality and plastic properties of bituminous coals as well as the recovery of froth flotation product in beneficiation. The increased oxygen content results in a lower Btu value, renders coal more susceptible to spontaneous combustion, and causes degrading in stockpiles. Oxidized coal negatively affects coal charging and handling in a coking operation and increases the difficulty of maintaining high-bulk density of coke-oven charges. Blast-furnace performance can also be adversely affected by the coke produced from oxidized coal. In addition, cokes produced from oxidized coals have lower physical strength and higher reactivity in CO_2 at elevated temperature, both of which are undesirable.

In some instances coal oxidation is advantageous. Oxidation destroys the agglomerating properties of bituminous coals. Therefore, processes that involve heating of coals on grates or in fluid beds where agglomeration is undesirable may benefit from the use of oxidized coal or a preoxidation step to inhibit particle sticking.

This chapter reviews a wide variety of tests involving carbonization, swelling, flotation-phase inversion, plasticity, chemistry, microscopy of coal and coke, infrared spectroscopy, alkali solubility, and electrostatic charge to detect naturally oxidized (weathered) coal. Induced oxidation is discussed only to contrast it with natural weathering.

6.2 WEATHERING OXIDATION

The earth scientist divides all Earth forces that change its surface into destructional forces that wear down the continents and constructional forces that build up or raise portions of the continents. The destructional forces include weathering and erosion whose energy is derived from the sun. The principal agents of weathering and erosion are the atmosphere, wind, wave action, rivers, and glaciers. The distinction between weathering and erosion is commonly made on the basis of whether or not the rocks fragmented by weathering are transported from the site. All forces, both chemical and physical, that breaks rock into fragments at or near the earth's surface but do not carry them away are considered weathering. The agents include the atmosphere and weather, as well as plants and animals.

Weathering is generally grouped into two types, mechanical and chemical. Mechanical weathering involves the break up of the rock into smaller pieces without compositional changes. Chemical weathering involves alteration or decay which break up the rock minerals into different substances. Some mechanical weathering is caused by temperature changes. Water expands about 10 percent by volume when it freezes. The action of frozen water in splitting rocks is called frost action or frost wedging. Dense rocks peel or scale by a process called exfoliation. Some believe water reacts with exposed minerals to cause their swelling or expansion which results in peeling. Corners or edges of rocks weather more rapidly due to greater exposure resulting in rounding. Differential expansion causes exfoliation. This is caused by the heating and cooling at the rock surface versus the more uniform internal temperature of exposed rocks. Removal of the overlying rocks resulting in reduced pressure may also cause exfoliation.

Plants such as moss and lichens as well as many other plants grow on rocks and their roots grow into cracks and pores. Acids formed by roots dissolve or react with rocks.

Animals contribute indirectly to weathering. Worms, ants, burrowing animals, and insects dig, exposing rock to air and water and bringing particles to the surface.

Chemical weathering of rock is generally a very slow process and involves geologic time—thousands, millions, or even billions of years. The chief reactions involve oxygen, water, and the atmosphere, which contains carbon dioxide. Small amounts of nitric acid and ammonia and other gases from the action of lightning with air as well as soil acids formed by plant and animal decay also contribute to chemical weathering.

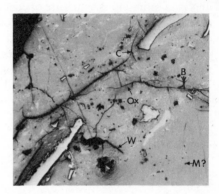
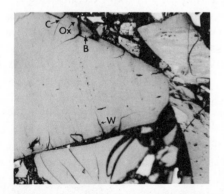
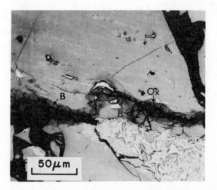

Figure 6.1 Photomigrographs showing the development of cracks and micropores, **M** due to coal weathering. Some cracks are wedge-shaped: **W**, while others are concoidal: **C**, and some bifurcate: **B**. Darker oxidized areas: **OX**, are softer producing polishing relief. Reflected light in oil.

The reaction of water and rock-forming minerals is called *hydration*. The union of oxygen with other substances is termed *oxidation*. The reaction of carbon dioxide with substances is termed *carbonation*. The acids act to dissolve substances. A warm humid climate favors weathering, particularly where repeated wetting and drying are common. The end product of rock weathering is soil.

When coals weather, the inherent bedding planes, cleats, and other cracks widen and the coal expands. The cleat in bituminous coals is generally at right angles to the bedding plane. Cleat generally occurs as two sets that intersect at right angles and each set consists of parallel fractures. The best developed cleat are called *face*; those that are less developed and occur at right angles to the face are called *butt*. As oxidation progresses, the crack surfaces become more extensive, and the initial cracks or fractures, commonly at right angles in bituminous coals, are opened. A new system of very irregular fractures develop between the initial cracks. Coal expands due to weathering and cracks develop at right angles to cleat and bedding plane fractures. These cracks are wedge shaped with the narrow end terminating in fresh coal. When extended, they tend to become conchoidal and bifurcated. Micropores also develop during weathering. The margins of weathered surfaces are lower reflecting (darker) and softer than fresh coal and create relief when polished for microscopic observation (Figure 6.1).

6.3 DAVID WHITE: HISTORIC WORKS ON COAL OXIDATION

White, in 1911,[4] performed and reported a very significant study concerning the effect of oxygen on coal properties. The purpose of the study was to (1) determine the importance of oxygen as an impurity in coal by comparing the ultimate analyses of approximately 250 coals of various ranks and (2) relate these findings both to coal origin, and the technological properties of coal. His most important conclusion was that oxygen is nearly as harmful as ash on the technological properties of coal. The findings of his study are summarized as follows:

1. Oxygen and ash have the same negative relationship to the calorific value of a coal. High oxygen and low ash produce the same effect as high ash and low oxygen as long as the carbon content is the same.
2. The negative effect of oxygen is the same whether it is bonded to carbon or hydrogen.
3. The calorific value of a coal is indicated by the ratio of total carbon to the sum of ash and oxygen. The C/(O + ash) calorific value relationship is fairly accurate for all ranks of coal.
4. If a coal is weathered, this ratio does not reliably predict the calorific value of that coal.

5. The weathering of coal is associated with a loss in calorific value especially for low-rank coals. They sometimes show a loss in their heating value from the mine to the test apparatus.

6. The coking properties of a coal can be predicted by the H:O ratio. A coal with a ratio of 59 percent will generally produce a good coke. Weathering will decrease this ratio, thus reducing a coal's coking power.

There is nothing astounding about these statements; however, one must consider their time frame. With relatively unsophisticated analytical equipment, White reached valid conclusions that are being "discovered" again with modern technology.

6.4 MICROSCOPY OF NATURALLY OXIDIZED COAL

Natural oxidation or weathering, as well as induced oxidation, cause changes in coal microstructure. These changes do not affect all macerals the same.

Franz Schulze,[5] found that the bulk of coal can be oxidized in nitric acid and potassium chlorate (Schulze solution). The oxidized coal can be dissolved in an alkali solution, a process called maceration. The residue from this procedure consists principally of spores, pollen, cuticle, inert organic macerals and mineral matter. Palynologists use this technique or some modification of the procedure to isolate fossil liptinite (spores, pollen, cuticle, resin, etc.) for studies to aid in dating, identifying, and correlating coal seams.

The appearance of bituminous coal macerals in reflected light in oil is shown in Figure 6.2. Vitrinite which is most affected by coal oxidation although some changes do occur in the liptinite, inertinite and mineral matter. This was demonstrated by Ferrari.[6] In his study of the origin of mine fires by spontaneous combustion, he showed vitrain oxidation was most important while other constituents absorb oxygen much less readily. This is expected with liptinite (exinite) since it is an incorporated maceral, preserved as a result of its resistance to chemical attack. Liptinite is more abundant in coals formed in acid conditions and tends to be less abundant in coals formed in fresh water or alkaline conditions. Alkali ash coals seldom have high liptinite contents unless the carbonates represent secondary deposition. This indicates liptinite is more resistant to acid attack. In fact, prolonged alkali treatment in the maceration process results in liptinite loss. Fusinite and other inertinite macerals such as semifusinite, micrinite, macrinite, and inertodetrinite are also incorporated coal macerals formed in the peat or biochemical stage of coalification by oxidizing processes which include thermal, chemical, and biological activity. Plumstead[7] attributes the high inertinite content of Gondwana coals to dry conditions with high oxidation in the peat stage. Since organic inerts are oxidation products of plant alteration, they are less susceptible to additional oxidation.

Marevich and Travin[8] found that bright coal (high in vitrinite) oxidized more rapidly than dull coal (high in inertinite, liptinite, and sometimes mineral

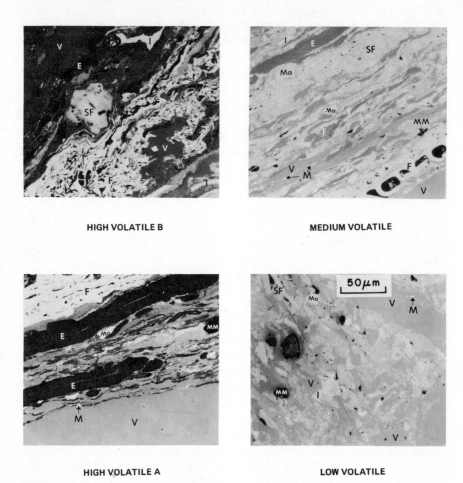

Figure 6.2 Photomicrographs showing coal macerals in high, medium, and low-volatile bituminous coals. Vitrinite: **V**, exinite: **E**, resinite: **R**, micrinite: **M**, macrinite: **MA**, intertodetrinite: **I**, semifusinite: **SF**, fusinite: **F**, and mineral matter: **MM**. Reflected light in oil.

matter). This persists up to the stage of advanced oxidation where oxidation of bright coal is retarded. Marevich and Travin[8] also found mylonitized (finely brecciated) coals were most oxidizable. Fusain in contact with vitrain was also found to be hazardous with respect to spontaneous combustion.

It is important to be able to detect oxidized coal in order to assess the effect of oxidation on the characterization of coals and their response to processing. Certain features of coal oxidation are readily recognized by microscopic features. Benedict and Berry,[9] Gray et al.,[1] Benedict et al.,[10] Crelling et al.,[3] Lowenhaupt and Gray,[11] and Marchioni[12] have described microscopic features characteristic of oxidized coal. Marchioni[12] gives an excellent review of the

subject. Lowenhaupt and Gray[11] assigned oxidized coal to three categories—
slightly, moderately, or badly oxidized based on microscopically distinguishable
oxidation features. Normal or unoxidized coal would constitute another
category. The features are the same for various ranks of coals.

In the petrographic procedure of Lowenhaupt and Gray,[11] the coal is
analyzed at a magnification of about 500 × with an oil-immersion objective. The
sample consists of a crushed coal (− 20 mesh), mixed with an epoxy binder,
briquetted, ground, and polished for microscopic examination. The sample is
analyzed using a point-counting procedure in which 5 points per field are
identified. The spacing between fields is about 500 μm, and a total of no less than
1000 points, excluding plastic, are identified. Particles below 5 μm are very

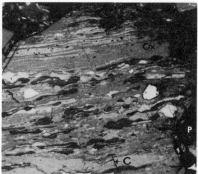

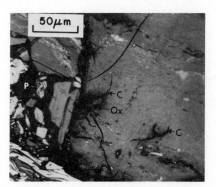

Figure 6.3 Photomicrographs showing weathered high-volatile bituminous coal from
Mingo County, West Virginia. Slitted structures: **S**, micropores: **M**, cracks: **C**, darkened
oxidized margins of particles and cracks: **OX**, and plastic mounting media: **P**, Reflected
light in oil.

difficult to assign to categories and their identification will introduce some error.

The weathering of coal produces changes principally in the physical and chemical characteristics of vitrinite and in the mineral matter. These changes are manifested in the development of degradation features and in the discoloration of vitrinite bordering on the cracks that occur naturally in coal and on the fracture development that accompanies natural oxidation. The extent and severity of the development of these features are used to allocate coal to the different levels of oxidation.

The features used in distinguishing the microscopic levels of oxidation[11] are illustrated in Figures 6.3 and 6.4. These features can be compared to those of unoxidized coals shown in Figure 6.2. The description of the microscopic characteristic features of oxidized coal are as follows:

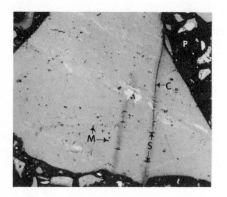
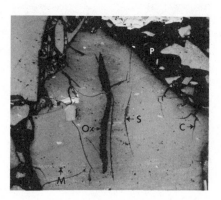
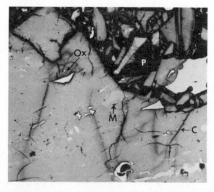
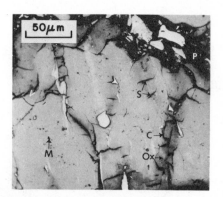

Figure 6.4 Photomicrographs showing weathered low volatile bituminous Pocahontas No. 4 coal from McDonald County, West Virginia. Slittled structures: **S**, micropores: **M**, cracks: **C**, darkened oxidized margins of particles and cracks: **OX**, and plastic mounting media: **P**. Reflected light in oil.

1. The vitrinite particles display cracking unrelated to cleat, sample preparation, or pseudovitrinite. This causes fines to develop. The extent of crack development is related to the rank and type of coal as well as the degree of oxidation.

2. A slight to severe discoloration (darker than the vitrinite) may occur along the edges and cracks of vitrinite particles. The degree of discoloration increases with increased oxidation.

3. The edges of the vitrinite become rounded.

4. A slight to severe alteration of FeS_2 (pyrite) accompanies oxidation. Clay minerals may exhibit swelling due to oxidation.

5. The cleat surfaces and edges of the vitrinite particles are often coated with secondary minerals in oxidized coals.

6. All these conditions may exist simultaneously or in any combination, but any one is an indicator of oxidized coal. The features become more pronounced as oxidation progresses and the number of oxidation features displayed also frequently increases with increased weathering.

Lowenhaupt and Gray[11] reported good agreement between petrographically determined coal oxidation and the oxidation of coal determined by an alkali-extraction tests (U.S. Steel procedure) as shown in Figure 6.5.

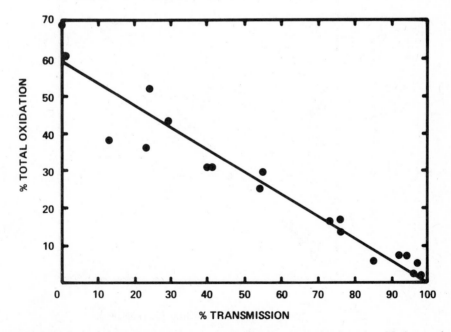

Figure 6.5 Relationship between petrographically determined coal oxidation and transmission values (after Lowenhaupt and Gray.[11]

Crelling, et al.[3] also showed good agreement between the petrographically determined weathered coal in samples and the actual percentage of weathered coal that was incrementally substituted for fresh coal (Figure 6.6).

One method of quantifying the extent of coal oxidation may lie in determining the thickness of oxidation zones and the reflectance of these zones relative to the centers or unaltered cores of coal particles. In general, studies of oxidized coal showed that the oxidation rings or borders of vitrinite have lower reflectance in weathered coals and higher reflectance in coals with induced oxidation. Chandra,[13] in studies of air-oxidized, naturally oxidized, and fresh coals observed no reflectance changes. In fact, Chandra[14] suggests that the reflectance of vitrinite in weathered coal could be used to predict the normal coal composition.

Ferrari,[6] Huntjens[15] Goodarzi and Murchison,[16] Nandi et al.[17] and Gray and Krupinski[2] all reported the formation of bright rims in response to laboratory-induced oxidation. Low reflecting rims have been reported due to natural weathering by Nakayanagi,[18] Gray et al.,[1] Marchioni,[12] and Crelling et al.[3] Benedict and Berry[9] reported on low temperature, laboratory-induced oxidation of −60 mesh coal. They found initial decrease of reflectance followed by an increase with very irregular overall results. Kojima and Ogashi[19] and Alpern and Maume[20] found a trend similar to Benedict and Berry.[9] Goodarzi

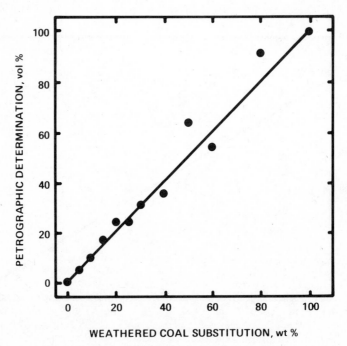

Figure 6.6 Petrographic determination of weathered coal in coal mixes (after Crelling et al.[3]

TABLE 6.1 Vitrinite Reflectance from Oxidized Centers and Oxidized Edges of Coal Particles

Vitrinite Reflectance, Ro% Mean Maximum	High-Volatile Coal			Low-Volatile Coal		
	Highwall	Intermediate	Outcrop	Highwall	Intermediate	Outcrop
All Vitrinites	0.91	0.93	0.91	1.48	1.50	1.46
Unoxidized Centers	0.96	0.98	0.92	1.46	1.53	1.48
Oxidized Rims	0.71	0.75	0.76	1.46	1.26	1.19

and Murchison[16] found vitrinite reflectivities increased after 64 days of oxidation at 150°C. They suggest that reflectivity in the blue region of the ultraviolet is best for distinguishing oxidized coal from fresh coal.

Gray's previously unreported result (Table 6.1) for samples of variously weathered high- and low-volatile coals show the vitrinite reflectance measured as mean maximum for randomly distributed vitrinite on unoxidized center of particles and on oxidized edges.

Figure 6.3 shows photomicrographs of the high-volatile coal and Figure 6.4 shows photomicrographs of the low-volatile coal samples. The detection of oxidized coal can be done microscopically. This technique can be used to determine the amount and the extent of oxidation. Measuring the reflectance of the oxidized portions of coal particles can be used to determine the quantitative differences between oxidized and unoxidized coal.

6.5 STAIN TEST FOR MICROSCOPIC DETECTION OF OXIDIZED COAL

It is very important to detect coal weathering during exploration and development for the following reasons:

1. To establish the depth of weathered coal,
2. To fix the mining line for coal recovery,
3. To predict fresh coal characteristics, and
4. To estimate the effects of weathered coal on recovery, beneficiation, handling, and use.

In cases where the properties of fresh coal are known, weathering can be detected and quantified by comparing the analytical data for the fresh and weathered samples. Rheological properties of coal such as Gieseler fluidity and free swelling index (FSI) deteriorate with coal oxidation and can be used for this comparison. There are very few tests that permit recognition of weathered coal directly from a sample. Microscopic examination of coal serves to disclose optical features that are characteristic of coal oxidation, both artificial and natural.

Gray et al.[1] discovered that bituminous coal can be rendered more amenable to quantitative microscopic analysis for oxidized coal by immersing a polished petrographic sample in potassium hydroxide solution (0.8 g NaOH per 25 ml water) for about 10 minutes. The etched sample is immersed in a saturated solution of saffranin O (a red stain) in alcohol. The technique has also been employed by Quick and Kneller.[20] The stained sample is washed in water or alcohol and blown dry with air. Unweathered coal of high-volatile A bituminous

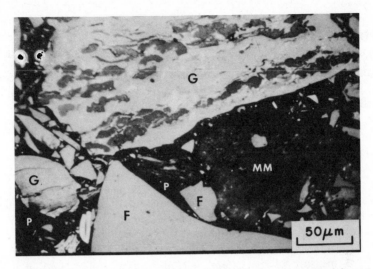

Figure 6.7 Photomicrographs showing coal that has been immersed in potassium hydroxide solution then washed and stained in a saturated solution of Saffranin 'O' (a red stain) in alcohol. Green coal "G" is oxidized, gray coal "F" is fresh, mineral matter "MM" is red, and plastic mounting media "P" is black. Reflected light in oil.

coal rank or higher is unaffected by the etch stain. However, oxidized areas are accentuated by preferential staining. The red stain turns the oxidized areas green as shown in Figure 6.7. Oxidation appears to be topochemical so that exposed surfaces appear to have a zoned area in which the intensity and depth (width of the zone) of staining increases as oxidation increases. Slightly oxidized, moderately oxidized, and highly oxidized coals become light green, yellow-green, and olive-green respectively. Lower rank coals stain even if they are fresh. However, oxidized zones stain more intensely. Fresh, high-rank coals (medium- and low-volatile) do not stain easily, but the oxidized areas do take on a stain. Lower rank coals that are oxidized dissolve in the hydroxide solution. Therefore, the potassium hydroxide treatment can be dispensed with in staining low-rank oxidized coals. The proportions of grains that have selectively taken up stain can be determined by a microscopic analysis and grain (point) counting of the coal samples.

Marchioni[12] in his study of detection of weathering on exploration samples of Canadian coals concluded that the staining with (saffranin-O) appeared to be the most sensitive petrographic parameter for detecting oxidized coal. He found it was sometimes more sensitive than rheological tests. In addition, Marchioni found that the staining test is suitable for detecting the weathered zone in a relatively unexplored area. It does not require a knowledge of the fresh coal's characteristics and is not influenced by variations in petrographic composition.

6.6 MINERAL MATTER AS AN INDICATOR OF COAL OXIDATION

A number of properties of oxidized coal may be used to indicate the degree of oxidation. Since the mineral matter that accompanies coal also undergoes oxidation, it has been investigated as an indicator of coal oxidation. Coals commonly contain clay minerals such as illite, kaolinite, and mixed silicates as well as lesser amounts of calcite, pyrite, and quartz.

Huggins et al.[22] used Mössbauer spectroscopy to investigate suites of oxidized coals from several strip mines in high-, medium-, and low-volatile bituminous coals. They found that the transformation of pyrite to FeOOH correlates with other parameters of oxidation such as oxygen from ultimate analysis, FSI, and petrographically determined coal oxidation.

Pearson and Kwong[23] found that the amount of $CaSO_4 \cdot 1/2 \ H_2O$ (bassanite) determined in x-ray diffraction patterns of low-temperature ash might be used as a measure of the extent of coal oxidation. This finding has been corroborated in Fourier Transform Infrared (FTIR) experiments done independently by Griffiths[24] and Painter.[25]

Very recent work by Lin et al.,[26] utilized diffuse reflectance infrared Fourier transform (DRIFT) spectroscopy, computer-controlled scanning electron microscopy (CCSEM),[27] Fe Mössbauer spectroscopy, Gieseler plastometer measurements, and other techniques to investigate laboratory-oxidized and naturally weathered coals. They found Gieseler plasticity to be highly sensitive to oxidation. The FTIR showed a reduction in intensity of aliphatic and aromatic bands for naturally oxidized coal. These changes were accompanied by an enhancement in intensity of certain carbonyl bands. The enhancement in carbonyl bands was not apparent in spectra of laboratory oxidized coals until oxidation was severe.

Lin et al.[26] found pyrite to be the most readily oxidized mineral in the coals investigated. In the naturally weathered samples, the principal pyrite oxidation product is iron oxyhydroxide, whereas the laboratory oxidized coals (constant humidity) initially formed iron sulfates, and then formed iron oxyhydroxide. The mechanism and rate of pyrite oxidation varied from coal to coal. No attempt was made to correlate the type of oxidation products or rate of oxidation with the pyrite form in coal, such as euhedral, framboidal, massive (secondary), and so on.

6.7 FREE SWELLING INDEX TEST FOR OXIDIZED COAL

Increasing demands for cleaner coal dictate more coal beneficiation to assure a uniform high-quality product. The properties of oxidized coals affect the washing characteristics, particularly the froth-flotation recovery. Oxidized coal recovered in mining is often lost to refuse in the froth-flotation step so that the cost of coal recovery and the additional cost of disposal of the coal to refuse adds to the coal cost. Thus, it is very important to detect oxidation in the coal recovery step so that the problems related to its beneficiation and use can be avoided.

The FSI of coal (ASTM Standard Method D720-67)[28] is often used as a screening test particularly in strip mine operations. This has the distinct advantage of being a rapid field test relatively sensitive to oxidized (weathered) coal as shown by Brisse and Richards.[29] This test is particularly useful if the FSI of the fresh coal is known. In the test a 1 g sample of −60 mesh coal (−250 μm) is rapidly heated to about 820°F in a closed crucible. The coherent button formed is graded by its comparison to standard shapes indexed between one-half and nine. The higher the number the greater the agglomerating and swelling characteristics of the coal. It is generally the bituminous coals that agglomerate. As agglomerating coals become oxidized, they lose their agglomerating characteristics.

In contour stripping operations, the FSI is sometimes used as a control variable. An auger sample is often used to recover samples of coal at about 10 ft

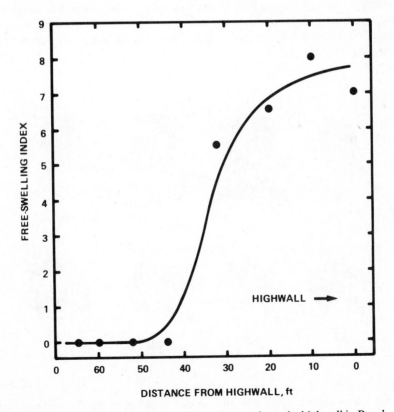

Figure 6.8 Relation of free-swelling index to distance from the highwall in Pocahontas No. 3 strip mine coal.

intervals between the coal outcrop and the highwall. The samples are then tested for FSI. If the fresh coal (highwall) will make an FSI button of about six to nine, then any coal less than about three will be considered oxidized. If the coal is intended for cokemaking, the coal that is considered oxidized should be rejected. Thus, the FSI is useful in establishing the mining line for coal recovery.

Coal samples from a low-volatile Pocahontas No. 3 from southeastern West Virginia were taken for FSI testing by Gray et al.[1] The graphic results are shown in Figures 6.8. Photographs of the FSI buttons for the low-volatile coal are shown in Figure 6.9. The FSI test data is very useful for establishing mining lines aimed at excluding oxidized coals. However, the test is not particularly sensitive and is best used as a comparative evaluation. Experience in sampling and in data interpretation of FSI results is required to assure maximum recovery of usable coal. An example of the changes in FSI in a vertical sequence and horizontal direction for high volatile A rank bituminous coal (Lower Cedar Grove) from southwestern West Virginia is shown in Figure 6.10. Oxidation or weathering takes place from the outcrop into the bed and from the surface down to the bed but is frequently tested only from the outcrop into the bed.

If any of the tests such as FSI, Gieseler or transmission density from the alkali extraction tests indicate oxidation, then other laboratory tests such as for ash, sulfur, and rheological properties may yield results which are atypical and misleading.

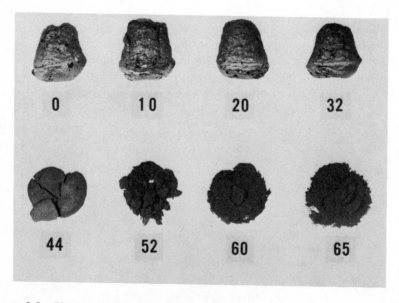

Figure 6.9 Photographs of FSI buttons of a Pocahontas No. 3 coal from various distances from the highwall.

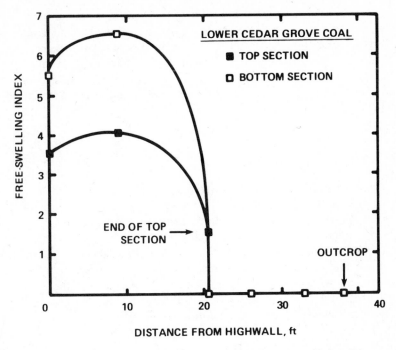

Figure 6.10   Free-swelling index as a function of distance from highwall lower cedar grove coal.

When the Gieseler and FSI data indicate lack of oxidation and the transmission tests indicates oxidation then the coal tested contains a small amount of highly oxidized coal. When all tests indicate oxidation then the entire coal is oxidized. When only the Gieseler indicates oxidation and the FSI and transmission test do not, then the coal is only slightly oxidized.

6.8   DETECTION OF OXIDIZED COAL: ALKALI-EXTRACTION TEST

The effects of weathering on metallurgical coal properties for cokemaking are obviously undesirable. Therefore, a rapid and reliable test is required to detect weathered coal to prevent its use in metallurgical coal blends. There are various tests that can be employed to detect weathered coal, as discussed previously. However, only one of these methods is reliable, rapid, and inexpensive to establish and operate. This is the alkali-extraction test for the detection of weathered (oxidized) metallurgical bituminous coal.

Over the years a certain amount of confusion has surrounded the oxidation test because its intent has been misinterpreted. The test is not a quantitative measurement of the change in oxygen content between fresh and oxidized coal but an empirical guide to alert users and sellers if a particular coal is suitable for metallurgical use. If the test is applied correctly, it is an inexpensive and efficient method of avoiding serious problems in coal preparation, cokemaking, and sales. With this in mind, a brief discussion of the role of alkali extraction in coal testing and the development of the alkali-extraction test is in order. The present alkali-extraction test and its applications to quality control will be addressed subsequently.

Schulze[5] discovered that spore coats in coals are resistant enough to allow chemical treatment to separate them from coal. He used nitric acid with potassium chlorate (Schulze's solution) to oxidize coal. After oxidation the acid is washed out and the vitrinized material is dispersed or rendered soluble by a base such as potassium hydroxide (KOH) or sodium hydroxide (NaOH). The alkali-soluble material is decanted, concentrating organic inerts, inorganic inerts, and spores. This maceration technique is widely used by palynologists. The spores, which concentrate in the residue, are studied to aid in dating, identifying, and correlating coal seams. When coals are oxidized or weathered, the treatment with Schulze's solution can be by-passed and the coal can be directly treated with alkali. Knowledge of the alkali solubility of oxidized coal has been the basis of many coal oxidation studies.[30]

Initial oxidation, according to Yohe[31] involves adsorption of oxygen without release of other compounds. However, severe oxidation involves chemical bonding of oxygen to form complex hydrocarbons commonly called humic acids and are characterized by their alkali solubility. Probably a number of different humic acids are produced and their relative proportions change with the degree of oxidation.

Atkinson and Hyslop[32] tested two methods of humic acid extraction of coal: (1) extraction with pyridine and (2) extraction with sodium hydroxide. With the pyridine extraction, the results were extremely variable but did indicate an initial *humic acid value* for each coal. This value was presumed to be related to the initial oxygen content of the coal and to some extent to the susceptibility of a coal to oxidation.

In the mid-1960s, considerable fluctuations in bulk density not explained by changes in moisture content of the coals, pulverization levels, or blend proportioning were observed at a U.S. Steel coke plant. Microscopic analyses of coal samples from periods of satisfactory and unsatisfactory bulk-density control indicated the presence of oxidized coal in the samples from the unsatisfactory period of bulk-density control. Additional studies established that weathered coal reduces oven charge bulk density and its immediate elimination from the coal blends became paramount. This meant detection and elimination at the source of oxidized coal traced to several stripping operations. The photomicrographs in Figure 6.11 show the changed appearances of coal from the highwall to the outcrop in a strip mine. However, there were no rapid analytical techniques available for detecting coal oxidation at the time. To solve this problem, R. J. Gray and others simply applied the knowledge that humic

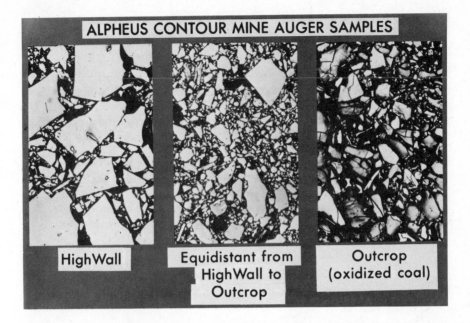

Figure 6.11   Photomicrographs showing the change in microscopy appearance of coal highwall to the outcrop in a coal strip mine.

acids will be extracted in the presence of a caustic and developed the alkali-extraction test, (under consideration as an ASTM standard).

The actual test procedure is straightforward and described along with the necessary equipment in Table 6.2. One gram of −60 mesh coal is immersed in 100 ml of a one normal NaOH solution and brought to a boil as depicted in Figure 6.12. If the coal has reached the oxidation stage where humic acids are present, these will be leached from the coal by the caustic solution and discolor the solution. This discoloration increases with increasing coal oxidation. It should be emphasized that some ranks of coal extract when not oxidized, but these coals are noncoking and generally are of little interest to coke producers. After the solution is filtered, the solution density is measured as percent light transmission at 520 nm relative to a standard of one normal NaOH representing 100 percent light transmission at 520 nm. The entire procedure takes no longer than half an hour and can be mastered rather quickly. Presently, a coal with a transmission value of less than 80 percent is rejected as too oxidized for metallurgical use in many by-product coking operations.

A study using bituminous coals of various ranks was performed by Lowen-haupt and Gray[11] that correlated results of the alkali-extraction test with petrographic, infrared, and oxygen analyses. The results are summarized in the following discussion.

TABLE 6.2   U.S. Steel Alkali-Extraction Test for Oxidized Coals

Scope

The test is a rapid, inexpensive method for determining if bituminous coal is oxidized and, if so, the relative extent of oxidation. Weathered (oxidized) bituminous coals produce a brown to black solution when treated with sodium hydroxide solution. The optical density of the solution is proportional to the degree of coal oxidation. Based on microscopically determined oxidation, this method detects oxidized coal present in amounts greater than 2 to 3 percent over the range of coking coals. Coke plants experience coal-handling problems and bulk-density control problems when the coals being processed have less than 80 percent transmittance. Bituminous coals, which have transmittance below this value, should not be considered to yield analytical data that is typical of fresh coal. This is particularly true for the rheological tests.

Reagents and Materials

1. In NaOH,
2. Anionic Tergitol (20 percent solution of tergitol TMN in ethanol),
3. 250 ml Erlenmeyer flasks or beakers,
4. 400 ml beakers,
5. Whatman No. 40 and No. 42 filter paper,
6. 60° glass funnels,
7. Spectrophotometer (Bausch and Lomb Spectronis 20 or equivalent),
8. Hot plate,
9. Glass stirring rods,
10. 17 mm ID (19 mm OD) round cuvette,
11. 100 ml graduated cylinders,
12. Thermometer, and
13. Boiling chips or glass beads.

Procedure

1. Weight out 1.0 g ($\pm$ 0.001 g) of -60 mesh coal.
2. Transfer the coal quantitatively to a 400 ml beaker containing 3 boiling chips or glass beads.
3. Add 100 ml in NaOH solution and one drop tergitol to the coal.
4. Bring the solution to a boil for 2 to 3 minutes, intermittently stirring with a glass rod, then remove the beaker and allow the contents to cool to room temperature.
5. Filter the slurry on No. 40–42 filter paper (No. 40 on top, No. 42 on bottom) into an Erlenmeyer flask. Transfer the filtrate into a graduated cylinder. There should be approximately 80 percent recovery. If not, the filtrate volume should be brought up to 80 ml using distilled water.
6. Allow spectrophotometer to warm up for about one-half hour to assure stability and set it to a wavelength of 520 nm.
7. Adjust the spectrophotometer to read zero percent transmittance, then fill a cuvette with a blank[a] solution, place it into the spectrophotometer and adjust the reading to 100 percent transmittance (making sure the meter has stabilized).

8. Using the same cuvette as used for the blank, test the transmittance of the coal alkali extract and report percent transmittance (%*T*).

Repeatability

Percent transmittance is repeatable within 2 percent *T* by the same operator in the same laboratory.

---

*a*Blank solution is prepared by adding one drop of tergitol to 100 ml NaOH solution, boiling for 2 to 3 minutes and filtering (i.e., the same procedure as outlined above is followed, except the beaker contains *no* coal).

Note: Periodically during the analysis, check the span (zero and 100 percent transmittance) and adjust for instrument *drift*.

All samples should be processed in duplicate.

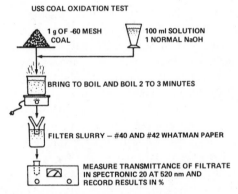

Figure 6.12 Schematic of the experimental procedure used in the alkali-extraction test.

6.8.1 Rank Study

To determine if coal rank affects the alkali-extraction test, samples representing different rank coals and levels of oxidation were collected. Several stripping operations were sampled from outcrop (most oxidized) to highwall (least oxidized to obtain an oxidation profile across the strip pit. The coals selected were high-, high- to medium-, medium-, and low-volatile in rank and represent coal seams from the Allegheny formation of Lower Pennsylvania series in Central Pennsylvania and from the Pottsville formation of the Lower Pennsylvanian series in the Cahaba coal basin in Alabama. The coals are listed in Table 6.3 according to their rank (expressed as dry, ash-free volatile matter) and location.

The amounts of oxidized coal in each sample were determined petrographically for correlation with the transmission values of each coal. Table 6.4 shows the percentage of slightly, moderately, and badly oxidized coal, and the

TABLE 6.3 Location and Rank of Coals Selected for Study

Coal	Location	Rank[a]
A	Central Pennsylvania	18.7
B	Central Pennsylvania	26.0
C	Central Pennsylvania	33.2
D	Central Alabama	36.2

[a]Expressed as dry, ash-free volatile matter.

corresponding transmission value for each sample. All three levels of oxidation (slight, moderate, and bad) show a linear trend with the transmission values for the individual coals. For simplification, only the total percentage of oxidized coal in each sample is correlated with the transmission value.

Figure 6.5 shows the relation of petrographically determined oxidized coal to the transmission value, which was previously discussed under microscopic detection of oxidized coal. Figure 6.5 is a graph of the data in Table 6.4. A regression analysis of the data produced a correlation coefficient of 0.964, which indicates a linear relationship. The graph shows that approximately 8 to 12 percent oxidized coal is present in these samples before the transmission values approach U.S. Steel's rejection level of 80 percent transmission. The graph also indicates that the alkali-extraction test has an initial detection level of 3 percent oxidized coal. When a coal is oxidized to 3 percent or less, the discoloration imparted to the extract will not significantly affect transmission values. This relationship between petrographically determined oxidized coal and transmission value indicates that the alkali-extraction test can be used successfully as a quality control parameter to avoid using oxidized coal in metallurgical coal blends.

Further evidence to support this statement is presented in Tables 6.5 and 6.6. Table 6.5 shows the transmission values and corresponding oxygen contents determined by neutron activation for the raw and washed strip samples from Pennsylvania. For each coal an increase in oxygen content is coupled with a decrease in transmission, with the outcrop sample (most oxidized) having the lowest transmission value and highest oxygen content for all three coals.

Table 6.6 lists the intensities of certain infrared spectral bands affected by coal oxidation, and the corresponding transmission values for the washed A, B, and C coals. Figures 6.13 and 6.14 are plots of these data and show that as oxidation increases (decreasing transmission value), the intensity of these spectral bands decreases. Thus, the changes in these infrared bands correlate with transmission values and degree of coal oxidation.

The rank study showed that the alkali-extraction test is in accord with petrographically determined amounts of oxidized coal for the coal ranks tested. The test results also correspond with oxygen content determined by neutron

TABLE 6.4 Percentages of Oxidized Coal and Corresponding Transmission Values for the Raw and Washed Coal Samples

Coal	Percent Oxidation								Percent Transmission	
	Raw				Washed					
	Slight	Moderate	Bad	Total	Slight	Moderate	Bad	Total	Raw	Washed
A										
Outcrop	19.9	16.3	32.8	69.0	21.0	8.8	30.8	60.6	0	1
Middle	20.5	3.0	2.0	25.5	27.0	2.4	0.4	29.8	53	55
Highwall	4.0	1.0	1.0	6.0	6.0	0.6	0.4	7.6	89	92
B										
Outcrop	36.0	10.0	6.0	52.0	5.6	11.8	25.6	43.0	24	29
Middle	21.0	8.0	2.0	31.0	21.0	5.2	4.8	31.0	41	40
Highwall	14.0	2.0	1.0	17.0	13.8	2.0	0.8	16.6	76	73
Unoxidized	2.4	—	—	2.4	2.0	0.2	—	2.2	96	98
C										
Outcrop	31.3	5.0	1.7	38.0	33.7	1.4	1.1	36.2	13	23
Middle	13.4	0.5	—	13.9	13.9	—	—	13.9	82	76
Highwall	5.9	—	—	5.9	7.6	—	—	7.6	97	94
D										
Acceptance	1.0	—	—	1.0	—	—	—	—	99	—
Unacceptable	11.0	—	—	11.0	—	—	—	—	76	—

^aDetermined petrographically.

TABLE 6.5 Oxygen Content and Corresponding Transmission Values for the Raw and Washed Coals

Coal	Percent Oxygen Content[a]		Percent Transmission	
	Raw	Washed	Raw	Washed
	A			
Outcrop	8.95	7.64	0	1
Middle	4.04	3.76	53	55
Highwall	2.64	2.37	85	92
	B			
Outcrop	4.90	3.90	24	29
Middle	3.80	3.37	41	40
Highwall	2.47	3.28	76	73
Unoxidized	0.85	2.24	96	98
	C			
Outcrop	6.24	7.11	13	23
Middle	5.24	5.45	76	76
Highwall	1.1	4.95	94	94

[a]Determined by neutron activation.

TABLE 6.6 Comparison of Peak Intensities and Transmission of A, B, and C Washed Coals

Coal	Peak Intensity (1/64 in.)		Percent Transmission
	2962 cm^{-1} (Aliphatic)	1450 cm^{-1} (CH$_3$)	
	A		
Highwall	37	75	92
Middle	25	49	55
Outcrop	11	28	1
	B		
Highwall	63	115	98
Middle	32	61	40
Outcrop	21	38	29
	C		
Highwall	79	138	94
Middle	62	115	76
Outcrop	42	82	23

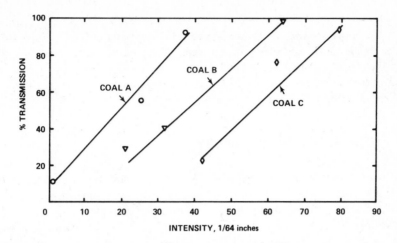

Figure 6.13 The decrease in intensity of spectral line 2962 cm^{-1} with increasing oxidation for indicated coals.

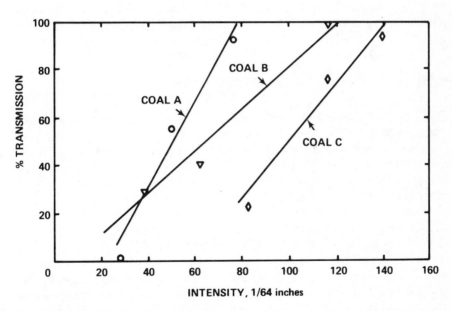

Figure 6.14 The decrease in intensity of spectral line 1450 cm^{-1} with increasing oxidation for indicated coals.

activation and certain spectral lines determined by FTIR. The detection limit of the test is approximately 3 percent petrographically determinable oxidized coal. Under this level, the discoloration imparted to the extract will not appreciably alter the transmission value from 100 percent. Nevertheless, the test is still superior to the FSI for detecting oxidized coal because it will detect amounts of oxidized coal for which the FSI test is insensitive.

6.9 PHASE INVERSION TEST

Coal weathering has an adverse effect on froth-flotation recovery of coal. The phase inversion test, described by Gray et al.,[1] is a simple method that can be used to make a rapid estimate of the floatability of coal prior to washing. This test utilizes 1 quart of water to which 2 tablespoons of −16 mesh coal and 5 drops of No. 2 fuel oil are added. The mixture is agitated for 2 minutes in an electric blender and then is poured into a graduated cylinder. If all the coal floats, the flotation will be good in beneficiation; if all the coal sinks, the flotation recovery is judged to be poor. The froth-flotation recovery of low-volatile coal for various degrees of oxidation is shown in Figure 6.14. Fresh coal is water repellent or hydrophobic, whereas weathered coal is hydroscopic. This hydroscopicity is attributed to the formation of acidic groups during oxidation.[33]

6.10 FLOTATION OF OXIDIZED COAL

Wen and Sun[34] studied the electrokinetic behavior of oxidized coals and their amine flotation relative to mineral matter. They found that increasing oxidation increased the negative value of the zeta-potential and decreased the isoelectric point. They also determined that flotation recovery increased with decreased pH values and that the optimum coal flotation in ferrous sulfate occurred at pH 6.5 with maximum pyrite depression.

Gayle et al.[35] found that oxidation reduced floatability in alcohol-type frothers. In the case of oil-type collectors, Sun[36] and Gayle et al.,[35] found that slight oxidation may increase coal's floatability. Cationic collectors may be more effective on oxidized coal flotation. The determination of electrokinetic potential is widely used to characterize surface characteristics of coal and minerals and their relation to adsorption to long-chain collectors. Campbell and Sun[37] found hydronium and hydroxyl ions were the determining ions for coal.

Wen and Sun[34] found the negative value of the zeta-potential of coals increased with decreasing coal rank. They found the isoelectric point (IEP) for anthracite, low-volatile bituminous coals, high-volatile coals, and lignites to be 5.0, 4.6, 4.5, and 2.3, respectively. They also found that increased oxidation increased the negative value of zeta-potential and that the zeta-potential variations due to oxidation were relatively small at pH's above 7 compared to pH's below 7. Wen and Sun[34] suggest that humic acids, the organic oxidation products on coal surfaces, are insoluble in acid but soluble in alkali and their solubility varies with molecular weight. Lower molecular weights dissolve at lower alkaline solution strengths.

Wen and Sun[34] studied the contact angle of oxidized coals and found that it decreased and the negative zeta-potential increased as oxidation increased. In general, as the contact angle of minerals increases, their floatability increases. They also found that most short-chain soluble organic acids from oxidized coal did not show much effect on the zeta-potential of unoxidized coal or the flotation of unoxidized coals with oily collectors. Brown[38] found that the floatability of coal is affected by the presence of soluble inorganic electrolytes. The major soluble electrolytes are Fe^{2+}, Ca^{2+}, Mg^{2+}, and Al^{3+}. Baker and Miller[39] found the hydrolyzed metal ions of Fe^{3+}, Al^{3+}, Ca^{3+}, and Cu^{2+} are all potential pyrite depressants. Wen and Sun[34] proposed that in the presence of electrolytes, the maximum hydrophobicity of coal and oxidized coal occurs near its IEP.

Wen and Sun[34] found that the adsorption of amine ions on oxidized coal is always greater than on quartz and pyrite. Amine ions do not adsorb on quartz between 6 and 8 pH at concentrations of 5×10^{-5} M to 10^{-4} M or less. Thus, selective flotation should be possible at this condition.

Gray et al.[1] tested the froth-flotation recovery of low-volatile coal samples that represented various degrees of oxidation (Figure 6.15). The tests were

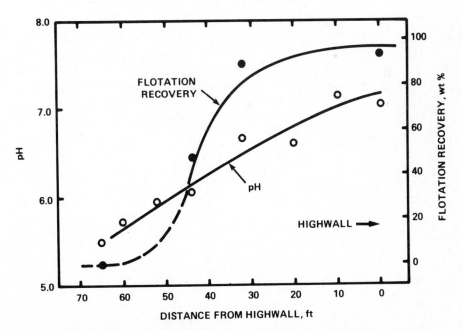

Figure 6.15 Relation of change in pH of coal slurry and froth-flotation yield to distance from the highwall in Pocahontas No. 3 strip mine coal.

conducted in a Wemco or Denver flotation cell with a coal–water slurry containing 5 percent solids and an equivalent of 0.28 to 0.30 lb of alcohol frother (No. 77) per ton of coal. Compressed air was used and the froth was skimmed, dried, weighed, and reported as recovered product. The flotation recovery decreased as the oxidation increased. However, examination of the various flotation products showed both oxidized and unoxidized coal in all samples tested. There is some indication that some coal particles have both oxidized and fresh surfaces and float in nonpolar and heteropolar reagents. Thus, flotation testing may not be very sensitive as an indicator of coal oxidation. However, Baranov et al.[40] have shown that flotation of 1.0 to 0.5 mm coal in a water slurry containing weakly oxidized coal with hexyl alcohol shows that floatability is impaired in even slightly oxidized coal, suggesting froth flotation is a very sensitive indicator of the degree of coal oxidation.

Gutierrez and Aplan[41] have correlated coal froth-flotation recovery with U.S. Steel's alkali-extraction test and found recovery increased as the coal rank increased and the degree of oxidation decreased. The transmittance value for the alkali-extraction tests for coals ranging in rank from anthracite to subbituminous C are shown in Table 6.7 together with transmittance data for two oxidized coals. The relation of transmittance to ultimate carbon (DAF) are shown in Figure 6.16 and the relation of transmittance to oxygen percent (DAF) are

TABLE 6.7   Results of the Alkali-Extraction Tests[a]

Sample No.	Rank	Transmittance Percent	
		Unoxidized Coal	Oxidized Coal
1	AN	98.0	—
2	LV	93.0	88.0
3	MV	94.0	—
5	HVA	92.0	—
6	HVA	88.0	—
7	HVA	85.0	—
8	HVB	71.0	48.0
12	HVC	77.0	—
13	Sub-A	18.0	—
14	Sub-B	9.0	—
15	Sub-B	1.0	—
17	Sub-C	0.0	—

Source: J. A. Gutierrez and F. F. Alpan, The effect of oxidation on the hydrophobicity and floatability of coal. *Colloids Surf.* (submitted for publication; the original work was Gutierrez's thesis at the Pennsylvania State University).

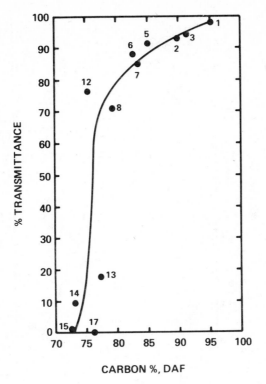

Figure 6.16 Relationship between percent transmittance and carbon content (DAF) of coal (after Gutierrez and Aplan[41]).

shown in Figure 6.17. The froth-flotation recovery for unoxidized and oxidized coals are shown in Table 6.8. Gutierrez and Aplan[41] also tested the effect of oxidation time at ambient temperature on the surface properties of coal as measured by the captive bubble technique (Figure 6.18).

Iskra and Laskowski[42] found that the pH of a coal–water slurry decreased and became more acidic as the coal became more oxidized. In their study, Gray et al.[1] added 1 g of −100 mesh coal to 125 ml of distilled water (pH 7) and measured the pH with a meter. A total of eight samples of low-volatile bituminous coal collected between the highwall and outcrop in a strip mine were used for these tests. The pH of the slurry for these samples decreased as the distance from highwall to outcrop increased as shown in Figure 6.15. Some later sampling at U.S. Steel showed that this relationship may not be valid for samples that are taken and experimentally washed before testing. This indicates that some of the acid surface coatings on oxidized coals are water soluble or soluble in the sink-float solutions used in coal experimental washing. In addition, some of the most oxidized coal may be mechanically removed in washing.

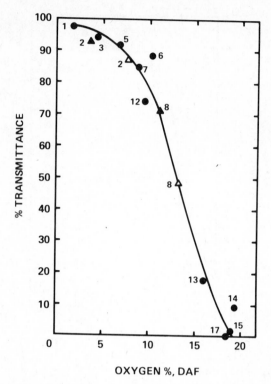

Figure 6.17 Relationship between percent transmittance and oxygen content (DAF) of coal. △ unoxidized value, ▲ oxidized value (after Gutierrez and Aplan[41]).

6.11 CHEMICAL CHARACTERISTICS OF OXIDIZED COAL

As would be expected, the oxygen content (by difference) from the ultimate analysis decreases from the outcrop (weathered coal) to the highwall (fresh coal) in a strip mine as shown for a low-volatile bituminous (Pocahontas #3 Seam) strip mine in southeastern West Virginia (Figure 6.19). There is also a change in DAF volatile matter with the degree of oxidation. The volatile-matter contents (DAF) for the Pocahontas incremental samples from a West Virginia strip mine are shown in Figure 6.20. The volatile matter increased as the coals sampled approached the outcrop. This indicates that the more oxidized coal also has a higher volatile-matter content. It is very likely that the composition of the volatiles from the fresh coal is different and contains more combustibles than the volatiles from the weathered coal.

In another group of samples from a strip mine in a high-volatile Alma Seam

TABLE 6.8   Froth-Flotation Recovery for Unoxidized and Oxidized Coals

Flotation Tests for Sample No. 5[a]

	Frother	lb/ton	Collector	lb/ton	Yield 1 min	Yield 2 min
			Nominal −28 mesh			
Unoxidized	MIBC	0.5	—	—	71.3	83.3
	MIBC	0.4	—	—	70.7	—
Oxidized	MIBC	0.5	—	—	67.9	80.2
	MIBC	0.2	Fuel oil	0.4	64.5	—

Flotation Tests for Sample No. 8[b]

	Frother	lb/ton	Collector	lb/ton	Yield 1 min	Yield 2 min
Unoxidized	MIBC	0.42	Fuel oil	2.0	55.1	64.9
	MIBC	0.42	Fuel oil	6.0	84.0	89.7
Oxidized	MIBC	0.42	Fuel oil	2.0	23.6	38.7

[a]Pittsburgh Seam Coal, HVA; 200 g Sample; pH 7.5; Fagergren Flotation Cell
[b]Illinois No. 6, St. Clair, HVB (Sample No. 8); Nominal −28 mesh; 200 g Sample; pH 7.5; Fagergren Flotation Cell.

coal and a low-volatile Pocahontas #4 coal, the following changes were noted in volatile matter relative to the position of the samples in the strip mine.

Location	Alma Seam Volatile Matter, (DAF)	Pocahontas Volatile Matter, (DAF)
Highwall	34.8	20.1
Intermediate	34.3	19.9
Outcrop	33.0	20.1

These data suggest a lack of trend between volatile matter (DAF) and degree of weathering. However, the individual samples were washed and each had a different yield; thus, a variable was introduced that might affect the results.

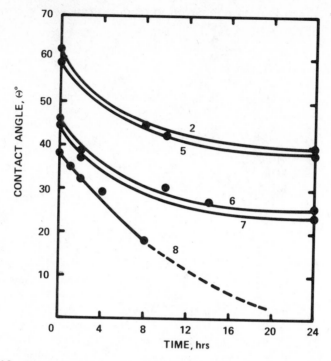

Figure 6.18 Effect of oxidation time at ambient temperature on the surface properties of coal as measured by the captive bubble technique (after Gutierrez and Aplan[41]).

6.12 INSTRUMENTAL DETECTION OF OXIDIZED COALS

Although the alkali extraction test is sufficient for identifying coals weathered to the humic acid stage, it fails to indicate earlier stages of coal weathering. This occurs because the alkali solution will not extract the functional oxygen groups produced in the beginning of coal weathering. These groups are primarily peroxides, hydroxyls, and carbonyls. It is extremely important to detect initial weathering because it can severely affect coal fluidity and thus the carbonization potential of coal. Further, the identification of these early stages will lead to a better understanding of the weathering mechanism. Recent advances in instrumentation applied to this problem have made significant progress in unraveling this weathering phenomenon.

Generally, the studies have approached the problem from opposite directions. One direction has concentrated on the changes in the concentration of the functional oxygen groups in the organic component of the coal, and these studies mainly employed infrared devices. The other direction concentrated on the chemical change in the mineral matter wrought by weathering, and these studies utilized Mössbauer equipment and scanning electron microscopes. Although the emphases are different, both approaches have produced significant results.

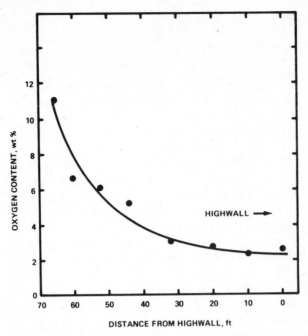

Figure 6.19 Relation of oxygen content to distance from the highwall in a Pocahontas No. 3 coal strip mine operation.

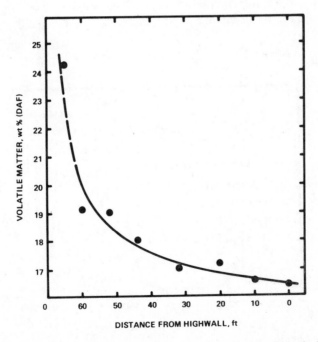

Figure 6.20 Relation of dry ash-free volatile matter to distance from the highwall in a Pocahontas No. 3 coal strip mine operation.

6.13 INFRARED STUDIES

Infrared techniques have been used to study coal for many years, and it was previously very difficult to interpret small changes in the coal spectra because of poor signal to noise ratios and the lack of suitable computing facilities. This has now been circumvented by the use of the rapid-scanning FTIR spectrometer that rapidly collects spectra and maintains high signal to noise ratios. This produces details not obtained by traditional infrared analyses. In addition, the coupling of the spectrometers with computers provides for the manipulation of spectra by highly sophisticated mathematical routines, allowing for subtraction of spectra with wave resolving to isolate small differences between spectra.

Two infrared studies on naturally weathered coal Griffiths et al.[24] Painter and Coleman[25] have been most enlightening and are discussed here. Before this discussion, however, a brief description of the basic infrared coal spectrum is appropriate.

Figure 6.21 is a typical infrared spectrum of a bituminous coal with the appropriate band assignments to the major peaks. There are three regions of major interest. The area between 2800 to 3200 wave numbers (cm^{-1}) is a rank indicator; that is the ratio of the aromatic C—H stretching mode to the aliphatic C—H stretching mode increases in the series of lignite to anthracite. The emphasis in this study was in the area between 1800 to $1200 \, cm^{-1}$ since it is most affected by weathering. Finally, the area between 1200 to $900 \, cm^{-1}$ indicates the types of mineral matter present. The intensities (concentrations) of the various species will change with coal rank, but the band assignments

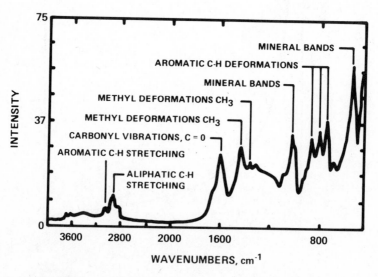

Figure 6.21 Band assignments for the unoxidized A coal.

generally remain the same. One can easily seen that the FTIR spectra of coals contain a considerable amount of important information.

Griffiths et al.[24] used a variation of FTIR called Diffuse Reflectance Infrared Fourier Transform Spectrometry (DRIFT). The method has all the advantages of FTIR spectrometry but avoids complicated sample preparation techniques by utilizing powdered samples. The impetus for this study arose from unusual experiences with the alkali-extraction test. Occasionally, the test indicates that a coal is unoxidized, but the coal behaves as if it were oxidizdd. To solve this problem, DRIFT experiments were addressed to the changes in the infrared spectra caused by weathering in strip pits from highwall to outcrop for three ranks of coal. These coals are listed in Table 6.9 with their corresponding rank and transmission values. All the coals showed the same basic peaks in their infrared spectra. The changes in the spectra of all three coals from highwall to outcrop are similar. Therefore, these changes will only be interpreted for the B coal (Figure 6.22). As oxidation increases, the out-of-line deformational variations of the aromatic C—H bands in the 900 to 700 cm^{-1} region decrease in intensity. This indicates that isolated hydrogen atoms bonded to aromatic nuclei are being oxidized. Also, the methylene peak at 1465 cm^{-1} and the methyl peak at 1450 cm^{-1} decrease in intensity with increasing oxidation.

The characteristic carbonyl (C=O) group absorptions in the 1850 to 1580 cm^{-1} region also showed some changes. A band at approximately 1600 cm^{-1} decreases with increasing oxidation. This band could be assigned to

TABLE 6.9   Rank and Transmission Values of the Coals Used in the DRIFT Study

Coal	Rank^a	Percent Transmission
	A	
Outcrop	18.7	1
Middle		52
Highwall		92
	B	
Outcrop	26.0	29
Middle		73
Highwall		98
	C	
Outcrop	33.2	23
Middle		76
Highwall		94

^{a}Dry, ash-free volatile matter.

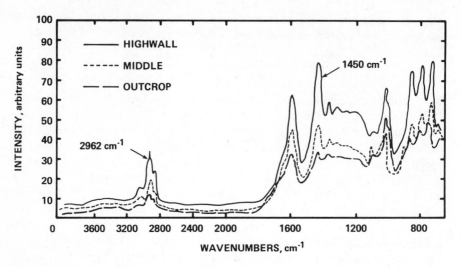

Figure 6.22 Change in infrared spectrum of B coal as oxidation increases, (highwall to outcrop).

aromatic ring vibrations that occur in this frequency region. A more plausible assignment is to carbonyl groups hydrogen-bonded to other functional groups. As a coal oxidizes, the bridging hydrogen is lost with the formation of additional unbonded groups. This explains the decrease in absorption observed at $1600\,\text{cm}^{-1}$ and the band broadening and formation of more pronounced shoulders on the high-frequency side of this band with increasing oxidation. This is evident in the oxidized spectra of all these coals. The intensities of the aromatic and aliphatic C—H stretching bands in the 2800 to $3100\,\text{cm}^{-1}$ region decrease with increasing oxygen. Thus, coal oxidation could involve the formation of carbonyl groups at the expense of both aliphatic and aromatic groups as well as bridging hydrogens in hydrogen-bonded functional groups.

The spectral changes of the B coals from highwall to outcrop are similar to those for the A and C coals. To confirm this, the intensities of spectral bands $1450\,\text{cm}^{-1}$ and $2962\,\text{cm}^{-1}$ were measured and plotted versus transmission values for all three coals (Table 6.6). Figures 6.13 and 6.14 indicate that as oxidation increases the intensity of these bands decreases. Thus, both spectral changes and transmission values correlate with the degree of coal oxidation.

This correlation indicated that the FTIR might have the sensitivity to detect oxidation in coals that passed the alkali-extraction test but failed bulk-density tests. To test this idea, DRIFT was used to analyze four coals with acceptable coal oxidation test values (>80 percent transmission) but with various bulk-density responses. The three coals with acceptable bulk-density responses had similar spectra, but the coal with the poor bulk-density response had a shoulder peak at $1735\,\text{cm}^{-1}$. This represents carbonyl groups that probably appear during the initial stages of coal oxidation. However, for coals in the initial stages of oxidation, the carbonyl groups are not bonded with groups acidic enough to be extracted by the alkali oxidation test. Therefore, the carbonyl would never be isolated as the cause of poor bulk-density response.

To determine the relationship between the carbonyl peak and poor bulk-density response, a series of unoxidized coals with varying bulk-density responses were analyzed by DRIFT for the presence of the carbonyl band at $1735\,cm^{-1}$. The spectra of these samples were converted to their second derivative spectra to resolve the carbonyl band at $1735\,cm^{-1}$ which appears as a shoulder in the primary spectra. To normalize the second derivative data, the ratio of the intensity of the 1735^{-1} band compared to that of the $1615\,cm^{-1}$ band, common to all coals was determined. This ratio was plotted against the change in bulk density for low-volatile bituminous coals and showed a reasonable correlation between the measured absorption ratio and the change in bulk density values (Figure 6.23). This result indicates that measurement of the second derivative of the DRIFT spectra of coals may be a more accurate method of monitoring the degree of weathering of mildly oxidized coals than the alkali-extraction test. A good correlation was not observed for high-volatile coals, however, indicating that a variety of factors contribute to the weathering and coking properties of bituminous coals. It is not entirely established that the peak at $1735\,cm^{-1}$ is a result of coal weathering or a coal rank or coal type phenomenon.

The work done by Painter and Coleman[25] is similar in nature but employed transmission FTIR rather than DRIFT. Painter and Coleman correctly believe that this technique is more quantitative than DRIFT, but it involves a laborious preparation of a potassium bromide pellet. Painter and Coleman's[24] results are nevertheless very similar to Griffeths et al.'s results.[24] Both found that the

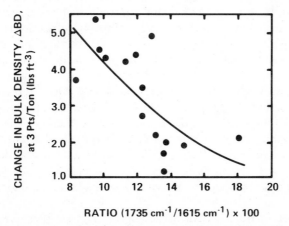

Figure 6.23 Plot of change in bulk density of fuel oil addition versus the intensity of the feature at $1735\,cm^{-1}$ in the second derivative spectra of several coals, normalized using the feature at $1615\,cm^{-1}$ for several low-volatile bituminous coals.

initial products of coal weathering are various carbonyl groups, followed by the formation of carboxyl groups. The FTIR has proven a most useful tool for substantiating supposed mechanisms of weathering that have escaped less sensitive equipment in the past. The use of FTIR equipment is restricted somewhat by cost and the specialized personnel required to interpret the results.

It should be mentioned that serious research is still being pursured in this area. Recently, Painter[43] has performed synthetic oxidation experiments and has shown, through FTIR studies, that the initial loss in coal fluidity is not necessarily a result of ether/ester oxygen crosslinks but a decrease in the aliphatic CH content during the initial stages of oxidation. Recent oxidation studies by Wu et al.[44] and Huggins et al.[45] show similar results.

6.14 MÖSSBAUER STUDIES

Studies of Huggins et al.[22] concentrated on the transformation of iron-bearing minerals in samples of weathered coals from strip mine operations. Mössbauer spectroscopy is one technique highly suited to detecting structural changes in iron-bearing minerals due to weathering (oxidation), and these transformations can be used to indicate the weathering stage of a coal. The same coals were used as in the Griffiths et al.[24] study. The ultimate analyses and data showing the degree of oxidation, as measured petrographically, and the related deterioration of coking properties are summarized in Table 6.10.

Typical Mössbauer spectra of a suite of oxidized coals are shown in Figure 6.24. The iron is generally distributed between two or more phases: pyrite (FeS_2), clay (illite), szmolnokite ($FeSO_4H_2O$), jarosite [$(Na,K)Fe_3(SO_4)_2(OH)_6$], goethite (α-FeOOH), and lepidocrocite (γ-FeOOH). The Mössbauer data on the distribution of iron among different minerals are summarized for all three coals in Table 6.11. It is restated that the most distinct trend with increasing weathering is the alteration of pyrite to iron oxyhydroxide as the samples proceed from highwall to outcrop for coals A and B. In the coal C suite, all the pyrite has been altered to iron oxyhydroxide either as the intermediate stage q-FeOOH or to the final stage α-FeOOH (goethite).

Because pyrite indicates unweathered coal and α-FeOOH is the final product of pyrite weathering, it is apparent that the Mössbauer ratio of α-FeOOH to pyrite could be used as an indicator of coal weathering. The trends with petrographic values of coal oxidation are substantially different for the three coals studied, however, (see Figure 6.25). Other factors such as coal rank and pyrite particle size would also have to be included to develop a widely acceptable quantitative measure of coal weathering based on the ratio of pyrite to FeOOH.

The presence of iron oxyhydroxide appears to be a sensitive indicator of weathering as it was detected in all the samples except the unoxidized sample of coal **B**. Therefore, it is possible to discern weathered or unweathered coal qualitatively based on the presence or absence of FeOOH. Because pyrite is

TABLE 6.10 Analytical Data for Oxidized Coal Samples

Sample	Ultimate Analysis (Weight Percent, Dry)						Carbon (DAF)	Petrographic Oxidation (%)	Swelling Index
	Carbon	Hydrogen	Nitrogen	Oxygen[a]	Sulfur	Ash			
Coal A									
Highwall	72.39	4.59	1.27	2.04	5.62	14.09	6	9.0	9.0
Middle	80.38	4.82	1.45	5.76	1.14	6.45	85.89	14	3.0
Outcrop	75.92	4.43	1.34	6.88	0.64	10.79	85.10	38	
Coal B									
Unoxidized	81.24	4.70	1.39	0.38	3.81	8.48	88.77	2	9.0
Highwall	80.59	4.51	1.34	2.30	2.00	9.26	88.81	17	9.0
Middle	81.73	4.51	1.40	3.98	1.28	7.10	87.98	31	6.5
Outcrop	81.85	4.35	1.41	4.58	0.76	7.05	88.06	52	2.0
Coal C									
Highwall	85.51	4.49	1.45	2.06	0.65	5.84	90.81	6	8.5
Middle	82.89	4.02	1.48	4.00	0.64	6.88	89.11	26	2.0
Outcrop	72.36	3.12	1.34	8.39	0.57	14.22	84.36	69	0.0

[a]By difference.

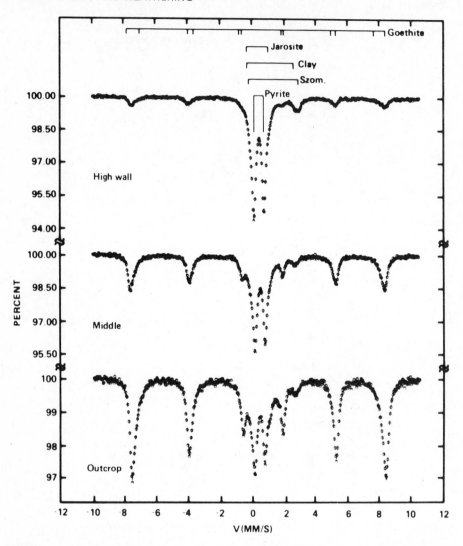

Figure 6.24   Typical Mossbauer spectra of a suite of oxidized coal samples from a strip mine operation.

present in almost all coals, the formation of FeOOH during coal oxidation can be expected to be widespread and any method based on FeOOH could be widely applied.

The results of this study are very informative and important. The various Fe-bearing minerals formed during weathering not only indicate the stage of weathering but may also affect other coal characteristics such as ash fusion temperatures, and possibly coal conversion.

TABLE 6.11 Mössbauer Data on Iron Distribution for Oxidized Coals

Sample	Iron Distribution as Percentage						Wt. Ratio[a]
	Pyrite	Clay	Szomolnokite	Jarosite	α-FeOOH	γ-FeOOH	FeOOH/(FeOOH + FeS$_2$)
Coal A							
Highwall	79	1	5	5	10	—	0.09
Middle	36	2	—	4	58	—	0.54
Outcrop	3[b]	2	—	—	95	b	>0.95
Coal B							
Unoxidized	82	5	4	10	—	—	0.00
Highwall	64	4	6	9	17	0	0.16
Middle	33[b]	5	—	11	51	b	>0.53
Outcrop	15	4	—	9	71	—	0.78
Coal C							
Highwall	—	8	—	—	48	44	~1.00
Middle	—	25	—	—	52	23	~1.00
Outcrop	—	23	—	—	66	12	~1.00

[a]Ration calculated as $\dfrac{(\%\text{Fe in FeOOH} \times \text{mol. wt. FeOOH})}{(\%\text{Fe in FeOOH} \times \text{mol. wt. FeOOH} + \%\text{Fe in FeS}_2 \times \text{mol. wt. FeS}_2)}$

[b]Pyrite may include minor γ-FeOOH contribution.

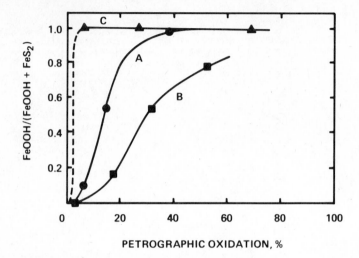

Figure 6.25 The relationship of the $FeOOH/(FeOOH + FeS_2)$ to petrographically determined oxidation for coal samples from three strip mines.

6.15 ELECTROSTATIC PROPERTIES OF OXIDIZED COAL

6.15.1 Static Charge

When oxidized coal impinges on a metal plate, it acquires an electrostatic charge.[46] The change in charge from positive to negative increases as oxidation increases. The British National Coal Board has designed and patented a convenient instrument for determining the degree of oxidation using this principle. Walton[47] describes the device in British Patent 733,00, July 6, 1955.

Gray et al.[1] developed a test for measuring the electrostatic properties of coal, shown graphically in Figure 6.26. Fifteen gram samples of -100 mesh dry coal are dropped from a hopper onto a grounded metal plate (stainless steel) with side guards inclined at about $45°$. The coal slides down the chute (as a thin layer) and is collected at the bottom in an insulated metal catchpan. A sensitive voltmeter equipped with a test probe measures the charge acquired. A series of eight samples of coal taken at 10 ft intervals across a strip pit of low-volatile Pocohontas No. 3 bituminous were tested in this device. The results are shown in Figure 6.27. The unoxidized coals are positively charged, whereas the oxidized coals acquired negative charges. There is a sharp break between the unoxidized (normal) and the oxidized (weathered) coals.

6.15.2 Resistivity

A test was conducted at U.S. Steel to determine if an electrical field separator could be used to separate oxidized from fresh coal, using Pocohontas No. 3 coal. A bench-scale nonuniform electrical field separator, consisting of two nonparal-

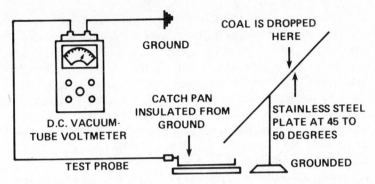

Figure 6.26 Generalized schematic drawing showing equipment used in determining electrostatic properties of coal.

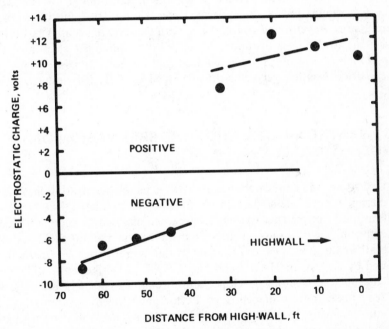

Figure 6.27 Relation of electrostatic charge developed in coal to distance from the highwall in a Pocahontas No. 3 coal strip mine operation.

lel metal plates held in plastic (lucite) frames and insulated from ground was the test unit. A positive potential of 25 kilovolts was applied to the top plate. The electrode spacing at the high-field side was 1/2 in. and at the low-field side was 2 in. Approximately 5 to 8 g of coal were placed in a narrow band about 1/4 in. from and extending to 20 mm beyond the wall at the high-intensity field side of the separator. The electrical field was applied for a period of 3 minutes. The coal that had moved from the initial band (20 mm from the wall) to the opposite wall or low-field intensity area was collected, weighed, and recorded as the amount of coal separated. The test results for three runs per sample were averaged. The separation of coal in a nonuniform electrostatic field separator relies on the inductive charging of some of the particles. Inductive charging is controlled by the particle resistivity, contact area with the electrode, particle size, and the electrical field strength. Charged particles are attracted to one electrode by electrical forces and travel to the other electrode when the process is repeated. The particles bounce back and forth gaining momentum as they move toward the area of lower field strength where they can be separated. Oxidized coal showed the most movement.

Coal samples consisting of 50 percent fresh and 50 percent oxidized coal were ground to different top sizes (8, 20, 40, and 60 mesh) and were prepared at different moistures (2, 4, 6, and 8 percent) for testing. The best separation of oxidized coal occurred at -40 mesh with 5 percent moisture. About 60 percent of the oxidized coal portion could be separated, indicating a low separation efficiency. Examination of the separated coal showed that fusain and some mineral matter traveled and concentrated with the oxidized coal. A test using a Carpco Electrodynamic Separator indicated that modified electrodes were needed to increase *pinning* force if oxidized coal was to be separated. However, the separation was not very efficient as judged by FSI. This test is not very sensitive.

6.16 NEUTRON-ACTIVATION DETERMINATION OF OXYGEN IN COAL

A fast, accurate, and completely nondestructive neutron-activation method for the determination of elemental oxygen in coal was evaluated. The method is based on the principle that oxygen atoms when exposed to neutrons become radioactive nitrogen atoms. A measurement of the radiation emitted is a quantitative indicator of the oxygen concentration in the coal. The method was first used by Hevesy and Levi[48] and had limited use until after World War II when the nuclear reactors developed during and immediately after the war provided ample sources of neutrons. Taylor and Havens[49] summarized the applications of neutron-activation analysis for this period. A neutron-activation method for the determination of oxygen content in coal was described by Veal and Cook[50] and also by Steele and Meinke[51] with many papers on the subject issued since that time.

The determination of oxygen depends on the nuclear reaction $O^{16}(n, p)N^{16}$. The nitrogen (N^{16}) formed is radioactive and decays to oxygen 16 by emitting a beta (β') particle. The half-life of the decay is 7.14 seconds. The emission of the β' is followed by emission of very high energy gamma rays in about 74 percent of the cases. The gamma rays with energies of 4.5 to 7.5 Mev are used for the quantitative measurements. A mylar sheet that contains 33.3 percent oxygen by weight is the primary standard.

An important aspect of analysis for oxygen by neutron activation is that the oxygen analysis is independent of the chemical state of the oxygen. In the case of coal, this is a disadvantage since the total oxygen includes the organically combined oxygen, the oxygen in the ash-forming minerals, and the oxygen bound as moisture. It is the organically combined oxygen that determines many of the coal's properties such as coking characteristics and heating value. The coals can be dried to eliminate moisture but the oxygen in the mineral matter still remains. Gray et al.[1] showed that a factor can be used to correct for the oxygen in ash.

The common method for oxygen determination consists of chemically analyzing for carbon, hydrogen, nitrogen, sulfur, and ash and subtracting the sum of these from 100 percent to get oxygen by difference. The relation of oxygen by difference to oxygen by neutron-activation is shown in Figure 6.28. all the data being for weathered coal. In a series of coals that were intentionally and rapidly oxidized in air, the neutron-activation method indicated a regular increase in oxygen with time of oxidation. The oxygen determined by difference also showed changes. The results are as follows:

Oxidizing Time, (hours)	Neutron Activation Oxygen (%)	Chemical (by Difference) Oxygen (%)
0	8.1	8.4
24	8.5	9.7
48	9.2	9.4
72	9.9	9.6

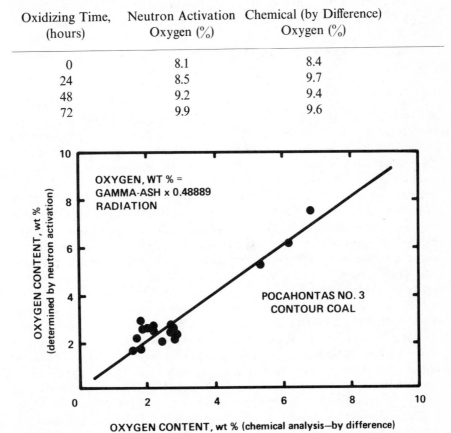

Figure 6.28 Relation of oxygen content from neutron activation studies and oxygen content from chemical analyses (by difference) in Pocahontas No. 3 contour mine coal.

Neutron-activation oxygen offers a rapid method of oxygen determination. This should have use in coal characterization studies since oxygen significantly affects the technological properties and should be determined preferably by a nondestructive, rapid and accurate method.

6.17 EFFECT OF COAL OXIDATION ON COKE CARBON FORMS AND COKE REACTIVITY

The study of the amounts and kinds of carbon forms in coke can be used as an analytical tool in assessing the following:

1. Blend proportions of various rank coals used to make the coke,
2. Relating coal maceral composite, rank, and grade parameters to coke microstructure,
3. Studying the response of various carbon forms to reactivity in CO_2 and with alkalies as well as reactions in the blast furnace or simulated blast furnace,
4. Relating coke porosity to coal characteristics and carbonization variables such as heating rate, coal size, and charge bulk density.

Gray and Schapiro[52] explained the thermal behavior of coal entities (macerals). They emphasized that coking coals consist of thermally inert organic and inorganic materials that act as filler in the coking process. They also showed that reactive macerals, such as vitrinoids, formed bond material during coking. The optical anisotropism of this bond carbon increased as the stage of coalification of the coals used to make the coke increased through high-, medium-, and low-volatile bituminous coals. They paved the way for relating coal petrography to coke petrography. Peters, Schapiro, and Gray[53] expanded on some of these concepts. In 1963, Schapiro and Gray[54] related coke structure and carbon forms to coke reactivity in CO_2 at about 1000°C. Schapiro, Gray, and Eusner[55] developed a system to predict coke strengths from petrographic data. Gray[56] consolidated these efforts with the publication of a complete system of coke petrography.

Different ranks of coal produce coke carbon forms that are optically distinguishable from one another even in coke produced from a blend of coals, provided the coals are of different rank. The photomicrographs in Figure 6.29 show coke carbon from marginal-coking high-volatile B rank bituminous coal, high-volatile A, medium-volatile, and low-volatile coal. The marginal coking high-volatile B coal produces optically isotropic carbon (noncrystalline or poorly ordered), whereas the high-volatile A coal displays incipient or a fine mosaic (pinpoint) anisotropism (slightly ordered). Coke from the medium-volatile coal displays lenticular or fibrous anisotropism (ordered), and coke from low-volatile coal displays large ribbon or leaflet anisotropism (highly ordered)

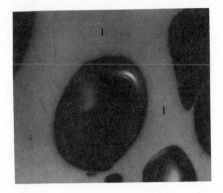

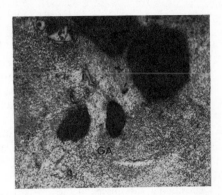

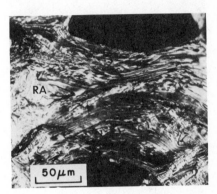

Figure 6.29 Photomicrographs showing carbon forms in coke produced from different ranks of bituminous coals. Marginal coking high-volatile B bituminous coals produce isotropic **I** carbon: high-volatile A coals produce granular **GA** anisotropic carbons; medium volatile coals produce lenticular **LA** anisotropic carbons; and low-volatile coals produce ribbon **RA** anisotropic carbons. Reflected light in oil.

domain. The anisotropy of the coke carbon increases as the rank and reflectance of the coal used to make the coke increases. The coke carbon forms and their relative reactivities are shown in Table 6.12. Generally, isotropic carbons are characteristic of physically weaker cokes that are highly reactive to CO_2 at elevated temperatures, whereas the more anisotropic carbons are characteristic of strong coke with lower reactivity. The ribbonlike anisotropic carbons are somewhat weaker and more reactive than the lenticular forms. The coarse inert macerals in all coking coals are more reactive than the anisotropic carbon forms.

When bituminous coals are oxidized thermally or naturally, the oxidized portions produce isotropic carbons or carbons with lower anisotropism than the carbon from the fresh coals as shown by the photomicrographs in Figure 6.30

TABLE 6.12   Coke Carbon Forms

U.S. Steel	Japanese	Reactivity
Isotropic	Isotropic	High
Incipient	Fine mosaic	Intermediate
Circular	Coarse mosaic	Intermediate to low
Lenticular	Fibrous	Low
Ribbon	Leaflet	Low to intermediate
Organic inerts	Inerts	High
Coarse	Fusite	
Fine	Fragment	
Depositional and miscellaneous	Miscellaneous	Variable

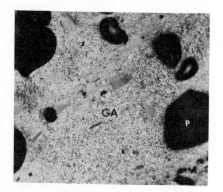
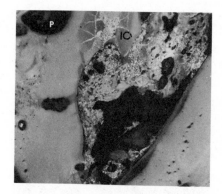
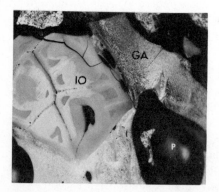

Figure 6.30 Photomicrographs showing granular isotropic **GA** carbon from high-volatile A bituminous coal and the isotropic **IO** carbons produced from oxidized (weathered) coal. Pores, **P**, are black. Reflected light in oil.

for the coke from high-volatile coal and in Figure 6.31 for the coke from low-volatile coal. Cokes from unoxidized and weathered high- and low-volatile coals are compared to show how the carbons of the cokes from oxidized and unoxidized areas of coal are distinguishable by their microstructures and degree of anisotropism.

Oxidation of coking coals renders them poor coking. The cokes are weaker and more reactive to CO_2. The reactivity of strip pit samples of high- and low-volatile coals of different degrees of oxidation have been coked and tested for reactivity. The results are given in Table 6.13. Weathered coals lose their ability to display an exothermic reaction during pyrolysis.

Oxidized coals produce less tar than fresh coals. A 1 percent addition of oxygen results in about 10 percent loss in tar yield and requires a greater heat of

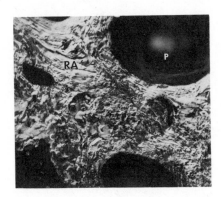
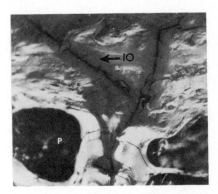

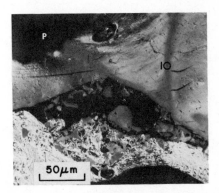

Figure 6.31   Photomicrographs showing ribbon anisotropic **RA** carbon from low-volatile bituminous coal and the isotropic **IO** carbons produced from oxidized (weathered) coal. Pores, **P** are black. Reflected light in oil.

TABLE 6.13 Coke Reactivity in CO_2 [a]

	High-Volatile Alma Seam, WVA			Low-Volatile Pocahontas No.4, WVA		
	Highwall	Intermediate	Outcrop	Highwall	Intermediate	Outcrop
Position of sample in strip mine Percent reacted	16.8	25.5	29.5	18.3	23.9	25.8

[a]Bethlem Reactivity Test.

carbonization. Including oxidized coals in blends results in green (smokey) pushes, which are unacceptable from an environmental standpoint. Oxidized coals also have different angles of repose and handling characteristics than the unoxidized. Most of the physical characteristics of oxidized coals are less desirable than their unweathered counterparts. Of course, oxidized coals have lower calorific value than fresh coals and are generally less desirable for combustion purposes. Some bituminous coals when oxidized show less tendency to produce smoke, probably because oxidized coals produce less tarry substances when heated. Oxidized coals generally have an adverse affect on coke making as well as combustion.

6.18 SPONTANEOUS COMBUSTION

The spontaneous combustion of coals has long fascinated coal scientists and mining engineers. Recent problems with "hot coal" at export terminals and subsequent ship fires have provoked further interest, Berry and Gosinski[57] and Mason.[58] Generally, the tendency towards spontaneous combustion is controlled by rank. The high-rank coals oxidize more slowly than low-rank coals. This is because high-rank coals have less active carbon sites and less internal surface area which controls the moisture holding capacity of the coal. Alternate wetting and drying is the single greatest cause of spontaneous combustion. A straight rank classification is not entirely sufficient to predict the tendency of coal to spontaneous combustion because the composition also has to be considered. Work by Chamberlain et al.[59] has shown that exinite has a much greater oxidation rate than vitrinite or inertinite from the same coal. Thus, coals of similar rank will have different tendencies for spontaneous combustion as their petrographic compositions vary.

Superimposed on these factors are the following coal quality factors that affect a coal's tendency to spontaneous combustion.

1. The amount and form of pyritic sulfur
2. Size
3. Moisture—equilibrium and surface
4. Grindability
5. Bulk density in storage and/or transport
6. Initial degree of weathering

Factors not related to coal quality also affect the spontaneous combustion tendency, viz:

1. Mining technique used to recover coal,
2. Type of product—raw or washed,

3. Weather conditions (rain, temperature, humidity) during storage or transport,

4. Length of time since coal was mined or in storage,

5. Temperature of coal in storage or transport,

6. Whether the product has been blended from several coals of different quality,

7. Storage pile configuration, drainage, and degree of compaction, and

8. Interfacing of coarse and fine coal to produce a chimney effect.

It is obvious that the *hot coal* problem is extremely complex.

There is no standard test available by which coal can be consistently categorized as to its susceptibility to spontaneous combustion. A number of tests do exist that can achieve this distinction with some success. They are:

1. Static isothermal method,

2. Chemical methods,

3. Determination of ignition temperatures,

4. Adiabatic method,

5. Paced adiabatic nonisothermal method, and

6. Product gas analysis during spontaneous heating.

Monitoring the gaseous products of spontaneous heating is the most successful and practical test for incipient combustion because carbon monoxide is always the first product of this process. Presently, monitoring for CO is used in ships and in underground mines to detect the early stages of the spontaneous combustion process.

Spontaneous combustion, an oxidation reaction, is rate controlled by temperature and oxygen concentration. Temperature has an exponential effect on the rate of this reaction and the reaction cannot proceed without oxygen. Therefore, a significant number of ocean going vessels are equipped with inert cargo holds, temperature sensors, and CO monitors. Many storage areas are now monitoring stockpile temperatures. Generally, temperatures over or near 120°F are generally considered dangerous and remedial action is necessary. The problem of "hot coal" is extremely complex with no simple solution. Therefore each coal must be evaluated separately to determine its potential for spontaneous combustion.

6.19 DESTRUCTION OF AGGLOMERATING PROPERTIES OF COAL BY THERMAL OXIDATION

In addition to natural oxidation or weathering, it is important to discuss induced oxidation and compare the microscopic and rheological properties that accompany each of these oxidation conditions to illustrate how they differ.

TABLE 6.14 Proximate and Sulfur Analysis, Weight Percent of Indicated Coal Seams

Coal Seams and Location	Rank Designation	Total Reactives Volume (%)	Total Inerts	Reflectance Mean Max, % in Oil	Volatile Matter (DAF)[a]	Volatile Matter	Fixed Carbon	Ash	Sulfur Content, Weight Percent (dry)
Adaville, Lincoln County, Wyoming	Subbituminous	90.2	9.8	0.42	47.7	45.6	49.9	4.5	0.60
Illinois No. 6, Franklin County, Illinois	Bituminous high volatile B	85.2	14.8	0.65	39.8	36.6	55.3	8.1	1.93
Pittsburgh, Greene County Pennsylvania	Bituminous high volatile A	78.6	21.4	0.91	37.9	34.9	57.2	7.9	1.52
Sewell, Nicholar County, West Virginia	Bituminous medium volatile	78.0	22.0	1.07	30.4	29.0	66.5	4.5	0.75
Pocahontas No. 3, McDowell County, West Virginia	Bituminous low volatile	61.9	38.1	1.61	18.6	17.3	75.8	6.9	0.71
Lower Alfred, South Africa	Semianthracite	52.6	47.4	2.31	11.6	10.8	82.3	6.9	0.72

[a]Dry, ash-free volatile matter.

TABLE 6.15   Gieseler Plasticity and Free Swelling Index of Indicated Coal Seams

	Maximum Fluidity		Plastic Temp. Range, °C		
	DDPM[a]	°C	Softening	Solidification	FSI
Adaville	—	—	—	—	0
Illinois No. 6	115	404	350	443	3
Pittsburgh	29,100	429	336	474	8.5
Sewell	7,200	433	373	478	+8
Pocahontas No. 3	70	465	408	498	8
Lower Alfred	—	—	—	—	—

[a]Dial divisions per minute.

In general, changes in the chemical and physical properties of coal due to oxidation cause problems in the characterization, handling, and use. Generally the problem of natural oxidation (weathering) is most frequently encountered. The major concerns are with the effect of oxidation on coke strength, combustion, behavior, and so on. In most cases, the oxidation of coal adversely affects coal properties. However, when bituminous coking coals are processed in gasifiers, the agglomeration of coal forms a mass that reduces gas flow and causes blockage in the reactors and feed lines. Agglomerating coals can also cause defluidization and blockage in fluid beds. In such instances, pretreatment of coals to oxidize them and decrease their agglomerating properties is beneficial.

Gray and Krupinski[2] thermally treated coals to lessen their agglomerating tendencies. They treated six coals ranging in rank from subbituminous through semianthracite in a 1 in. ID miniature fluidized bed in air for 10 minutes at 177°C and in air for 30 minutes at 400°C. Before they were oxidized, the samples were analyzed for petrography, proximate and sulfur (Table 6.14) Gieseler plasticity and FSI (Table 6.15). Photomicrographs showing the appearance of untreated coal are shown in Figure 6.32. After induced oxidation, each coal sample was analyzed for FSI, Table 6.16 examined microscopically and photographed for microstructural changes (Figures 6.33 and 6.34). The reflectance and thickness of the oxidized zone of the particles were also measured (Table 6.17). In addition, a 0.2 g sample of 35 by 60 mesh coal was tested with the same size char in the fluid bed at 750°C in nitrogen with a gas velocity of 0.3 ft/sec. This was done to determine the amount of the added char of the same size that was agglomerated by the coal and to develop an agglomerating index for coal in fluid beds. Particles larger than 20 mesh were considered agglomerates, and the agglomeration index for this test is the number of grams of agglomerates formed per gram of coal added. The analytical data for the thermally treated coals are shown in Table 6.18.

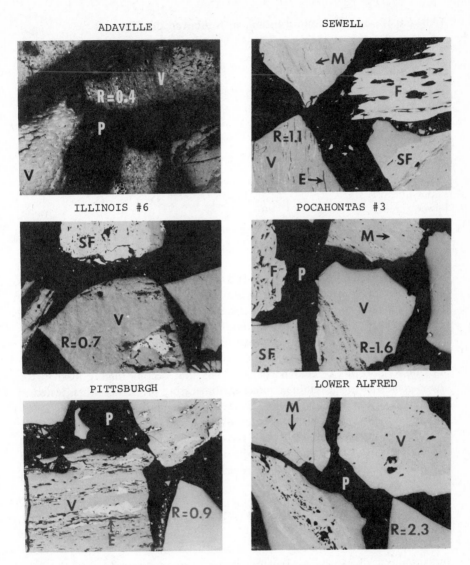

Figure 6.32 Photomicrographs of entities in indicated coals. Reflectance **R** increases with increase in rank. Key: vitrinoid **V**, exinoids **E**, micrinoids **M**, semifusinoids **SF**, fusinoids **F**, plastic **P**. Reflected light.

Gray and Krupinski's work[2] showed that coal rank, as measured by vitrinite reflectance, correlated with the plastic properties of the coals. The vitrinite was the main coal maceral affected by thermal oxidation, whereas fusinite, semi-fusinite, and micrinite were virtually unaffected at the treatment temperatures tested. Coals with lower-rank vitrinite (<0.6 percent R_o) and those with higher-rank vitrinite (>1.9 percent R_o) than bituminous coals decompose during or

TABLE 6.16　Free Swelling Index for Indicated Coal Seams

	Untreated		Treated at 117°C		Treated at 400°C	
	35 × 60 Mesh	−60 Mesh	35 × 60 Mesh	−60 Mesh	35 × 60 Mesh	−60 Mesh
Adaville	0	0	0	0	0	0
Illinois No. 6	2	2	1	2	0	0
Pittsburgh	8	8.5	8.5	9	2.5	4
Sewell	7.5	8	8	8	2	2
Pocahontas No. 3	7	7.5	7	7	1	1
Lower Alfred	0	0	0	0	0	0

*a*Standard sample for FSI determination is ground to pass a No. 60 sieve.

before softening and are nonagglomerating. The vitrinite of bituminous coals passes through a plastic stage during heating (Figure 6.34). Oxidation decreases the plasticity and agglomerating tendencies of bituminous coals (Tables 6.16 and 6.17). However, the thermal oxidation of all coals examined proceeded topochemically (Figure 6.34). This has also been shown in a calorimetric study by Kaji et al.[60] Discernible rings of higher reflecting material developed around the unoxidized portions of coal particles. Coals that become highly fluid in the Gieseler plasticity test retained some agglomerating properties when tested in the miniature fluid bed and in the FSI test even after oxidation at 400°C for 30 minutes. When these coals are heated above the temperatures at which they were oxidized, they swell and break the oxidation shell. The gas generated in heating extrudes the plastic mass from inside the particles and the extruded material causes agglomerating (Figure 6.35).

6.20　RELATION OF COAL MICROHARDNESS TO OXIDATION

Indention microhardness is a sensitive indicator of oxidation. Most of the work on this subject, done by Schuyer et al., Alpern,[61] Handa and Sanada,[62] and Schapiro and Gray,[60] was aimed at relating either the Knoop or Vickers microhardness index to the rank of coal (carbon content, volatile matter, or reflectance). The relation of indention hardness to rank is shown in Figure 6.36.[63] Indention hardness is measured by pressing an indenter made of some hard substance such as diamond into the material to be tested. The force applied is specified and the time of indentation may also be specified. The Vickers indenter is shaped like a pyramid and the Knoop is shaped like a pyramid elongated in one direction. Molt[27] published a very good text on *Micro-Indention Hardness Testing*.

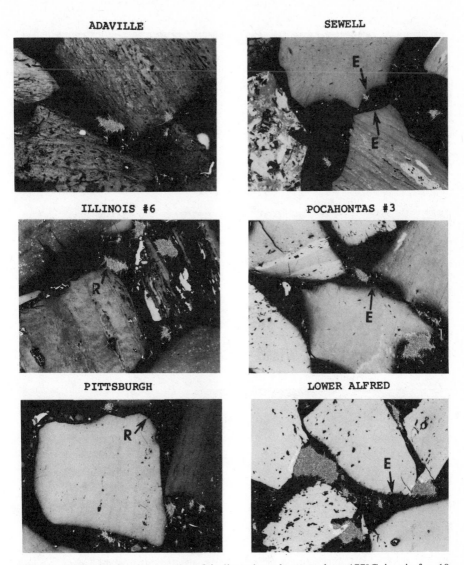

Figure 6.33 Photomicrographs of indicated coals treated at 177°C in air for 10 minutes. Treatment results in increased vitrinoid reflectance and some particle rounding **R** and edge relief **E**. Reflected light.

The size of the indention for a given load (pressure) is a measure of hardness. Das[64] and Nandi et al.[17] found different types of impressions. The impressions formed by the indentor may have smooth edges indicating the material is plastic. Some materials recover after the indenter is removed leaving an outline that is shallow and indistinct, indicating the material is elastic. Figure 6.37 shows the outline of indentions in unoxidized and oxidized (weathered) coal.

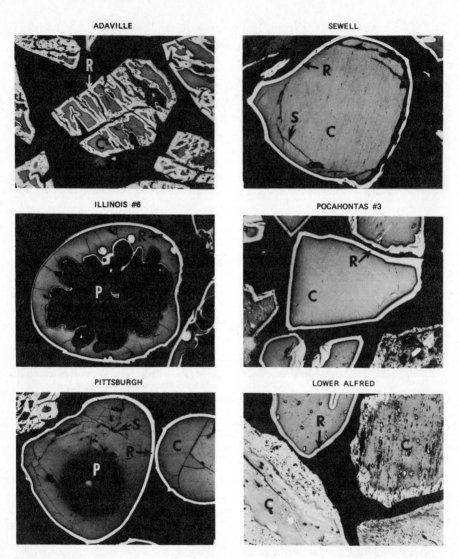

Figure 6.34 Photomicrographs of indicated coals treated at 400°C in air for 30 minutes. Key: oxidation layer **R**, unaltered coal **C**, pores **P**, shrinkage cracks **S**. Reflected light.

When the vitrinite in coal is oxidized, it is changed from the plastic to the elastic state. Nandi et al.[17] studied the variation of the microhardness and reflectance of coal under conditions of oxidation simulating weathering. Coals above 92 percent carbon content, such as anthracite, show elastic behavior, whereas coals between 70 and 92 percent carbon show plastic behavior. Nandi et al.[17] showed that as oxygen penetrates into coal particles the physical state of the maceral vitrinite changes from the plastic to elastic. They found that the

TABLE 6.17 Reflectance of Edge and Center of Coal Particles and Oxidation-Layer Thickness for Indicated Coal Samples

| Coal Seam | Reflectance in Oil, Percent | | | | Average Rim Thickness in Microns |
| | Treated at 177°C in Air | | Treated at 400°C in Air | | Treated at 400°C in Air |
	Center	Rim	Center	Rim	
Adaville	0.51	0.58	1.09	2.02	9.0
Illinois No. 6	0.65	0.67	1.17	2.12	8.0
Pittsburgh	0.91	0.95	1.05	2.07	7.5
Sewell	1.13	1.13	1.17	2.21	8.8
Pocahontas No. 3	1.64	1.62	1.70	2.45	8.8
Lower Alfred	2.33	2.40	2.32	2.95	10.5

TABLE 6.18 Agglomeration Index Determined by Rapid Addition of Coal to a Hot Fluid Bed (1 in. unit)

| | Agglomeration Index[a] | | |
	Untreated	Treated at 177°C	Treated at 400°C
Adaville	0.05	0.05	0.05
Illinois No. 6	7.1	4.8	0.10
Pittsburgh	5.8	3.4	5.8
Sewell	5.4	5.9	4.9
Lower Alfred	0.15	0.10	0.05

[a]Agglomeration Index = grams of agglomerates formed per gram of coal added.

extent of oxygen penetration can be determined from the marked change in state that occurs from the unoxidized centers of coal particles to their oxidized margins. The unoxidized coal displays lower reflectance and plastic deformation; the oxidized coal has higher reflectance (induced oxidation) and elastic deformation. The transformation occurs rapidly in high-volatile coals, but more severe oxidation is needed to produce this change in low-volatile coals. By using reflectance and microhardness, Nandi et al.[17] were able to determine the boundary between unoxidized and oxidized coal and determine the rate of oxygen penetration into coal particles.

The study by Nandi et al.[17] deals exclusively with coal that has been oxidized in air at 100°C. Work at U.S. Steel in microhardness testing of unoxidized and

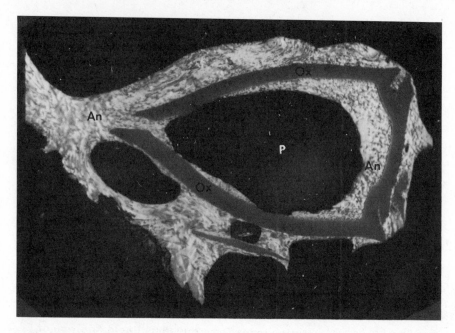

Figure 6.35 Photomicrographs of carbon forms from FSI test. Key: oxidized layer **R**, anisotropic carbon from fluid coal **F**, isotropic carbon from nonagglomerating and inert coal entities **I**, pores **P**. Reflected light.

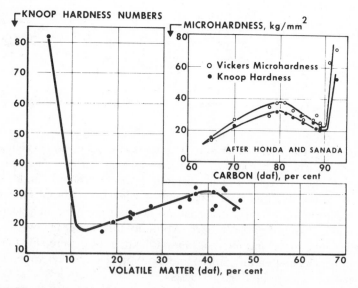

Figure 6.36 Relationship between hardness of vitrinite and rank in coals (after Schapiro and Gray[63]).

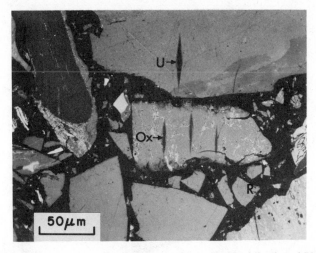

Figure 6.37 Micro-idention hardness impressions in unoxidized coal **U** have distinct margins while those in oxidized coal **OX** tend to close. Reflected light in oil.

oxidized coals indicated that the hardness changes from the plastic to the elastic state in going from fresh to naturally weathered coal. The reflectance of oxidized vitrinite is lower in the weathered portion of coal particles.

6.21   EFFECT OF OXIDIZED COAL ON COKE STRENGTH

The inclusion of a high-volatile bituminous coal from a West Virginia strip mine in a coking blend for use in an eastern steel mill coincided with a blast-furnace production decrease primarily due to downtime from burned tuyeres. The subject Alma Seam coal from the Mingo–Logan counties area in West Virginia was suspected of being oxidized.

In recovering the coal, the FSI was used as a control parameter. Auger samples of coal taken at 10 ft intervals from the outcrop to the highwall were tested for FSI. The mining line was set at an FSI of 1, but some of the recovered coal could still contain more than 60 percent oxidized coal. Thus, the practice of setting a mining line at an FSI of 1 was inadequate for excluding oxidized coals.

To test the affect of oxidized coal on coke properties, samples of oxidized coal with an FSI of zero and a Gieseler fluidity of zero were taken from the coal outcrop, and samples of unoxidized coal with an FSI of $7\frac{1}{2}$ and a Gieseler fluidity of 6300 were taken near the highwall and washed. The washed oxidized and unoxidized coals were coked separately and in blends to determine the effect on coke strength. In addition, the blends of oxidized and unoxidized high-volatile coals were blended with low-volatile bituminous Pocahontas No. 3 coal from West Virginia and coked to determine the effect of oxidized coal on the resultant coke strength.

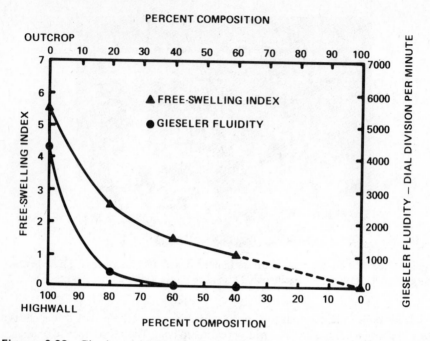

Figure 6.38 Gieseler and FSI versus amount of outcrop coal in highwall coal— Lower Cedar Grove seam bulk samples.

The FSI and Gieseler properties of the outcrop coal (oxidized), highwall coal (unoxidized), and various mixtures of the two were determined, and the results are shown in Figure 6.38. The fluidity of the unoxidized coal is significantly reduced by the addition of as little as 20 percent of the oxidized coal; no fluidity (1.0 DDPM) is obtained with 40 percent addition of oxidized coal. The effect of oxidized coal did not drastically affect the FSI values of the blends, 60 percent or more of oxidized coal was required to reduce the blend FSI to below 1.

The coking properties of the unoxidized coal, oxidized coal, and blends of the two were determined by carbonizing them in a 30 lb capacity pressure-test oven to a final temperature of 1850°F over a period of $3\frac{1}{2}$ hours. Ten pounds of the 1 by $1\frac{1}{2}$ in. coke were tumbled for 500 revolutions and the $+\frac{3}{4}$ in. and $+\frac{1}{4}$ in. fractions (corresponding to the stability and hardness factors) were determined. The relationship of the modified tumbler index (percent $+\frac{3}{4}$ in.) to the ASTM (percent $+1$ in.) stability is shown in Figure 6.39. Stability data for the individual coals and blends of unoxidized and oxidized high-volatile coal are shown in Figure 6.40. They indicate that addition of more than 20 percent oxidized coal to the unoxidized coal significantly reduces the tumbler strength of the coke.

Since the high-volatile coal is used commercially in blends with low-volatile coal, the unoxidized coal with the same amounts of oxidized coal (as used in the previous tests) was blended with 20 and 30 percent Pocahontas No. 3 low-

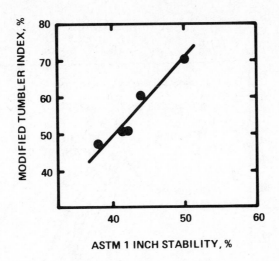

Figure 6.39 Relationship of modified tumbler index to ASTM stability.

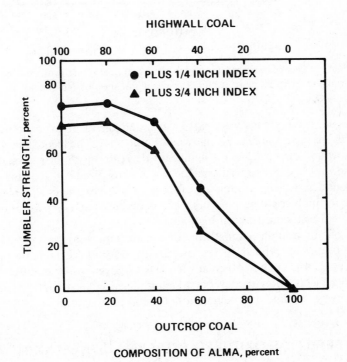

Figure 6.40 Effect of highly oxidized coal additions on the coking properties of Alma coal.

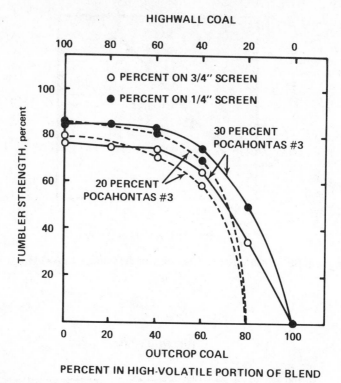

Figure 6.41 Effect of amount of outcrop (oxidized) coal in highwall coal on tumbler strength in blends with 20 percent and 30 percent Pocahontas No. 3 coal.

volatile coal. These blends were carbonized, and the resultant coke tested for strength properties. The data are shown in Figure 6.41. The blend containing 30 percent low-volatile coal has no significant decrease in tumbler strength until over 40 percent of oxidized coal is added to the unoxidized high-volatile portion of the blend. In a blend containing 20 percent low-volatile coal, a significant reduction in tumbler strength occurs when between 20 and 40 percent oxidized coal is added to the unoxidized high-volatile coal.

Although the tumbler strength of coke is an important parameter of coke quality and closely related to the operating performance of blast furnaces, oxidized coal also creates other adverse effects. Even when oxidized coal is present in a blend below the level at which it affects coke strength, its removal reduced tuyere burning and increased hot-metal production.

6.22 EFFECT OF OXIDIZED COAL ON COKEMAKING

Badly oxidized coal is unsuitable for coke making because of the loss of rheological properties and the resultant loss of coke strength. Controls are frequently exercised to avoid mining coal sufficiently oxidized to detract from the coke strength. About 16 years ago a major coke plant (Clairton Works, U.S. Steel Corp.) experienced frequent upsets in the control of pulverized coal bulk density, despite the use of preventive mining controls. McGinnis and Gray[65] reported on a study of the problem.

Severely oxidized coal was generally avoided to prevent upsets in preparation plant circuits, such as froth flotation, and to prevent loss in coke strength. At that time, little was known of the pronounced effect that partially oxidized coal exerts, not only on coal bulk density control, but also on charging delays, heating irregularities, carbon formation, oven pushing problems, coke production, tar and pitch quality, and the control of smoke emission.

Coke plants strive to control coal bulk density since it affects coke productivity. Commonly understood factors such as moisture content and size consist affect bulk density. Bulk density decreases as the size consist decreases. The bulk density also decreases as the moisture increases from very low values. Additional increases in moisture, above about 8 percent, increase the bulk density. Generally, oil additions to the coal charge are used by coke plant operators to control bulk density. Normally, three pints of oil addition per ton of pulverized coal raises the coal bulk density by three pounds per cubic foot, as measured by the 6-foot drop test for bulk density determination. The oil lubricates the coal and allows it to flow freely and pack better. This increase in coal bulk density (about $3\,lb/ft^3$) due to oil additions is important because it results in approximately a 6 percent increase in the weight of coal charged. However, on occasions the oil addition was less effective than normal in increasing bulk density. Bulk density response to oil additions for normal and unsatisfactory operating periods are shown in Figure 6.42. A statistical analysis

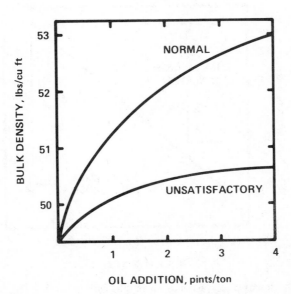

Figure 6.42   Relative increase in coal bulk density for normal and unsatisfactory coals.

of test data taken during normal and problem periods for bulk density control revealed that only 40 percent of the variability in coal bulk density could be related to changes in coal moisture, coal size, and oil addition. The large remaining proportion of unexplained variability confirmed the existence of other factors affecting coal bulk density.

Other problems accompany poor bulk density control. For example, when the amount of coal in an oven is decreased without opportunity for a corresponding cut in underfiring gas, the oven charge is overheated. This increases energy consumption, results in overheated ovens, and contributes to the accumulation of hard carbon deposits on oven walls and roofs. These deposits provide an obstacle to the pushing of coke from the oven at the end of the coking cycle. Production delays, excessive pull-back of coke by the pusher ram onto the pusher side bench, and damage to oven walls and machinery result. In addition, there are increased tar solids, resulting in increased pitch solids. These can detract from the quality of electrode binder pitches, impregnating pitches, and pipe enamels. Periodic overheating of coke also increases the time required for quenching, causing higher and more variable moisture contents.

A chronological correlation of the dips in coal bulk density with coke output per oven led to some interesting findings. As expected, low coal bulk densities reduce coke output since less coal per unit of oven volume is charged. Actual production losses were much greater than the losses based on the bulk density changes, Figure 6.43. For instance, it was found that a 5 percent decrease in bulk density correlated with 13 percent drop in oven productivity. The additional 8

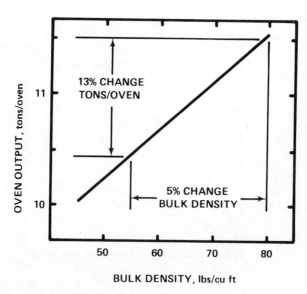

Figure 6.43 Relative increase in oven output with increasing coal bulk density.

percent of coke loss over the 5 percent loss assignable to lower coal bulk density, was caused by lower and irregular height of the coal charge. When coal bulk density decreased, the coal became sticky and assumed a steeper angle of repose in the oven. This created excessive valleys between charging holes and low coke-side ends. At other times coal would pack into the charging holes, blocking the tunnel head, and would create charging emissions. The problem was compounded when larry car volumetrics were raised. This later resulted in excessive coal height in the oven when coal bulk density increased back to normal. In addition, coal arched and rat-holed in the coal bunkers and larry cars, causing delays in charging.

The cause of these problems was isolated when the periods of low coal bulk density were related to the time of unloading a specific low-volatile coal. However, when conventional laboratory analyses for sulfur, ash, and volatile matter were made, there was no evidence of corresponding changes in these coal properties. The most difficult periods of bulk density control correspond to periods of low-coal moisture in the product coal. This was confusing because lower coal moisture normally results in higher, rather than lower, coal bulk density. Further investigations at the coal preparation plant showed that the low-coal moistures were caused by yield losses in the flotation recovery of extremely fine and wet coal particles. Following this finding studies were conducted to find the causes of bulk density control problems.

When samples of both satisfactory and unsatisfactory low-volatile coals were compared under a microscope, characteristic differences emerged. The coarse particles of unsatisfactory coal were irregular. They were characterized by the presence of fissures, predominantly in the vitrinite particles and by the presence of fines resulting from degradation of the particle surfaces, Figure 6.44, on the surfaces of coarser particles. A film of extreme fines develops on the coarse particle surfaces even if the fine particles are removed once by water washing (Figure 6.44). Additional degradation of coal surface occurs in subsequent coal handling operations, since the removal of extreme fines in the coal washing operation is not permanent.

It is likely that this film formation increases the frictional forces among coal particles during compaction and results in low bulk densities. Figure 6.44 is a photomicrograph showing displaced, fine coal and generation of fine particles on oxidized coal surfaces.

Oil addition is also less effective in increasing the bulk density of unsatisfactory coal because of the tendency for oil to penetrate into the surface fissures of oxidized coal and not be available at the particle surfaces to lubricate and aid compaction, Figure 6.44.

Samples were taken at the source of unsatisfactory coals at various distances from the highwall of the strip mine. These samples were studied for physical and chemical characteristics relating to the reported problems with control of coal bulk density. Photomicrographs of samples of coal from the highwall, equidistant from highwall to outcrop, and from the outcrop are shown in Figure 6.45. As the samples approached the outcrop, the particles took on a negative electrosta-

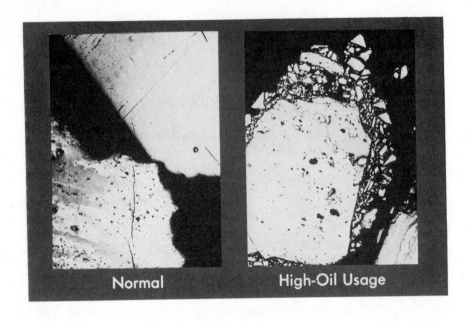

Normal

High-Oil Usage

FINE COAL PARTICLES AT SURFACE OF
COARSE COAL PARTICLES

Figure 6.44 Photomicrographs in a show wet screened coal from a normal oil usage period and from a high oil usage period. Displaced fine coal coating coarse coal particles is characteristic of the high oil usage period. Photomicrographs in B show fine coal forming at the surface of oxidized coal particles.

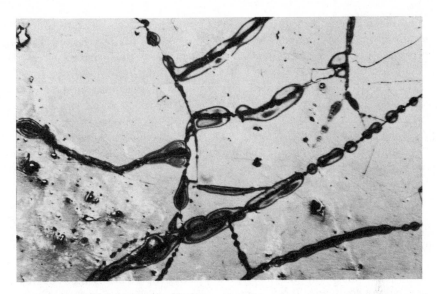

Figure 6.45 Photomicrographs showing oil penetration of cracks in oxidized coal.

tic charge, lost rheological properties (FSI and Gieseler), increased in oxygen, the pH values became more acid, and recovery from froth flotation decreased. All of these changes were found to be characteristic of oxidized coal.

Since the results from Saffranin O correlate closely with data from phase inversion testing at some mines, the phase inversion test is also used routinely for detection of coal oxidation. The unoxidized coal is water repellent and tends to float. Oxidized coal is hydroscopic and tends to sink.

As part of the study, additional plant tests were conducted to further explore the effects of coal moisture and oxidation on operating problems. The analyses of pulverized coal samples for bulk density and oxidation were correlated to establish quantitative relationships. This information indicated the coal bulk density decreased approximately $3\,lb/ft^3$ when the increase in coal oxidation caused a drop of 35 percent in light transmittance. The correlation line also shifted toward a lower bulk density at higher levels of coal moisture. From this information, it was evident that the light transmittance value for the coal blend must be 90 percent or higher to maintain a control bulk density of $53\,lb/ft^3$ for coal ranging as high as 7.5 percent moisture. Light transmittance values approaching 100 percent are required when coal peaks at 8.5 percent moisture.

Other tests were conducted to relate the significance of decreased bulk density to the coal flow problems during the charging of the coke ovens. It was found that when the coal was partially oxidized and the bulk density decreased, more time was required for the discharge of a fixed amount of coal from a test hopper. The discharge time was also increased when the coal was wetter. From this analysis, it became apparent that the severe coal handling problems were really the combined effect of coal oxidation and high coal moisture, following periods of local heavy rain. Because efforts to control peak coal moisture have been largely unsuccessful, attention was focused on the correction of coal oxidation problems.

The caustic solubility test is now being used at the mines supplying U.S. Steel's coke plants. Sources of raw coal are regularly tested prior to washing, and the raw coal is rejected if it fails to meet standards. Prospective new coals must pass oxidation tests for acceptance. As a result, productivity and coke quality have improved, operational problems have diminished, and smoke emissions are under control. Conditions on top of the battery improved with the return of free coal flow from larry cars and coal bunkers.

6.23 RHEOLOGICAL PROPERTIES OF UNOXIDIZED AND OXIDIZED COALS

The principal interest in the softening and plastic properties of heated coal relate to the coking coals since this is a unique property of the coking coals. Thus, an examination of some of the theories of why coals 'coke' should precede a discussion of the rheological (thermoplastic) properties of coal and how these properties are affected by coal weathering.

The plastic properties of heated coal have been widely investigated as a means of characterizing coking coals. The various methods used to investigate plastic properties include: absolute, agglutinating, and free-swelling tests, dilatation, torsional, gas flow permeability, extrusion, and penetration. These tests are empirical in nature but serve to measure this important property of coal. A few of the more widely accepted tests are described below.

The Audibert-Arnu dilatometer is a popular rheological test widely used in Europe. A finely ground 2 g coal sample is moistened and compressed in a mold to form a tapered cylindrical pencil 60 mm high and 6.5 mm in diameter. The pencil is placed in a metal tube with an inside diameter of 8 mm and a closed bottom. The sample and container are placed in a vertical electric tube furnace, and a 7.8 mm diameter rod is placed on top of the sample. The rod is connected to an indicator pointer that moves along a graduated scale to indicate volume changes during pyrolysis at 3°C/minute. The total weight on the sample is 150 g. Movement of the plunger measures softening (contraction) and swelling (expansion) of the heated coal. The maximum dilatation percentage is characteristic of a coal's rank or type. Oxidation decreases dilatation. Chapter 4 by Loison et al.[66] gives a very good description on the plastic properties of coal.

The free-swelling test is another commonly used procedure to determine the agglomerating and swelling characteristics of coal. A 1 g sample of -60 mesh coal is placed in a covered crucible and heated to 820°C in $2\frac{1}{2}$ minutes. The height of the botton produced is graded from the lowest button of 1 to that which fills the crucible, graded 9. In agglutinating tests, such as the Roga Test, inerts are mixed with coal and then the strength of the agglomerate on heating is measured.

Sopoznikhov has developed a penetrometer for measuring the strength and thickness of the plastic layer developed during carbonization. This instrument is widely used in eastern Europe. Reportedly there are correlations between the thickness of the plastic layer, coal type, and volatile matter.

One of the most widely used instruments for measuring the plastic properties of coal is the constant torque Gieseler plastometer. This instrument is popular in the United States and Japan. In this test, a 5 g sample of −40 mesh coal in a retort crucible is tightly packed around a rod with rabble arms. A constant torque of 40 g/in. is maintained on the stirrer throughout the test. When the coal is heated from an initial temperature of 300°C at 3.0°C per minute, the coal softens, becomes plastic, reaches a maximum plasticity, and then becomes less plastic as it is transformed to a semicoke solid. The important data measured are the temperature of initial softening, maximum fluidity, solidification, and the maximum fluidity in dial divisions per minute (the stirrer rotations or 100 DDPMs = 1 revolution). Typical Gieseler curves for high-, medium-, and low-volatile coals are shown in Figure 6.46. For most coking coals the plastic range, which extends from softening to solidification, falls within the temperatures of 350 to 550°C. Generally, the results of these individual tests correlate with one another and with other coal tests.

The maximum fluidity of coal varies considerably with rank as is shown in Figure 6.46. Coals less than high-volatile B in rank have little or no fluidity, high-volatile A coals generally have the highest fluidity, low-volatile coals have low fluidity, and coals higher in rank than low-volatile coal have no fluidity.

The Japanese use the individual fluidity values of coals to blend them for coke making. This works well for the Japanese because the fluidity of coals and coal

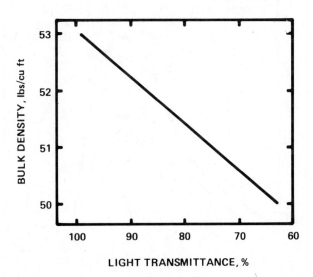

Figure 6.46 Relationship between bulk density and light transmittance resulting from coal oxidation.

blends can be related to other characteristics of coals important in cokemaking. Weathering reduces fluidity so it is important to know the weathering state of a coal.

The effect of weathering on coal fluidity is seen in Figure 6.38, which shows the rapid decrease in fluidity for a coal from highwall (least weathered) to the outcrop in a strip pit (maximum weathering). This occurs because weathering mainly alters the chemistry of the maceral vitrinite, which is the primary source of fluidity in a coal. The general reaction governing this phenomenon is the replacement of linking hydrogens between aromatic clusters with oxygen reducing the available hydrogen necessary for fluidity. Larsen et al.[64] and Gethner[65] discuss this weathering reaction and others as possible mechanisms that destroy coal fluidity. This reduction in fluidity is most dramatic in highly fluid high-volatile coal, although the net effect is the same for low-fluid coals. In terms of rank, the addition of oxygen to vitrinite virtually reduces the rank of the vitrinite, transforming it to an essentially inert material.

Thermal oxidation reduces the fluidity of a coal as shown in Figure 6.47. This coal was heated at 105°C for different time intervals. After only 6 hours of heating, the coal's fluidity was reduced to 1100 DDPM from an initial fluidity of nearly 10,000 DDPM. Continued heating eventually reduced the fluidity to almost zero. Thus, even heating in a stockpile can have profound effects on coal fluidity.

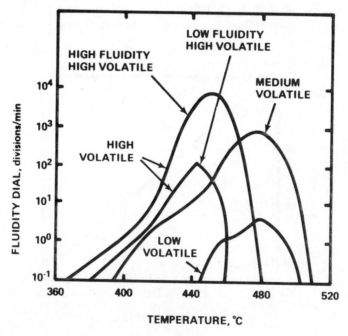

Figure 6.47

The FSI of coal is also affected by weathered coal, but not as severely. Thus, this test is not a particularly sensitive indicator of the weathering state of a coal.

It is evident that the plastic properties of coal are critical for producing high-quality coke. However, one must know what weathering state these properties reflect or incorrect conclusions will be drawn concerning the coking potential of a particular coal.

6.24 WEATHERING CHARACTERISTICS OF COAL IN STORAGE

Marginal coking high-volatile B-rank coal is known to be extremely sensitive to oxidation. U.S. Steel in an internal study conducted outdoor and accelerated oxidation tests on as-received size and $-\frac{1}{4}$ in. coal. The coal storage was in Utah. The coal consisted of 79 percent Sunnyside Seam coal from Utah and 21 percent Somerset B and C seam from Colorado. The coal was sampled monthly for testing and the test period was 1 year. Tests included size analysis, proximate, ultimate, petrographic, plasticity, FSI, microhardness, electrical resistivity, direct oxygen (infrared and neutron activation), calorific value, and carbonization. The results of the 1 year outdoor-storage test showed that the oxygen content of the coal determined by difference of the coal increased from about 9 to about 11 percent or nearly 20 percent, based on data from ultimate analyses (Figure 6.48). The increased oxygen in the coal corresponded to a decrease in maximum fluidity, FSI, and carbon-to-oxygen ratio during the 1 year test period. The decrease in maximum fluidity and FSI was determined to be about 50 percent. In contrast, the heating value or Btu decreased only 2 percent over the test period. The change in maximum fluidity, Btu, and FSI with time of storage are shown in Figure 6.49. Carbonization test results on as-received size and coal pulverized 100 percent $-\frac{1}{4}$ in. showed that the strength of the coke produced remained unchanged for the first 6 months of storage but decreased rapidly after 8 months. The carbonization data for the $-\frac{1}{4}$ in. high-volatile coal and a blend with 20 percent medium-volatile coal are shown in Figure 6.50. This rapid decrease corresponded to a coal oxygen content of 10.5 percent and a carbon-to-oxygen ratio of 7.2. Coking strength of the blend as measured in a small (30 lb) coke oven did not show the decrease in coke strength that was noted for the high-volatile coal when coked alone.

Microscopic observations of coal have shown that oxidized coal other than mildly oxidized can be detected easily. Unfortunately, no major change in microscopic appearance occurred as storage time increased. Oxidation rings did not develop and crack development was not pronounced. No change in vitrinite reflectance or microindention hardness occurred.

Small cokes prepared from fresh and oxidized coal were examined for changes in coke microstructure. About 50 g of coal was coked at 300°C per hour from 1300 to 1850°F and soaked for 2 hours. The resultant coke was cooled in nitrogen and prepared for microscopic examination. The coke from the oxidized

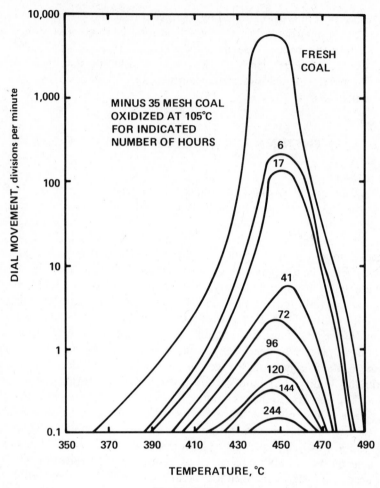

Figure 6.48

coal is much more granular than that from the fresh coal. This is due to better fusion between the coked particles from the fresh coal compared with poorer fusion between the coke from oxidized coal particles. In the course of this study it was found that drying the coal in a forced-air oven altered the coal.

Recommendations for storage of coal to minimize the alteration in coking characteristics due to weathering include the following:

1. Compact coal in 1 ft layers to a bulk density of 70 lb/ft^3.
2. Maintain gently sloping slides.
3. Compact sloping slides.
4. Smooth final surface.

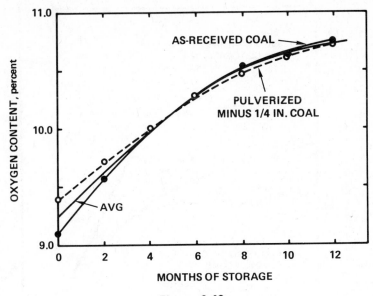

Figure 6.49

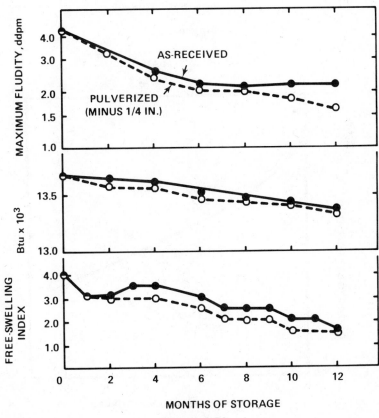

Figure 6.50

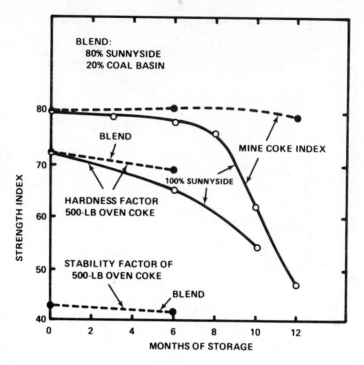

Figure 6.51

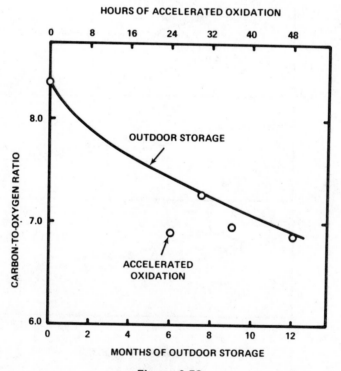

Figure 6.52

Accelerated oxidation tests using 1200 lb of $-\frac{1}{4}$ inch coal were also conducted. The coal was treated with oxygen (120 scf per hour) in a blender/dryer at 200°F. A small sample was removed and tested for oxygen content every 6 hours for 48 hours. The results are shown in Figure 6.51. As expected, the accelerated tests resulted in the oxidation of the coal at a faster rate than in a storage pile because of the higher temperatures and oxygen concentration used in the accelerated test. However, the carbon-to-oxygen ratios are similar to those from outdoor storage tests when the 4 hour accelerated oxidation test data is compared to the data for the 1 month outdoor storage sample, Figure 6.52.

REFERENCES

1. R. J. Gray, A. H. Rhoades, and D. T. King, Detection of oxidized coal and the effect of oxidation on the technological properties. Trans—*Soc. Min. Eng.* AIME, **260**, 330–341 (1976).

2. R. J. Gray and K. C. Krupinski, Use of microscopic procedures to determine the extent of destruction of agglomerating properties in coal. *Proc. Coal Agglom. Convers. Symp.*, 1975 (1976).

3. J. C. Crelling, R. H. Schrader, and L. G. Benedict, Effects of weathered coal on coking properties and coke quality. *Fuel* **58**, 542–546 (1979).

4. D. White, Coal oxidation. *Bull.—U.S. Bur. Mines* **29** (1911).

5. F. Schulze, Uber das Virkommen Wehlerholtener Zellulose in Braunkohle und Steinkohle. *Ber. N. Akad. Wiss (Berlin)* pp. 676–678 (1855).

6. B. Ferrari, Die Entstehung von Grubenbranden nach Untersuchunger auf Kollen-Petrographischer Grundlage. *Glueckauf* **74**, 765–774 (1938).

7. E. P. Plumstead, The permo-carboniferous coal measures of the Transvall, South Africa—An example of the contrasting stratigraphy in the southern and northern hemispheres. *C.R.4, Congr. Int. Strat., Geol. Carbonifere, Masst.* **2**, 540–550 (1962).

8. N. V. Marevich and A. B. Travin, Tendency toward spontaneous combustion of petrographic types of coal from Prokop'evsk deposits of Kuzbass. *Izv. Akad. Nauk USSR, Otd. Tekh. Nauk* pp. 110–117 (1953).

9. L. G. Benedict and W. F. Berry, Recognition and measurement of coal oxidation. Presented to Coal Group, Geological Society of America, Miami, Bituminous Coal Research Inc., Monroeville, PA, p. 41 (1964).

10. L. G. Benedict, R. R. Thompson, J. J. Shigo, III, and R. P. Aikman, Pseudovitrinite in Appalachian coking coals. *Fuel* **47**, 125–143 (1968).

11. D. E. Lowenhaupt and R. J. Gray, The alkali-extraction test as a reliable method of detecting oxidized metallurgical coal." *Int. J. Coal Geol.* **1**, 63–73 (1980).

12. D. L. Marchioni, The detection of weathering in coal by petrographic, rheologic and chemical methods. *Int. J. Coal Geol.* **2**(3), 231–259 (February 1983).

13. D. Chandra, Reflectance and microstructures of weathered coals. *Fuel* **41**, 185–193 (1962).

14. D. Chandra, Reflectance of oxidized coals. *Econ. Geol.* **53**, 102–108 (1958).

15. F. J. Huntjens, X. X. Doormans, and D. V. Van Krevelin, unpublished data (1950); taken from D. W. Van Krevelin and J. Schuyer, *Coal Science*, p. 314 Elsevier, 1957.

16. F. Goodarzi and D. G. Murchison, Petrography and anisotropy of carbonized pre-oxidized coals. *Fuel* **55**, 141–147 (1976).

17. B. N. Nandi, L. A. Ciavaglia, and D. S. Montgomery, The variation of the microhardness and reflectance of coal under conditions of oxidation simulating weathering. *J. Microsc. (Oxford)* **109**(1), 93–103 (1977).

18. Y. Nakayanagi, Microscopic study of outcrop coal—A microscopic study of the grades of coalification on the weathered coal sampled from outcrops. *J. Coal Res. Inst. (Tokyo)* **1**, 225–232 (1956).

19. K. Kojima and H. Ogoshi, Petrographic consideration of weathered coal. *J. Fuel Soc. Jpn.* **52**(56), 885–895 (1973).

20. B. Alpern and F. Maume, Etude petrographique de l'oxydation naturelle et artificielle des houilles. *Rev. Ind. Miner—Mines* **51**(11), 979–997 (1969).

21. J. C. Quick and W. A. Kneller, The use of dyes as an aid to coal petrology. *Int. J. Coal Geol.* **7**(1), 51–69 (1987).

22. F. E. Huggins, G. P. Huffman, D. A. Kosmack, and D. E. Lowenhaupt, Mossbauer detection of geothite (αFeOOH) in coal and its potential as an indicator of coal oxidation. *Int. J. Coal Geol.* **1**, 75–81 (1980).

23. D. E. Pearson and H. Kwong, Mineral matter as a measure of oxidation of a coking coal. *Fuel* **58**, 63–66 (1979).

24. M. P. Fuller, I. M. Hamadeh, P. R. Griffiths, and D. E. Lowenhaupt, *Fuel* **61**, 529–536 (1982); *Proc.—Ironmaking Conf.* **41** (1982).

25. P. C. Painter and M. M. Coleman, *The Application of FTIR to the Determination of Mineral Matter in Coal and Oil Shale*, Notes, No. 35, pp. 1–33. DIGILAB, 1980.

26. M. C. Lin, G. P. Huffman, F. E. Huggins, G. R. Dunmyre, and D. E. Lowenhaupt, *A Correlative Investigation of the Effects of Oxidation on the Minerals, Macerals, and Technological Properties of Coal*, Annual report February 1, 1982 to January 31, 1983. Prepared by U.S. Steel Corporation for Gas Research Institute, Contract No. 5081-261-0557, (1983).

27. B. W. Molt, *Micro-Indentation Hardness Testing*, pp. 1–272. Butterworth, London, 1956.

28. ASTM, Standard Method D-720-67: Free-swelling index of coal; gaseous fuels; coal and coke; atmospheric analysis. *Annu. Book ASTM Stand.* Part 26, pp. 268–274 (1983).

29. A. H. Brisse and P. J. Richards, Coal quality control at the mine by rapid carbonization. *Blast Furn., Coke Oven, Raw Mater., Proc.* pp. 265–278 (1955).

30. V. S. Engleman, Wet oxidation applications to coal utilization. *Proc. 2nd Annu. Pittsburgh Coal Conf., 1985* pp. 769–775 (1985).

31. G. R. Yohe, Oxygen absorption of coal—Some observations on oil treatment. *Trans. Ill. State Acad. Sci.* **62**(1) (1969).

32. H. B. Atkinson and W. Hyslop, The assessment of incipient oxidation in Durham coking coals. *Coke Gas* **23**, 102–106 (1961).

33. B. C. T. Labuschagne, The variation of coal hydrophobicity with rank: Hydrophobic virus hydrophilic parameters. *Proc. 4th Annu. Pittsburgh Coal Conf.*, pp. 861–876 (1987).

34. W. W. Wen and S. C. Sun, An electrokinetic study of the amine flotation of oxidized coal. *Trans.—Soc. Min. Eng. AIME* **262**, 177–180 (1977).

35. J. B. Gayle, W. H. Eddy, and R. Q. Shotts, Laboratory investigation of the effect of oxidation on coal flotation. *Rep. Invest.—U.S., Bur. Mines* **RI6620** (1065).

36. S. C. Sun, Effect of oxidation of coals on their flotation properties. *Trans. Am. Inst. Min. Metall. Eng.* **6**(4), 396–401 (1954).

37. J. A. L. Campbell and S. C. Sun, Bituminous electrokinetics. *Trans.—Soc. Min. Eng. AIME*, **247**, 111–114 (1970).

38. D. J. Brown, Coal flotation. *Froth-Flotations—50th Anniversary Volume* (D. W. Fuerstenau, ed.), p. 518. AIME, New York, 1962.

39. A. F. Baker and K. J. Miller, Hydrolyzed metal ions as pyrite depressants in coal flotation: A laboratory study. *Rep. Invest.—U.S. Bur. Mines* **RI7518** (1971).

40. L. A. Baranov et al., Assessing the oxidation of coal by a flotation procedure. *Coke Chem. USSR* (*Engl. Transl.*) **10**, 5–18 (1970).

41. J. A. Gutierrez and F. F. Aplan, The effect of oxidation on the hydrophobicity and floatability of coal. *Colloids Surf.* (submitted for publication; the original work was Gutierrez's thesis at the Pennsylvania State University).

42. J. Iskra and J. Laskowski, New possibilities for investigating air-oxidation of coal surfaces at low temperatures. *Fuel* **46**(1), 5–12 (1977).

43. P. C. Painter, personal communication.

44. M. M. Wu, G. A. Robbins, R. A. Winschel, and F. P. Burke, Low temperature coal weathering: Its chemical nature and effect on coal properties. *Energy Fuels* **2**(2), 19–26 (1988).

45. Huggins, F. E., G. E. Huffman, G. R. Dunmyre, and M. C. Lin, *A Correlative Investigation of the Effects of Oxidation on the Minerals, Macerals, and Technological Properties of Coal*, Final Rep., GRI Contrast 5081-261-0557. 1985.

46. D. G. A. Thomas, *Br. J. Appl. Phys.* **4**, 555 (1953); Anonymous, *Iron Coal Trades Rev.* **166**, 941–942 (1953).

47. W. H. Walton et al., Complete specification—Method and apparatus for measuring the degree of oxidation of coal. British Patent 733 (1955).

48. F. Hevesy and H. Levi, The action of neutrons on rare earth elements. *Mat-Fys. Medd.—K. Dan. Vidensk., Selsk.* **14**(5) (1936).

49. T. I. Taylor and W. W. Havens, Jr., Neutron spectroscopy and neutron interactions in chemical analysis. *Phys. Methods Chem. Anal.* **3** (1956).

50. D. J. Veal and C. F. Cook, A rapid method for the direct determination of elemental oxygen by activation with fast neutrons. *Anal. Chem.* **34**(2), 178 (1962).

51. E. L. Steele and W. W. Meinke, Determination of oxygen by activation analysis with fast neutrons using a low-cost portable neutron generator. *Anal. Chem.* **34**(2), 185 (1962).

52. R. J. Gray and N. Schapiro, *Petrographic Composition and Coking Characteristics of Sunnyside Coal From Utah*, Bull. No. 80, Central Utah Coals: A Guidebook Prepared for the Geological Society of America and Associated Societies of Mines and Mineral Ind., University of Utah, 1966.

53. J. T. Peters, N. Schapiro, and R. J. Gray, Know your coal. *Trans. Am. Coc. Min., Eng.*, pp. 1–6 (1962).

54. N. Schapiro and R. J. Gray, Relation of coke structure to reactivity. *Blast Furn. Steel Plant* pp. 273–280 (1963).

55. N. Schapiro, R. J. Gray, and G. R. Eusner, Recent developments in coal petrography. *Blast Furn., Coke Oven Raw Mater., Proc.* **20**, 89–112 (1961).

56. R. J. Gray, A system of coke petrography. *Proc. Ill. Min. Inst., Annu. Meet.* pp. 20–47 (1976).

57. W. F. Berry and J. S. Goscinski, Hot coal: Causes and remedies. *Bulk Syst. Int.* Oct., pp. 33–36 (1982).

58. R. Mason, Hot coal: The problems and precautions. *J. Coal Qual. Fall*, pp. 15–20 (1982).

59. E. A. C. Chamberlain and D. A. Hall, The liability of coals to spontaneous combustion. *Colliery Guardian* Feb., pp. 65–72 (1973).

60. R. Kaji, Y. Hishinuma, and Y. Nakamura, Low temperature oxidation of coals, a calorimetric study. *Fuel* **66**, 154–157 (1987).

61. B. Alpern, Microdurite des chargons et dos cokes in function due degree de houllification. *C.R. Hebd. Seances Acad. Sci.* **242**, 653 (1956).

62. H. Handa and Y. Sanada, Hardness of coal, *Fuel* **35**, 451 (1956).

63. ASTM, Standard Method D-2797-72: Standard method of preparing coal samples for microscopical analysis of reflected light. See note from N. Schapiro and R. J. Gray, p. 377 (1980).

64. B. Das, A study of the microhardness impression of coal. *Trans. Inst. Min. Metall., Ostrava Min. Geol. Ser.* **17**(2), 95 (1971).

65. J. P. McGinness and R. J. Gray, Effect of oxidized coal on coke plant operations. Paper presented at Winter Meeting of the Eastern States Blast Furnace and Coke Oven Association (1976).

66. R. Loison, A. Peytavy, A. F. Boyer, and R. Grillot, The plastic properties of coal. In *Chemistry of Coal Utilization* Ed. H. H. Lowery, ed., Suppl. Vol. 1, Chapter 4, pp. 150–201. Wiley, New York, 1963.

67. L. W. Larsen, D. Lee, T. Schmidt, and A. Grint, Multiple mechanisms for the loss of coking properties caused by mild air oxidation. *Fuel* **65**, 595–596 (1986).

68. L. S. Gethner, Kinetic study of the oxidation of allinois No. 6 coal at low temperatures. *Fuel* **66**, 1091–1096 (1987).

Commentary

Environmental concerns in the use of fossil fuels are highlighted by the acid rain problem. This has led to an increasing emphasis on coal beneficiation whereby certain constituents of the coal such as sulfur bearing minerals may be selectively separated. Beneficiation accounts for about 9% of the average cost of coal in the U.S.; beneficiation is an important factor in the coal economy. The subjects of oxidation and aging are continued in the following chapter where the effects of aging on beneficiation are examined.

7

AGING AND BENEFICIATION

P. Somasundaran
and
C. E. Roberts, Jr.
Henry Krumb School of Mines
Columbia University, New York

7.1 INTRODUCTION

Processing of coal by techniques such as flotation, agglomeration, dewatering, and filtration has gained increasing importance recently because of the requirement for the removal of sulfur and ash and the desirability of recovering valuable fines that were previously discarded as waste. The removal of pyrite and mineral matter requires finer grinding than in the past. Also, the amount of coal fines generated has been increasing in recent years due to the use of mechanized and continuous mining techniques. Almost half of the anticipated annual 1 billion tons of domestic coal will be subjected to some type of cleaning process in the future,[1] and yet coal processing techniques are inefficient and not well understood. This results partly from the complex physical and chemical nature of coal and, more seriously, from the variation in its properties from mine to mine, seam to seam, and bin to bin. Most interesting variations occur in the surface properties of these coals with marked effects on its wettability characteristics and, hence, on its behavior during processing using flotation, dewatering, and other beneficiation procedures (see Figure 7.1 for a flotation plant flowsheet). These variations are considered to be the result mainly of the oxidation of the coal during its weathering. Weatherized coal is known to exhibit inferior flotation properties compared to freshly mined samples. Also, in general, strip-mined coals do not float as well as deep-mined coals from the same seam.[2] Whereas the latter type might need only a frother reagent for flotation, the former often require, in addition to the frother, an oil such as kerosene.

Variations in flotability or hydrophobicity of coal even within the same seam could result from direct atmospheric oxidation in the surface and subsurface

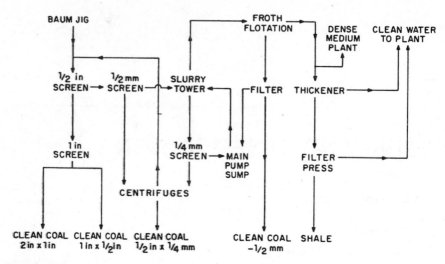

Figure 7.1 Flowsheet of a coal flotation plant (after Brown[6] reprinted with permission).

region as well as chemical alterations during groundwater percolation through the coal bed. The atmospheric oxidation of coal results in physical and chemical adsorption of oxygen from the air on the surface of the coal, followed by the formation of various acidic groups, peroxides, and phenols. In addition, oxidation of such secondary constituents as pyrites in the coal leads to the generation of various soluble inorganics that, when released into the slurry, can adsorb on the coal particles and affect the efficiency of subsequent beneficiation processes since most of these processes are dependent on the surface properties of these coal particles. Although it is known that aging affects the performance of various coal processing techniques, only a limited amount of work has been done to understand the basic reasons for it. A number of investigators have attempted to simulate the oxidation process in the laboratory by gaseous and aqueous chemical treatment. Even though no conclusive evidence for the direct effect of oxidation *alone* has been offered in any of these tests, both methods have been observed to yield the same effects on coal flotability as weathering.

In this chapter, flotation principles of coal and the mechanisms of aging and oxidation effects are discussed, and some recent results are analyzed to isolate and identify the effects of various pretreatments on the behavior of coal during such processes as flotation.

7.2 PRINCIPLES OF FLOTATION

Froth flotation of minerals results from the attachment of air bubbles to hydrophobic particles and their subsequent levitation while they are suspended in a stirred pulp (slurry of mineral particles in water). The attachment of air bubbles during their contact with the solid particles is governed by, among other things, the interfacial properties of minerals and bubbles, and the changes in them brought about by the addition of various chemicals. Selective separation of minerals from each other is possible when one of them is naturally hydrophobic or acquires hydrophobicity owing to selective adsorption of surfactants, called collectors. Minerals such as coal and sulfur have some natural hydrophobicity and should not require collectors. However, if these minerals have undergone any surface chemical alteration, they may require addition of oil-type collectors to enhance their hydrophobicity.

Adsorption of collector surfactants on the mineral surface results from a number of interactive forces individually or in combination with each other. The forces that can contribute to the adsorption process include those due to electrostatic attraction, covalent bonding, hydrogen bonding, and nonpolar bonding between the surfactant and the particle surface species, van der Waals and steric interactions among the adsorbed species, and solvation or de-solvation of adsorbate and adsorbent species.[3]

In the case of coal, both the nonpolar bonding of the collector with the hydrophobic sites on the particle and the electrostatic bonding of the ionic or polar flotation reagents with the oxidized and other reacted sites will be the major adsorption forces. On the other hand, pyrite and other mineral matter can adsorb the reagents by means of other mechanisms as well, depending on the pretreatment and aging the various particles might have undergone. The role played by different mechanisms in adsorption is usually studied by monitoring changes in surface properties such as contact angle and zeta-potential upon adsorption.

Surfactants will also adsorb on the bubble surface, and this will have its own influence on the attachment of particles to bubbles.[4] This can be seen from Young's equation, which defines the surface thermodynamic condition for the three phase contact:

$$\gamma_{SG} = \gamma_{SL} + \gamma_{LG} \cos \theta \qquad (7.1)$$

where γ_{SG}, γ_{SL}, and γ_{LG} are the interfacial tensions of the solid/gas, solid/liquid, and liquid/gas interfaces, respectively, and θ is the contact angle. The change in free energy accompanying the replacement of a unit area of the solid/liquid interface by solid/gas interface is given by Dupre's equation:

$$\Delta G = \gamma_{SG} - (\gamma_{SL} + \gamma_{LG}) \qquad (7.2)$$

Combining the above two equations gives:

$$\Delta G = \gamma_{LG}(\cos \theta - 1) \qquad (7.3)$$

Free energy decrease upon attachment of a mineral particle to an air bubble will occur if the contact angle has a finite value. It is clear from the above equations that surfactant adsorption at any of the interfaces can affect the bubble/particle attachment process. From Equation 7.2, it is seen that $\gamma_{SG} - \gamma_{SL}$

should be $<\gamma_{LG}$ for spontaneous attachment, and an ideal surfactant should help the system meet the above criterion for the mineral to be floated. In addition, in the case of coal, the rough porous nature of its surface will cause marked differences between contact angles of an advancing and a receding liquid film. For the flotation phenomenon, receding contact angle is considered to be critical, and the effect of aging and pretreatment on this angle has to be understood. The degree of porosity of the coal must indeed be taken into consideration while studying the flotation behavior of various coals. However, there is no reliable information on the effect of this important property.

7.2.1 Coal Flotation

The surface properties of coal reponsible for its flotation depend on a variety of geochemical and processing factors. First, since coal is a complex mixture of slow degradation products of organic and mineral matter, its properties will depend not only on the nature of the original vegetation and mineral deposits but also on the extent of the chemical and physical changes that have occurred. As described in Chapter 1, the process of coalification begins with the formation of peat and proceeds to the formation of lignite, bituminous coal, and finally anthracite. These transformations occur under conditions of high temperatures and pressures. The carbon content and the rank of the coal increase throughout the coalification process, whereas the oxygen content decreases.[5,6] Also, in addition to different aromatic and aliphatic groups, the coal will contain various acidic groups, volatile matter, moisture, and mineral ash, all varying in composition depending on the coalification rank and origin. The presence of all these materials in coal makes it a very heterogeneous substance. This heterogeneity, particularly on its surface, is the basic reason for numerous problems in coal beneficiation since the techniques employed depend to a great extent on the surface characteristics of the mineral.

The heterogeneous mineralogical composition of the coal provides it with hydrophobic and hydrophilic sites, the proportion of each varying with both the rank and the weathering. Because of such variations in the hydrophobic/hydrophilic character of the coal, the flotability also varies with the rank and aging.

Three theories have been proposed to explain the flotation of coal, but an examination of these theories reveals their limited applicability and instances where flotation behaviors predicted from theory are contrary to actual occurrences. The carbon/hydrogen ratio theory proposed by Taggart[7] suggests that the difference in flotability between anthracite and bituminous coal is due to the variation in carbon-to-hydrogen ratio. However, this theory is not found to be applicable to the actual flotation behavior of lower-rank coals. For example, the carbon/hydrogen ratios of lignitic coals are often similar to those of some bituminous coals and smaller than those of anthracite coals. Yet, bituminous coal is better floating than anthracite, and lignitic coals exhibit the most inferior flotation properties.[5] Another theory proposed by Wilkins[8] that relates flotation directly to the carbon content of the coal also does not fully account for the actual behavior of coals Anthracite, which has a higher carbon content than the bitumious coals are, as previously noted, less floatable than the latter. The third theory developed by Sun[9] considers coal to be made up of floatable and nonfloatable groups of materials and the balance indicates the flotation behavior of a sample. Based on the chemical analysis of the coal in terms of C, H, O, S, N, moisture, and mineral matter, a formula was proposed for the flotability of the coal:

$$F_c = x(N/2.0796) + y(C/12 - H/2.0796)$$
$$+ 0.4(S/32.06) - z(M/18) - 3.4(O/16) - (N/14)$$

where F_c is the calculated flotability based on a balance between floatable and non floatable materials. H, C, S, O, N, and M are, respectively, the weight percent of hydrogen, carbon, sulfur, oxygen, nitrogen, and moisture. x, y, z, are factors governed, respectively, by the amount of ash, carbon, or moisture remaining in the sample.

This relationship can account for the various natural processes that coals undergo. As the oxygen and moisture contents build up, the flotation is predicted to decrease. A major deficiency of this theory is that it does not take into account marked changes in flotation with the use of reactive flotation reagents.

Reagents Used in Coal Flotation. The reagents most frequently used for coal flotation include cresylic acids, pine oil, short-chain alcohols, kerosene, creosote, and fuel oil. The first three types are used as frothers and the latter three as collectors. Some reagents have both frothing and collecting properties, whereas others have either an activating (inducing flotation), depressing (reducing flotation), dispersing, or a combination of effects, depending on their concentrations and solution chemistry.

Collectors used for coal are generally nonpolar reagents, such as oils, with low solubility in water. These oils may be added to the pulp and dispersed during the conditioning process, or they may be added as an emulsion. They act by coating the coal particles and making them more hydrophobic. High-ranked bituminous coals need little or no collector for good flotation recovery; lower-ranked bituminous and lignitic coals and weathered, aged coals require increasingly larger amounts of collector. Whereas the reagents will adhere strongly to the hydrophobic high-rank coal and to vitrains with a low degree of oxidation, other reagents might be needed to activate the adsorption of the nonpolar agents on the low-rank coal.

In general, frothing agents possess both polar and nonpolar parts. They are usually short-chain alcohols of up to 10 carbon atoms per molecule, the most popular being methylisobutylcarbinol.[5,6] Other frothers include cresylic acid, pine oil, xylenols, and even salt solutions. These reagents are considered to act by changing the surface tension of water in order to facilitate the production of stable froths. Some frothers are adsorbed onto the surface of the coal and are thus removed from solution. This leads to a decrease in the frother concentration and, thereby, to a decrease in the flotation recovery. Note that the magnitude of this effect on flotation will depend on the type of coal and its weathered state since adsorption of polar reagents will be governed also by the polar/nonpolar nature of the surface of the particles.

The depression of either coal or its mineral gangue (nonvaluable portion of the ore) can be brought about by the use of starches, inorganic salts, or oxidizing agents. Starches and oxidizing agents are depressants for coal, whereas, inorganic salts, such as sodium silicate, are depressents for siliceous gangues.[6,10] Lime or sodium and potassium cyanides are used as depressants for pyrite. Flotability of coal is also dependent on pH, pulp density, particle size, aeration and agitation, and mixing conditions during reagentizing.

7.3   AGING EFFECTS ON COAL FLOTATION

The flotability of coals varies considerably because of their rank and petrographic constituents. However, coals of even the same rank can exhibit significant variation in flotability, depending on whether they are freshly mined or have been exposed to the atmosphere for a few hours or days. Also, flotability can vary depending on whether the coal is deep mined or surface mined. Yarar and Leja[11] have shown that coal mined at distances greater than 24.7 m were naturally hydrophobic and floated more readily throughout a pH range of 2–11 than coal samples from the surface. The surface samples were weathered and floated with collectors only if they were deslimed and the pulp pH was near 2.

These variations in flotability have been attributed to the surface oxidation of the coals that takes place even at atmospheric temperatures (see Chapter 5). The effects of oxidation have been observed on coals stored under argon and nitrogen.[12,13] The oxidation process results in the formation of hydrophilic groups at the coal surface, resulting in a reduction in its flotability. In this connection, it has been noted by Brown that exposure of freshly cleaved surfaces to air for a short time may even increase the flotability.[6] This has been considered to be due to the drying of the coal surfaces, leaving them difficult to wet and thus more hydrophobic. In addition to chemical alteration, there can be changes in the physical structure of coal upon weathering with major consequences in its processing. For example, formation of micropores and fissures observed to occur as a result of repeated drying and wetting of the outcrop coal[14,15] can lead to alterations in its wettability and thereby in its flotation and dewatering characteristics.

The products formed when coal is oxidized include organic acids, phenols, peroxides, and humic acids along with gaseous materials. The role of these products on the flotability of coal has not been adequately investigated, yet all these products can be expected to markedly affect the interfacial properties responsible for flotation. The effects of these products have always been considered to be simply the effects of "oxidation" on the flotation of coal. But the oxidation process itself has been reported to proceed in three main stages: (1) the physicochemical adsorption of oxygen on the surface of the coal, (2) the formation of humic acids, and (3) the production of soluble acids.

The physicochemical adsorption of oxygen on the coal surface is considered to proceed through, first, physical adsorption where oxygen molecules form a complex with the coal surface, and second, chemical adsorption where oxygen molecules bond with various C—H groups to form peroxides. Consequently, there is a decrease in CH_3, CH_2, and CH groups and an increase in —OH and —CO functional groups during this first stage of oxidation. As the oxidation process proceeds, much of the organic material in the coal becomes converted into alkali-soluble, acid-insoluble substances called humic acids. These acids vary considerably in molecular weight and are known to contain both carboxyl and phenolic groups. However, Gillet[16] has reported that the empirical formula of most of these humic acids is $C_{20}H_8O_8$, the main function groups being —CO, —OH, —COOH, and OCH_3.

Further oxidation will degrade the humic acids and cause the production of other acids that are soluble in either acid, neutral, or alkaline aqueous media. Wen[17] reported that these soluble acids vary from aliphatic types such as acetic acid (CH_3COOH), formic acid (HCOOH), and oxalic acid (HOOC—COOH) to aromatic types such as benzoic acid (C_6H_5COOH) and phthalic acid ($1,2-C_6H_4[COOH]_2$). Oxygen-containing functional groups control coal wettability through a balance of hydrophobic/hydrophilic sites and flotation kinetics by influencing the surface charge.[18]

The extent of surface oxidation during aging has been suggested by Vargha-Butler et al.[19] to depend strongly on the inorganic matter content as well as the dryness of the coal. These authors determined the surface tension of coal particles by the freezing point technique. They found that aging or drying did not affect the surface tension of wet-screened samples but caused significant changes in dry-screened samples. It was suggested that a water layer that remains adsorbed on the coal particles after wet-screening, owing to the inorganic matter in it, protects the wet-screened coal from being oxidized. Indeed, a wet-screened sample of one coal with a lower inorganic content did show measurable changes in its surface tension upon aging. The effect of aging on dry-screened samples was attributed to possible diffusion of low molecular weight hydrocarbons to the surface, making the particles more hydrophobic during the aging process. Drying of these aged samples was found to restore the hydrophilicity, apparently due to the removal of the hydrocarbons from the surface.

In the examination of the effects of oxidation on coal flotability, investigators have used various methods to effect surface oxidation. Sun[20] used an electrically heated glass tube, at temperatures of 150°C and above, with an inlet for compressed air, to oxidize his coal samples. He noted that the effects were related to the change of the coal surface from hydrophobic to hydrophilic in nature as the oxidation proceeded. Superficial oxidation at low temperatures was found to have little effect on the flotability of the coal when a neutral oil was used as the collector. In fact, the flotability has been reported to be improved occasionally by such low temperature oxidation. However, as the oxidation became more extensive the flotability gradually decreased (see Figure 7.2). Sun[20] attributed this effect to the accumulation in the slurry of the water-insoluble oxidation products of the coal. Certain cationic reagents were also found to be effective in the flotation of oxidized coals and coke. The flotability of oxidized coals in the presence of oily collectors as well as cationic collectors generally increased with decrease in pH. Selective flotation with sodium oleate and pine oil was also reported possible, selectivity being the highest in this case at pH 6.5. Mixtures of hydrocarbons and pine oil have been reported to be the best agent for surface-oxidized coal. Separation by agglomeration using oil can work with some fine coals. Even though the recovery using oil agglomeration can be high and moisture content low, the oil consumption of the process makes it unattractive unless the fuel value of the entrapped oil can be fully recovered in downstream operation.

Gayle et al. have also observed effects of oxidation or weathering on flotation.[21] They observed that flotation using aliphatic alcohol type frothers was adversely affected by oxidation under all conditions, whereas flotation using kerosene and pine oil was initially improved and later deteriorated. Baranov *et*

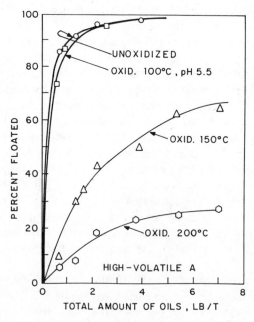

Figure 7.2   Effect of air oxidation on the flotation of −65 mesh high-volatile A coal using purified petroleum light oil (after Sun[20] reprinted with permission).

al.[22] observed that the flotation of oxidized coal decreased with increasing presence of phenolic and carboxylic hydroxyl groups in the coal. Moore[23] found flotation using methylisobutylcarbinol and kerosene to be suppressed by even a small amount of oxidation. Several investigators have reported oxidized coals to adsorb significantly higher quantities of frothers such as aliphatic alcohols; since such adsorption leads to depletion of frother required to maintain the froth stability, the flotation recovery will be poor with such coal.[21,24]

In an extensive study of the effect of gaseous oxidation, Wen[17] used a glass chamber and an oven at 125°C, with flowing oxygen, to oxidize his coal samples. He reported that an increase in the degree of oxidation increases the negative value of the zeta-potential (effective surface potential or the potential at the slip-plane of the interface) and decreases the isoelectric point (point at which the electrophoretic mobility of the particle is zero). The zeta-potentials of oxidized and unoxidized samples of vitrain, fusain, and durain are shown in Figure 7.3 as a function of pH.[17] All the samples are seen to become more negatively charged upon oxidation, the effect being most pronounced on vitrain. The effect on zeta-potential due to oxidation was found to increase with both the time and temperature of oxidation. Wen also examined the effect of oxidation on the contact angle of the coal, and he observed a good correlation between the zeta-potential and the wettability of the oxidized samples (see Figure 7.4).

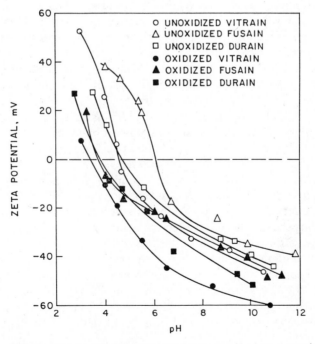

Figure 7.3 Diagram illustrating the effect of oxidation on the zeta-potential of vitrain, fusain, and durain (after Wen[17] reprinted with permission).

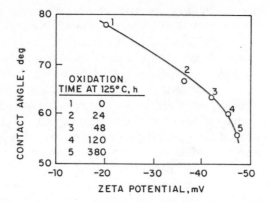

Figure 7.4 Correlation between zeta-potential and contact angle of oxidized coal samples (after Wen[17] reprinted with permission).

Most investigators, in their attempts to determine the effects of the oxidation process, subjected their samples to various chemical and gaseous means of oxidation. However, the techniques used often did not isolate and identify the effects. For example, Sun[20] and Wen[17] heated the coal samples at high temperatures in flowing air and determined the properties of the samples subsequently. Note that the heating process could conceivably produce an effect due to the removal of moisture or volatile matter.[25] Also soluble inorganics, resulting from oxidation, can produce their own characteristic effects in the flotation process.

Recently, Roberts and Somasundaran[26] have isolated the effects of oxidation by conducting control experiments in the presence of nitrogen as an inert (oxygen-free) gas. Results obtained for the flotation of coal after exposure to various gas mixtures are given in Figure 7.5. With nitrogen, only a slight increase in flotation was obtained. This was probably due to the removal of some adsorbed water from the surface of the coal. The exposure of the coal to air at room temperature had no significant effect on its flotability. At 90°C, however, there was a slight increase in flotability in the beginning and then a substantial decrease. The initial effect is possibly a result of the release of gases from the interior of the coal with a concomitant increase in hydrophobicity. As the oxidation proceeds, the direct effects of oxidation dominate those of heating. Interestingly, an oxygen–nitrogen mixture produced a drastic decrease in flotation from the beginning. This is the result of many reactive oxygen molecules adsorbing on the coal surface to produce a hydrophilic group. The difference in the effects produced by an oxygen–nitrogen mixture (O_2/N_2 ratio adjusted to approximately that of air) and air is most interesting and suggests a possible avenue to explore the mechanisms of oxidation effects on coal surface properties.

Chemical oxidizing agents have also been used to treat coal. Potassium permanganate and hydrogen peroxide are commonly used. The flotability of

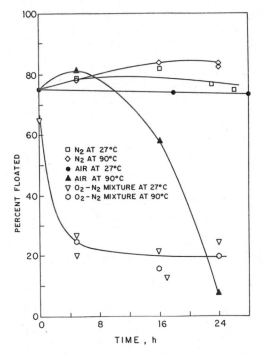

Figure 7.5 Flotation of $-65+100$ Tyler mesh high-volatile bituminous coal treated with various gas mixtures at 27°C and 90°C.

coals can be reduced or even completely destroyed by these chemicals, depending on their concentrations and the pH of the pulp. In this case also, there can be secondary effects produced by the adsorption of the inorganic species on the surface of the coal, or, in the case of hydrogen peroxide treatment, additional effects produced by gas evolution and entrapment.

Celik and Somasundaran[27] have reported that treating the coal with either an oxidizing or reducing agent, or even with an acid or alkali in concentrations in excess of 10^{-2} kmol/m^3, can bring about similar decreases in coal flotability (see Figures 7.6 and 7.7). Evidently it is not possible to attribute the above flotation depressions totally to the oxidation effects of the reagents on the coal samples. Chemicals such as $KMnO_4$ and $SnCl_2$ can be expected to exert an additional influence owing to possible adsorption or precipitation of various inorganic species on the particle surface. In fact, in this study, depression of coal with $SnCl_2$ was observed in the pH range where precipitation was found to occur. Flotation depression obtained above pH 12 can be similarly accounted for, since multivalent ions, such as calcium or iron, present in the mineral pulp should be expected to precipitate in this pH range. In this study, no simple relationship between the surface charge of the coal and its flotation was evident. It can be seen from Figures 7.6 and 7.8 that the isoelectric point of pH 2.5

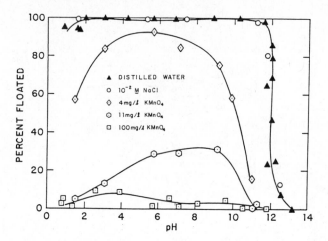

Figure 7.6 Flotation of untreated and $KMnO_4$-pretreated coal ($-50 + 100$ U.S. mesh bituminous A type) as a function of pH with methylisobutylcarbinol[9]. Reprinted with permission.

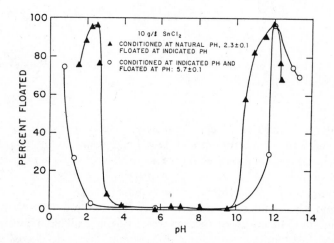

Figure 7.7 Flotation of $SnCl_2$ pretreated coal ($-50 + 100$ U.S. mesh bituminous A type) with methylisobutylcarbinol[29], Reprinted with permission).

obtained for the $KMnO_4$-treated coal is located away from the pH range where it floated most. Diaz and Hernanz[28] studied the relation between surface oxidation, flotation, and sulfur selectivity using creosote as a flotation agent and reported electrokinetic potential and contact angle to be poor indicators.

It is imporant to note that various dissolved inorganics such as Fe^{2+}, Fe^{3+}, Al^{3+}, Ca^{2+}, Mg^{2+}, Na^+, SO_4^{2-}, and CO_3^{2-} will be present in coal pulps and all these species can affect the surface properties of the particles in the pulp.

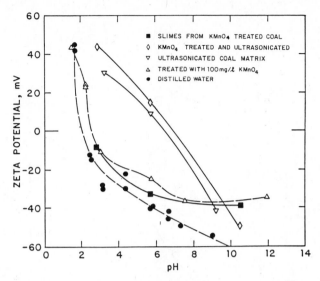

Figure 7.8 Zeta-potential of various untreated and treated coals[29] Reprinted with permission.

Flotation of coals have been shown to be affected significantly by dissolved products (See Figure 7.9).[26,29] Other minerals such as pyrite in the pulp can also be oxidized. In the case of pyrite, the soluble oxidation product is ferrous sulfate. It can adsorb on the coal particles and produce marked alterations in interfacial behavior. The magnitude of the effect of ferrous and other such ions is dependent, among other things, on the formation of their hydroxy complexes, which in turn is dependent on the pH of the pulp. Under appropriate conditions, the dissolved metallics can precipitate as hydroxides; and, if the precipitation occurs on particles of coal and other minerals, all the particles can be expected to behave as particles of metal hydroxide. Under such conditions, *selective* adsorption of surfactants on coal will not take place, and flotation separation will not be easily accomplished. In the case of amine, which adsorbs more on oxidized coal than on pyrite and silica, equal adsorption has been noted on all particles at pH values below 5 and above 7 at a level of dissolved iron likely to occur naturally in oxidized coal systems. This was considered to be the result of precipitation of iron hydroxide on all particles, causing them to behave in an identical manner. Under these conditions, oxidized and unoxidized coals also can be expected to behave similarly.

Salts, in general, have been reported to assist the flotation of coal. For example, coal can be floated in sea water without the addition of a frother.[6] Oxidized coals, however, float poorly in salt solutions.[30]

To test the possibility that the depression of chemically treated coal is due to the formation of the inorganic layers on the surface of coal, Celik and Somasundaran[27] removed the surface layers by ultrasonication and determined

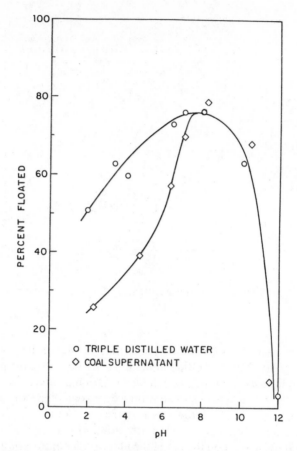

Figure 7.9 Effect of dissolved products from treated coals on the flotation of volatile bituminous untreated coal using methylisobutylcarbinol.[22]

the properties of the fines produced. Even though the isoelectric point of the fines obtained by ultrasonication of $KMnO_4$-treated coal did not correlate with that of MnO_2-treated coal, cleaning of both the $KMnO_4$-treated and untreated coal matrix by ultrasonication raised their isoelectric points to about 6.5 (see Figure 7.8); also, the flotability of the coal could be completely restored by ultrasonication. These results suggested that the contamination (not excluding oxidation) of the coal surface by the dissolved chemical is responsible for its negative surface character as well as depressed flotation.

Gogitidze and Plaksin[31] have examined the effect of exposure to air, oxygen under pressure, H_2O_2, alkaline, and neutral $KMnO_4$. Although all the treatments eventually produced a decrease in flotation, the $KMnO_4$ treatment caused a complete depression of the flotation. The flotability could be restored by boiling in water, heating in vacuum, or chemical reduction in 1 percent NaOH solution.

In the case of oxidized coals, if the oxidation is mild and limited to the surface, the normal properties can be regenerated by removing the surface layers. If the entire bulk has been oxidized by many years of weathering, as in the case of outcrop coal, the samples will respond poorly to most treatment.[15] For example, Wen[17] failed to improve the flotability of outcrop coal by grinding.

7.4 OXIDATION EFFECTS ON SEDIMENTATION AND DEWATERING

Sedimentation of fines in colloidal suspension and resultant dewatering is governed by interactions between the fine particulates, which in turn are determined by their surface properties. Therefore, aging can be expected to affect the dewatering of coal slurries. Recently, Fadaly et al.[32] have shown oxidation of coal to strongly affect the sedimentation rate and the ultimate compaction density (see Figure 7.10). These authors also observed a strong correlation between the compaction density and the zeta-potential of these coals. The effect of oxidation on the electrokinetic behavior of coal is marked, particularly during the early stages of oxidation. These changes have been attributed to an increase in acidic oxygen-containing functional groups. The effects of oxidation on sedimentation itself is suggested to result from the tendency of the system to form a clustered structure, entrapping liquid within the cluster.

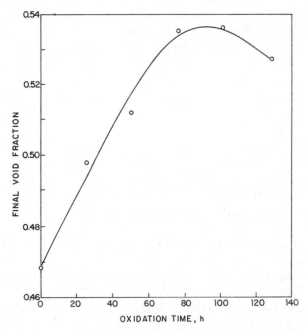

Figure 7.10 Diagram illustrating the effect of oxidation of coal on the sedimentation rate and ultimate compaction density as given by final void fraction (after Fadaly et al.[32]).

7.5 CONCLUDING REMARKS

It is clear that oxidation, wetting, and contact with dissolved inorganics during aging can drastically alter the behavior of coal during flotation, dewatering, and sedimentation. Surface properties such as zeta-potential and contact angle have also been found to undergo marked changes upon such treatment of the coal samples. Coal samples become less hydrophobic and more negatively charged upon oxidation and this is considered to be the result of its conversion to humic acid with —CO, OH, —COOH and —OCH$_3$ as the main functional groups. Chemicals are considered to oxidize the coal in a similar manner. However, such effects could be expected by simple adsorption or precipitation of dissolved inorganics on the coal surfaces. Pyrite and other minerals in the coal will also undergo oxidation and produce such species as Fe^{2+}, which can adsorb or precipitate on all particles in the pulp leading to a complete loss of selectivity in operations such as flotation.

The exact mechanism by which chemicals produce the above effects is not well established. Also, there is no agreement in the literature regarding the role of the electrochemical properties such as zeta-potential in flotation. In addition, in the case of severely weathered coal, the fissures, cracks and pores on the samples may affect the flotation process due to both the direct effect on contact angle and wettability and adsorption of frothers and collectors leading to their depletion. The precise role of the above factors in flotation has not been studied, and it is clear that systematic studies of the effect of chemical and physical pretreatment on all relevant surface properties of coal along with simultaneous tests on its behavior in such processes as flotation, sedimentation, and dewatering are necessary to elucidate the mechanisms involved in producing these effects.

REFERENCES

1. J. S. Browning, Removal of sulfur from coal by a combination of gravity and flotation methods. *SME-AIME Fall Meet.* SME Prepr. 81–318 (1981).

2. K. J. Miller, Flotation study of refractory coals. *Rep. Invest.—U.S., Bur. Mines* **8224** (1977).

3. P. Somasundaran, Interfacial chemistry of particulate flotation. *AIChE Symp. Ser.* **71**, IC (1975).

4. P. Somasundaran, The relationship between adsorption at interfaces and flotation behavior. *Trans.—Soc. Min. Eng. AIME* **241**, 105 (1968).

5. F. F. Aplan, Coal flotation. In *Coal Flotation* (M. C. Fuerstenau, ed.), Vol. 2, p. 1235. AIME, New York, 1976.

6. D. J. Brown, Coal flotation. In *Froth Flotation—50th Anniversary Volume* (D. W. Fuerstenau, ed.), p. 518. AIME, New York, 1962.

7. A. F. Taggard et al., Oil-air separation of non-sulfide and non-metal minerals. *Trans. Am. Inst. Min. Metall. Eng.* **134**, 180 (1939).

8. E. T. Wilkins, Coal preparation: Some development to pulverized practice. *Conf. Pulverized Fuel, Proc. Conf., 1974* p. 398 (1974).

9. S. C. Sun, Hypothesis for different flotabilities of coals, carbons, and hydrocarbon minerals. *Min. Eng. (Littleton, Colo.)* **6**(1), 67 (1954).

10. J. W. Leonard, *Coal Preparation*, 4th ed. AIME, New York, 1979.

11. B. Yarar and J. Leja, Correlation of zeta-potential and flotability of weathered coal. *Soc. Min. Eng., Prepr.* **82-44** (1982).

12. F. E. Huggins, G. R. Dunmyre, M.-C. Lin, G. P. Huffman, Storage of bituminous coals under argon. *Fuel* **64**(3), 350 (1985).

13. J. W. Larsen and T. E. Schmidt, Proper lab storage of coals. *Fuels* **65**(9), 1310–1312 (1986).

14. D. Chandra, Reflectance and microstructure of weathered coals. *Fuel* **41**(2), 185 (1962).

15. R. J. Gray, A. H. Rhoades, and D. T. King, Detection of oxidized coal and the effect of oxidation on the technological properties. *Soc. Min. Eng., Prepr.* **74-F-63** (1974).

16. A. Gillet, The solubility-confering oxidation of coal. *Fuel* **30**, 181–187 (19   ).

17. W. W. Wen, Electrokinetic behavior and flotation of oxidized coal. Ph.D. Thesis, Pennsylvania State University, University Park, 1977.

18. D. W. Fuerstenau, J. M. Rosenbaum, and J. Laskowski, Effect of surface functional groups on the flotation of coal. *Colloids Surf.* **8**(2), 153–173 (1983).

19. E. I. Vargha-Butler, T. K. Zubovits, R. P. Smith, I. K. L. Tsim, H. A. Hamza, and A. W. Neumann, Effect of aging on the surface characteristics of coal powders. *Colloids Surf.* **8**, 231–247 (1984).

20. S. C. Sun, Effects of oxidation of coals on their flotation properties. *Trans. Am. Inst. Min. Metall. Eng.* April, pp. 396–401 (1954).

21. J. B. Gayle, W. H. Eddy, and R. G. Shotts, Laboratory investigation of the effect of oxidation on coal flotation. *Rep. Invest.— U.S., Bur. Mines* **RI-6620** (1965).

22. L. A. Baranov, O. V. Budrina, Sh. Sh. Gubaidulin, and M. V. Romanovskaya, Assessing the oxidation of coal by a flotation procedure. *Coke Chem. U.S.S.R.* **10**, 15 (1970).

23. E. Moore, cited by Wen,[17] p. 39.

24. N. S. Vlasova, V. G. Kushnereve, and A. Ya. Barsukova, Influence of the degree of oxidation of coal on the adsorption of n-aliphatic alcohols. *Coke Chem. USSR (Engl. Transl.)* **6**, 5 (1973).

25. J. S. Gethner, Thermal and oxidation chemistry of coal at low temperatures. *Fuel* **63**(10), 1443–1446 (1985).

26. C. E. Roberts, Jr. and P. Somasundaran, unpublished results.

27. M. S. Celik and P. Somasundaran, Effect of pretreatments on flotation and electrokinetic properties of coal. *Colloids Surf.* **1**, 121–124 (1980).

28. J. M. Diaz and J. F. Hernanz, Flotation and sulfur selectivity of a bituminous coal with different degrees of oxidation. *Fuel* **63**(10), 1466–1468 (1984).

29. M. S. Celik and P. Somasundaran, Effect of multivalent ions on the flotation of coal. *1984 AIME Annu. Meet.* February (1984).

30. A. F. Baker and K. J. Miller, Zeta potential control: Its application in coal preparation. *Min. Congr. J.* **54**, 43–44 (1968).

31. T. A. Gogitidze and I. N. Plaksin, The influence of the surface oxidation of some bituminous coals upon their flotability. *Chem. Abstr.* **51**; 126a (1957); cited by Wen,[17] p. 35.

32. O. A. Fadaly, J. C. T. Kwak, H. A. Hamza, and A. M. Al Taweel, The role of surface forces in the compaction and dewatering of coal fines. *Colloids Surf.* (to be published).

Commentary

Aging and weathering of coal produce undesirable consequences in processes such as coking and beneficiation, as detailed in Chapters 6 and 7. This survey is extended in Chapter 8 by Robert Miller, who examines the impact of storage and treatment on chemical processes such as coal conversion to liquids. It is clear that aging and weathering, characterized by both physical and chemical changes in coal properties, have generally deleterious effects on coal utilization processes. An assessment of the coal processing–aging correlation is a requisite for making appropriate choices in attempting to optimize factors such as storage, sample selection, and pretreatment.

8

EFFECTS OF PREPARATION, PRETREATMENT, AND OXIDATION ON COAL CONVERSION

R. N. Miller
Air Products and Chemicals Inc.

8.1 INTRODUCTION

8.1.1 Theoretical Basis of Coal Conversion

Hydrogen Transfer. The ultimate aim in coal liquefaction is the attainment of a high-yield, high-quality distillate product obtained under minimal operating pressure and with selective and thereby minimal hydrogen consumption. Thermal conversion of coal to low-molecular-weight products depends on the efficiency of supplying hydrogen to passivate reactive sites formed during high-temperature treatment. As pyrolysis temperatures (typically 350–450°C) are approached, coal undergoes an initial depolymerization; bonds are thermally cleaved and reactive sites are produced. The transfer of hydrogen to these sites at a rate fast enough to avoid repolymerization is the fundamental problem of coal liquefaction. If sufficient hydrogen is not available to passivate the reactive sites, retrograde reactions involving heteroatoms or carbon–carbon reactions predominate. This results in the formation of condensed products of high-molecular weight at the expense of distillate products. If the conditions are too severe, the reactions result in an unnecessarily high consumption of hydrogen to form high yields of saturates accompanied by a high production of low-valued hydrocarbon gases.

Understanding and controlling the hydrogen transfer reactions has been of fundamental concern in the attempt to develop novel coal conversion processes. Indeed, many process configurations have been proposed and have incorporated a broad range of reactions and reaction conditions to improve conversion. The most pronounced of these efforts at optimization has been focused on catalysis, for the presence of a catalyst significantly lowers the severity of hydrogen-transfer reactions and increases the specificity of the hydrogen transfer. Higher oil yields are achieved in the presence of the proper catalyst.

The efficiency of catalysts in liquefaction is limited by the inability to obtain an effective dispersion of the catalytic material into the coal structure. Prior to 1965, exhaustive investigation of coal/catalyst systems had been done, and as was pointed out in a review by Hawk and Hiteshue,[1] "almost every element in the Periodic Table has been tried somewhere as a catalyst or as a component of a catalyst in coal liquefaction." Despite all of this effort in catalytic liquefaction the hydrogenation reactions were largely being controlled by the type of coal and not necessarily by the type of catalyst. In these earlier studies of coal, hydrogenation emphasis was generally placed on the *kind* of catalyst and its *concentration* rather than on how it was added—most catalytic materials were simply applied by dry-mixing them with the coal. When the coal is heated to pyrolysis temperatures, the dry-mixed catalyst is unable to participate efficiently in the initial hydrogen-transfer chemistry.

The hypothesis that the catalytic material should be brought into intimate contact on a molecular scale with the reactive moieties within the coal structure is now recognized as fundamentally important. Catalysts supported on silica- and/or alumina-based substrates or catalysts as salts of metals that are simply mixed with the coal exhibit their maximum effect only after relatively long residence times or at exceptionally high, uneconomical concentrations. Mixed as solids with the coal, they are not able to diffuse within the solid matrix of the coal structure prior to the onset of the primary liquefaction reactions. Consequently these types of catalysts will not have a significant effect on the kinetics of hydrogen transfer at the onset of liquefaction when the hydrogen must be transferred rapidly to the reactive sites to minimize competing reactions that lead to insoluble carbonaceous residues.

More recently, catalyst addition via impregnation methods to achieve better dispersion has been the basis of attention in catalytic liquefaction. Molecular dispersion or a dispersion of the catalytic material deep into the coal matrix will increase the probability for catalyst participation in primary reactions and enhance the conversion efficiency. Even dispersion of a catalyst precursor (that must first react with another substrate to form the catalytically active species) will show improvement compared to addition of standard supported catalysts. Repeated studies[2,3] over the last decade have conclusively supported these ideas.

It would follow that future research in coal liquefaction should continue to seek improved methods for achieving efficient dispersion of both hydrogen-transfer reagents and catalysts. Fundamental to studies of this nature is the need to develop a better understanding of the relationship of the physical and chemical features of the coal during liquefaction.

Modification of Coal Properties. Our understanding of the complex nature of coal and its behavior at high temperatures has progressed significantly over the last several decades. It has become increasingly evident that coal structural characteristics, including the presence of inorganic constituents, largely control liquefaction potential. Because of the diverse heterogeneous nature of coals, liquefaction yields have been directly related to generic origin, and these have allowed a broad categorization of coal behavior in hydrogen-transfer reactions.[4-9,10] Individual liquefaction yield structures have been attributed either directly or indirectly to a number of critical fundamental properties, as summarized in Table 8.1. For minimizing hydrogen consumption while maximizing conversation efficiency, Duriel[11] has stated that the most desirable characteristics are a high content of reactive macerals having a high hydrogen content, a high content of aliphatic material, and relatively simple aromatic structures, low oxygen content, and low ash, though others[9,10] would also add a high amount of pyrite and sulfur to the list. High-sulfur, low-rank bituminous coals are especially favorable for liquefaction and generally give the highest yield of distillates.[12]

Treatment of coal by a chemical or physical process intended to modify one or more critical properties will have a pronounced effect on the performance in liquefaction. Consequently, pretreatment technology aimed at modifying coal characteristics for both thermal and catalytic coal conversion have received much attention. Structural studies have principally focused on obtaining better dispersion and higher activities of both catalysts and hydrogen-transfer solvents. The most important techniques involving chemical pretreatment and physical beneficiation to effect positive changes in liquefaction behavior is a major aspect of this review.

Modification of critical properties detrimental to coal conversion can also take place during the routine handling and processing of coal. Of special concern are changes related to the process of dehydration, air oxidation, weathering, pulverization, storage, and beneficiation. These factors were discussed in earlier chapters. This chapter considers how and to what effect coal

TABLE 8.1   Fundamental Properties Important in Coal Conversion

Property	Desired Level	Effect on
Ash content	low	operations and handling
Pyrite content and distribution	high; fine particles	catalysis
Oxygen content	low	hydrogen consumption; gas make
Reactive macerals	high	liquid yield
Aliphatic character	high	liquid yield and quality
Extractables	high	liquid yield and quality
Particle size	very fine	operations
Moisture content	low	thermal efficiency
Hydrogen content	high	liquid yield and hydrogen consumption
Reflectance	medium rank	liquid yield

processing affects conversion and how important the effects of long-term storage are. Another important aspect, the detrimental effects on conversion resulting from physical and chemical manipulation of coal, is also discussed.

8.1.2 Coal Pretreatment Technologies

In a broad context, the technology of pretreatment can be defined as that which is done to a coal in a preliminary stage of a conversion process and is intended to directly improve yield structure and/or conversion efficiency. The technology of *coal pretreatment* rightly includes both chemical and physical modification, though pretreatment is often thought of only with regard to physical modification. In not all cases does the pretreatment technology produce beneficial effects on the conversion; the parallel attempt to increase the efficiency sometimes results in detrimental structural changes or loss of the reactive constituents. Within this broader context, pretreatment of coal can be conveniently categorized into the three following groups: physical coal beneficiation, chemical pretreatment, and structural modification and oxidation.

Physical Coal Beneficiation. Beneficiation by physical methods is the most common pretreatment technology considered in coal conversion and many of the techniques are already commercially established in coal processing. By utilizing a variety of coal-cleaning methods, physical beneficiation can produce significant changes in petrologic characteristics, which in turn can have a substantial effect on liquefaction. It is possible, using physical methods such as density separation, floatation, and sizing, to achieve a fractionation of both the inorganic constituents (the minerals) as well as the organic constituents (the macerals) that will result in marked changes in conversion. The most noted is the removal of the catalytically active component, the pyrite.

Concentrating the most reactive organic entities (the vitrinites and the exinites) and thereby eliminating the inert constituents (the minerals and the inertinite macerals) from a feedstock via physical methods can have a potentially large effect on the overall process economics. These effects include improving overall mass throughputs at higher yields, operating with smaller reactors, and reducing downstream solids handling and waste disposal needs.

Chemical Pretreatment. Chemical methods of beneficiation have also been considered as a means of improving coal conversion. Most applicable chemical technologies do not share the same level of commercial development as the physical processes and have been applied only in fundamental conversion studies. Chemical pretreatment has been used to manipulate both the inorganic as well as the organic constituents of coal, though the major thrust of past research has been on inorganic matter alteration and catalyst impregnation. Emphasis has also been placed on chemical treatment methods that utilize the interactions of various organic solvents with coal. The relevant chemical pretreatments covered are listed in Table 8.2.

**TABLE 8.2 Pretreatment Technologies
for Coal Liquefaction**

Physical Beneficiation

Fractionation of minerals and macerals
Moisture removal
Comminution

Chemical Pretreatment

Thermal pretreatment
Acidic-gas treatment
Solvent presoaking
Solvent extraction
Coal modification for catalyst impregnation
Mineral acid pretreatment

Structural Modification

Oxidative pretreatment
Functional group reactions
Irradiation

Structural Modification and Oxidation. When the chemical treatment produces an irreversible change in the three-dimensional structure, it is possible that the change will have a large impact on coal conversion. As evidenced by the extensive literature, coal scientists in their attempts to understand the structure of coal have been studying the reactions of coal with many types of organic reagents for years. Despite this effort, very little has been directly applied to coal conversion nor have any reactions proven to be practical for liquefaction. Functional group modification through oxidative or reductive techniques has received the most attention in attempts to improve conversion and/or hydrogenation. In certain cases, the chemical reactions are so severe as to nearly completely degrade the coal and leave it essentially soluble. Such treatments are beyond the scope of this chapter.

Preoxidation has been viewed with the most interest because of its potentially detrimental effect on coal conversion. This area includes studies of weathering, aging, oxidative effects through preparation, and various preoxidation techniques.

8.1.3 Defining Coal Conversion

It is apparent in surveying the literature of coal liquefaction that there is no uniformity or consistency among different research groups in evaluating conversion, and it is very difficult to interpret many of the results. Nearly every group has independently developed its own methodology for determining conversion using many different types of liquefaction reactors, experimental conditions, product work-up methods, and manner of reporting conversion data. The criteria most commonly employed to evaluate conversion yields are distillability and solubility. A number of distillation techniques are available, though the most common are those that have been standardized by ASTM. Conversion yields based on distillation are usually reported as the fraction distilled between two temperatures with end points for maximum oil yields ranging from 650°F to as high as 1000°F. If *light* liquids are the principal products from the liquefaction, then distillation serves as the best measure of oil yield.

Because of the relative simplicity of the tests and because a substantial fraction of coal conversion products are nondistillable, coal conversion is more often measured by solubility of the products in various solvents. The problem is that the choice of solvents has never been standardized. Most investigators consider solubility in pentane, hexane, heptane, or even cyclohexane as approximating distillate oil yields, though the actual yields may vary between laboratories depending on how the solvent extraction is conducted. Solubility of the nonoil or nondistillable components in benzene is by definition the asphaltene yield and defined as such in many studies. However, benzene solubility is more often used as a measure of total conversion, where the benzene-insoluble residue is used to conduct the stoichiometry. Ethylacetate and toluene are sometimes used to report coal conversion in place of benzene.

Pyridine solubility is the most common approach used to measure total conversion yields, though many other solvents such as cresol, tetrahydrofuran, methanol/MEK, and benzene/ethanol have been considered. In any case, the differential solubility of the coal liquids and subtle variations in the methods of extraction often result in major differences in conversion yields between solvents. In discussing the effects of pretreatment on conversion in this chapter, caution has been exercised in drawing parallels or differences where conversion criteria differed. Where possible, conversion data are accompanied by the type of method.

8.2 PHYSICAL BENEFICIATION

8.2.1 Physical Coal Cleaning for Coal Conversion Processes: Fractionation of Macerals and Minerals

Current techniques employed for mining coal often produce a run-of-mine (ROM) product that does not meet product specifications such as size, ash, or sulfur content. Additional mineral processing techniques such as crushing and sizing or physical separation are necessary to prepare the product for use. In these processes, the objective traditionally has been to remove the mineral matter to produce a feedstock that is higher in heating value, lower in ash, and lower in sulfur content and, therefore, higher valued as a combustion or metallurgical coke feedstock.

Recent work[13] has clearly shows that the preferred feedstocks to coal liquefaction plants will necessarily be physically beneficiated. The benefits include lower mineral-matter throughput with associated reduction in abrasion, reduced loading on solid–liquid separation systems, less sulfur to handle and thereby lower hydrogen demand, and a more consistent feed. Gray et al.[14] have recently studied the economic trade-offs of coal cleaning for direct liquefaction and concluded that the economic benefits of cleaned coal more than offset the costs of the coal preparation. From both an economic and an operations standpoint, cleaned coals appear to be the preferred feeds in coal liquefaction.

In considering cleaned coals for liquefaction, the question arises as to whether the reactivity would be significantly diminished as a result of removing catalytic minerals from the feed. It has already been shown by Anderson[15] and Garg et al.[3] that the addition of pyrite or other particulate iron minerals at high levels (above 3 percent total iron on coal) to low iron-containing eastern bituminous cleaned coals (less than 1 percent total iron on coal) has a significant effect on improving the distillate yields. Therefore, one would predict that removal of pyrite during cleaning operations would be detrimental to liquefaction. Killmeyer[16] recently reviewed some of the research efforts aimed at defining this effect. Experiments at the Pittsburgh Energy Technology Center[17] and at Hydrocarbon Research Inc.[18] on a run-of-mine and cleaned Kentucky #11 coal showed comparatively higher liquid yields and some improvement in product quality when using ROM coal. Following these works, Mazzocco et al.[19] ran a continuous experiment under SRC-II conditions using the same coal but added pyrite to the cleaned coal to bring it up to the level of that of the original ROM coal. The results showed that the product slate for this coal was similar to the ROM coal.

Hoover[9] investigated the influence of pyrite levels in cleaned coals for SRC-I processing. He showed that the mode of occurrence rather than the total amount of the pyrite was of greater importance. The beneficial effects of the dispersion of iron catalysts have also been confirmed by Anderson[15] and Garg et al.,[2,3] who demonstrated that the addition of iron as either pyrite or its oxidized (weathered) forms, the iron sulfates, had little significance on the liquefaction compared to how the iron compound was dispersed in the system.

Garg et al.[3] also reported the effect that different types of coal cleaning had on the liquefaction. They tested a coal cleaned by oil agglomeration to a moderate ash level (10.3 percent) but with a high level of retained pyrite (5.1 percent). The oil-agglomerated coal showed superior liquefaction reactivity to either the ROM sample or a cleaned coal with pyrite added. Their data suggest that it is possible to use selective processing techniques to tailor a desired product slate.

Coal cleaning acts to shift the maceral composition in addition to changing the inorganic-matter content. It was demonstrated by Benedict[20] and Bayer[21] that reactive macerals are concentrated by gravity techniques into the cleaner coal fractions, whereas inert macerals concentrate in refuse or middlings

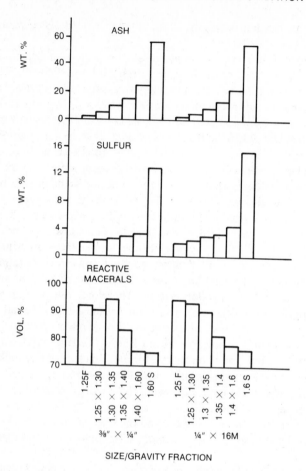

Figure 8.1 Physical beneficiation of components important in liquefaction of an Illinois #6 coal.

fractions. Hoover and Davis[22] showed by automated microscopy on a diverse set of coals that vitrinite content was inversely proportional to the specific gravity of separation. As an example of the magnitude of the effect, washability data from Spackman et al.[23] for an Illinois #6 bituminous coal have been plotted in Figure 8.1. It is evident that both particle size and density separation produce significant variation in maceral and mineral content. Hoover and Givens[24] have shown that the concentrating effect is commercially feasible. Conventional western Kentucky cleaning plants already enrich the clean coal by as much as 2–3 percent total reactives.

It is clear that physical beneficiation can be used to prepare coal liquefaction feedstocks using current technologies for modifying both the organic and the inorganic characteristics. The washability properties of the coal, as well as the specific technology used, will have an impact on coal liquefaction.

8.2.2   Moisture Removal

The preparation of coal prior to liquefaction calls for a reduction in the as-received moisture content to levels as low as 3–5 percent. The presence of large amounts of moisture in coal considerably lowers the overall thermal efficiency of a process because of the energy requirements to remove moisture from either the feed or product streams. Coals with high seam moisture such as lignites are not desirable liquefaction feedstocks. The presence of a high water-vapor pressure during the operation of a liquefaction reactor dilutes the hydrogen, thereby requiring a much higher pressure operation to maintain the optimal hydrogen partial pressure. The removal of moisture to as low a value as economically possible prior to liquefaction undoubtedly will improve a process, but removal of moisture can also be accompanied by oxidative effects detrimental to coal conversion. Proper conditions for drying the coal must be maintained.

Coals are dried prior to liquefaction using commercially available thermal drying technology often integrated with pulverizing and sizing operations. The coal is heated to temperatures ranging between 50–200°C in the presence of either air, steam, flue gases, or in some cases under a nitrogen or inert gas blanket. When coals are dried in air or any oxygen-containing environment, there is ample evidence to show that thermal drying procedures appreciably change critical physical and chemical characteristics. The magnitude of the effect is directly dependent on the temperature. These changes in turn have been shown to seriously affect liquefaction behavior.

The detrimental effects due to thermal drying are most serious for high moisture-containing, low-rank coals of more than 20 percent moisture content. They must be dried substantially before they are suitable as liquefaction feedstocks. Hence, most of the research has focused on the effects on drying of low-rank coals. Coronauer et al.[25–28] studied the effects of thermal drying (see Figure 8.2) on a Wyoming subbituminous coal (Bell-Ayre). They reported that drying in air resulted in a significant reduction in liquefaction product yield when compared with the liquefaction of as-received coal. The detrimental effects of air oxidation accompanying the drying operation are well known and are discussed in more detail. It was also found that drying resulted in a decrease in conversion even when the coal was predried in nitrogen (Figure 8.2).

Studies on the physical constitution of coal by Gauger[29] showed that the gel-like structure of low-rank coals is irreversibly altered by drying (Figure 8.3). Dried low-rank coal has been shown[30] to exhibit a significantly decreased surface area and a comparatively lower equilibrium moisture content than freshly mined coal, indicating a partial collapse of the pore structure. As illustrated by the work of Gorbaty,[31] Figure 8.4, the effect of drying can seriously retard absorptive properties. Such changes would contribute to the observed decrease in liquefaction yields as the dispersion of the donor-solvent and/or catalyst would be subsequently affected.

In actual pilot plant operation of the EDS Process[32] using both Wyodak subbituminous and Texas lignite, lower than expected coal conversions were

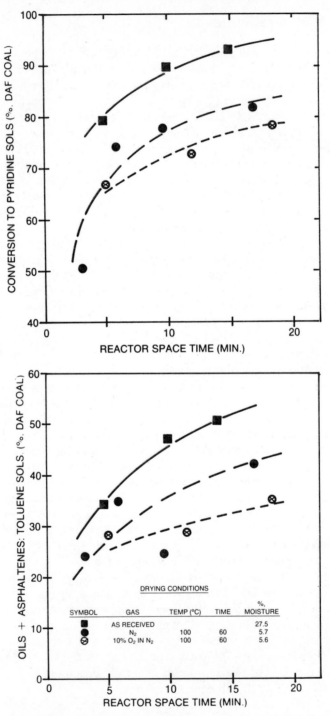

LIQUEFACTION AT 450°C; 1750 PSIA H₂; COAL:SOLVENT = 1:1.5

Figure 8.2 Effect of drying subbituminous coal on conversion.

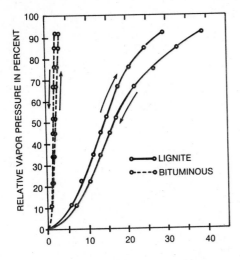

Figure 8.3 Hysteresis effect of water adsorption/desorption in low-rank coal.

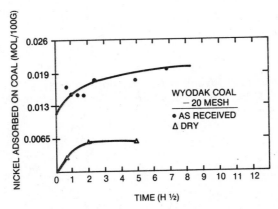

Figure 8.4 Change in the metal adsorption capacity of subbituminous coal caused by drying.

observed when the coals had been dried in a gas-swept mill prior to feeding to the liquefaction process. Subsequent studies were conducted in bench-scale reactors to test the magnitude and variation of effects due to predrying; drying temperatures were varied between 0 and 315°C and times were varied from 0 to 10 minutes at 315°C. The Exxon bench-scale data, Figure 8.5, demonstrates a linear conversion debit with temperature that begins to appear between 90–15°C. In a catalytic liquefaction experiment using brown coal with Fe/Mo, Konyashima and Nikiforova[33] found that an increase in oxygen during high-temperature steam drying produced a major decrease in the yield of liquefaction products.

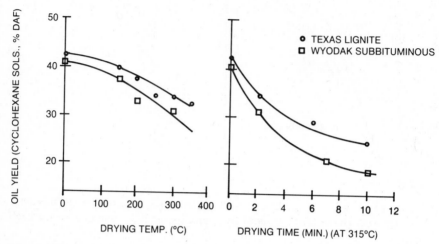

450°C; 40 min.; SOLVENT; 1500 PSIG H$_2$

Figure 8.5 Influence of drying conditions on the liquid yields from liquefaction of low-rank coals.

TABLE 8.3 Effect of Air-Drying Kentucky #6/11 Pyro Mine Bituminous Coal on Coal Liquefaction[a] Performance

Percent DAF Coal	Drying Conditions	
	Nitrogen/Argon	Air[b]
Total Conversion	90.1	91.9
Product distribution		
Gases	6.8	6.2
Water make	1.6	3.0
Oils	7.6	9.3
Asphaltenes	36.2	31.9
Preasphaltenes	38.0	41.5
IOM	9.9	8.1

[a]Liquefaction at 454°C; 2000 psig H$_2$; 30 minutes; hydrogenated creosote oil: coal = 3:2.
[b]Ground in commercial GSM at approximately 82°C.
Source: R. W. Skinner and E. N. Givens, *Effects of Accumulated Solids and Solvent IBP on Dissolver Performance, SRC-I Q. Tech. Rep., DOE/OR/03054-4. Vol. 2. U.S. Department of Transportation, Washington, D.C., 1981.*

Studies to compare liquefaction of dried versus as-received coals have not been done for bituminous coals. It may be presumed that the effects would be negligible for the higher-ranked coals. Skinner and Givens[34] report data for liquefaction of predried Kentucky bituminous coal to compare the effects of air and nitrogen drying. Their data, reassembled in Table 8.3, indicate no significant effect on total conversion to pyridine soluble material. They note that the SRC product from the air-dried coal contained more oxygen. In their experiments, a multiplicity of causes are superimposed (comminution, drying, and oxidation); thus, it is difficult to attribute the effects to drying alone. Moreover, their relatively severe reaction conditions (488°C, 2 hours) produced conversions in excess of 90 percent where the effects of drying would not be discernible.

Alternate methods of removing moisture from low-rank coal prior to liquefaction have been considered. One is the drying of the coal after it has been contacted with a hydrogen donor solvent to avoid the physical-state changes that occur during simple thermal drying. Accordingly, several novel drying processes have been described.[35-37] Wet coal is mixed with high-temperature organic solvents to form an oil-slurry from which the excess moisture is evaporated. The remaining dry coal-solvent slurry is passed on to liquefaction processing. In studying this dehydrating technique, Komiyama et al.[37] showed that the degree of liquefaction of an Australian brown coal treated in this manner was about 5 percent higher than that of air-dried coal. They suggested that the effect was largely due to a reduction in surface oxidation because of the intimate presence of the donor solvent prior to thermal drying, and the drying operation allowed ample time for the solvent to penetrate before initiating the liquefaction reaction.

Another technology for preparing dried low-rank coal for liquefaction without detrimental effect utilizes the principle of azeotropic distillation.[38-41] Wet coal is slurried with a miscible organic solvent, and the bulk of the water is displaced by evaporation at the azeotropic boiling point of the water/solvent system. For example, Dolkmeyer et al.[39] dried low-rank coal to less than 1 percent moisture content by mixing the coal with a recycle solvent and xylene. Following evaporative displacement of the water/xylene mixture, the dried coal/recycle-solvent slurry is then passed onto liquefaction. In another example, Greene[41] dried lignite to less than 3 percent moisture by using an azeotrope of benzene and water.

8.2.3 Coal Comminution as a Pretreatment for Liquefaction

Coal Pulverization Studies. Coal comminution is widely practiced in the preparation of coals for liquefaction, not only to improve transport properties of the coal/solvent slurry but to prevent operational problems. Coal pulverization has also been used as a pretreatment to improve the dissolution behavior in certain liquefaction reactors (or preheaters). To the surprise of a number of investigators, the efficiency of coal liquefaction was found not to depend on differences in the particle-size distribution. Whitehurst et al.[42] observed no significant mass-transport limitations between typical feedstock particle sizes ranging from 45–75 μm (200–325 mesh) to as coarse as 425–600 μm (30–40 mesh) in short contact time liquefaction of Illinois #6 bituminous coal. In a much earlier study, Curran et al.[43] drew a similar conclusion using Pittsburgh bituminous coal. Even as early as 1944, Rank[44] reported only marginal effects of fine grinding on liquefaction.

In other work, Yoshida et al.[45] proposed that particle size effects were negligible in liquefaction because of coal plasticity. That is, if the coal becomes plastic and swells when heated, the integrity of the physical structure is destroyed and particle-size effects diminish. In lower-rank coals where this behavior is absent, the effect of particle size and internal surface area on liquefaction is noticeable even though small. In contrast to pulverizing coal in the common range of 70 μm (200 mesh) very little is known of the effects of ultrafine grinding down to colloidal size (2 μm). However in a 1970 patent, Riedl et al.[46] claim that ultrafine grinding does markedly enhance conversion yields.

Detrimental Effects of Grinding on Coal Conversion. Possible detrimental effects on conversion can be brought about as a result of pulverizing, the most important being oxidation of the freshly generated surfaces. The consensus of those who have studied the effects of oxidation is that fresh, unoxidized surfaces generated during fine grinding readily adsorb oxygen from air. If the temperature is sufficiently high, the adsorption may lead to irreversible chemical changes in the coal structure. This may lead to significant losses in conversion and extra hydrogen consumption to remove the oxygen if the oxidation is severe enough.

In a carefully controlled liquefaction experiment (by this author) designed to test the effect of pulverizing coal in air at 180°C against pulverizing the coal under an inert condition (helium atmosphere at 0°C), coarse samples of Illinois #6 coal were pulverized under the above conditions for 60 minutes in a pressurized grinding bomb prior to liquefaction. The extent of oxygen adsorption during air grinding (as determined by a direct gas analysis of the atmosphere in the bomb during the grinding step) is shown in Figure 8.6. Following grinding, liquefaction solvent was added to both the air- and helium-ground samples. They were then liquefied according to the conditions in Table 8.4. The air-ground coal showed a marked comparative decrease in both the oil yield (as pentane solubles) and the total conversion (to MEK/methanol solubles). It was concluded that the uptake of oxygen was alone responsible for the observed differences. This was further substantiated by the *negative* oil calculated by difference, a negative yield indicating that solvent had been incorporated with the coal. Hence, it is a strong presumption that oxidation was the inherent cause of the solvent incorporation.

To minimize the effects of surface oxidation during grinding and to further enhance the surface reactivity toward hydrogenation, other investigators have

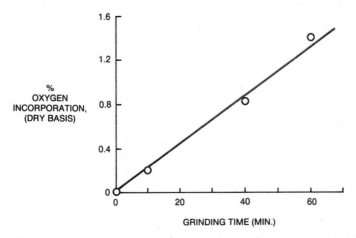

Figure 8.6   Incorporation of oxygen from grinding Illinois #6 Burning Star coal in the presence of air at 180°C (3 g, 20 × 100 mesh coal).

TABLE 8.4   Effect of Air-Grinding at Elevated Temperature on the Liquefaction[a] of Illinois #6 Burning Star Coal

Percent, DAF Coal	Grinding Conditions[b]	
	Helium	Air
Total conversion	78.3	66.5
Product distribution		
Gases	6.8	7.4
Oils	13.3	−3.9
SRC	58.1	63.1
IOM	21.7	33.5

[a]425°C; 1 hour; 850 psig H_2 cold; solvent : coal = 2 : 1.
[b]Helium grind at 0°C; air at 120 psi and 82°C.

pulverized coals under a variety of conditions, including cryogenic nitrogen[47] and inert gases.[48] Most of these techniques are too costly to be considered in commercial coal conversion applications.

Solvent-aided Grinding.   Wolk[49] considered process solvents as comminuting vehicles and discussed the use of a pretreatment reactor for soaking coarse coal prior to liquefaction in ebullated bed catalytic systems. It was claimed that the solvent acted to reduce the particle size within the slurry and therefore facilitated control of the ebullated bed. Skidmore and Kanya[50] also studied fragmentation of coal wherein the coal was pretreated at 200–300°C

with a hydrogen-donor solvent. They established that the efficiency of comminuting the coal in this manner approached that of mechanical comminution.

Nelson[51] and Chang et al.[52] discussed a speculative basis for distinct
chemical changes that could occur during coal liquefaction following solvent-
aided grinding. In a pulverizer operated under conditions of high kinetic energy,
it is possible that weak chemical bonds may rupture along the planes of shear in
the fracturing of a particle and thus provide reactive sites for radical acceptors.
Because of improved solvent contacting and exclusion of oxygen (a radical
acceptor molecule), there could be an improved transfer of hydrogen. Chang et
al.[52] indeed demonstrated in tubing-bomb experiments that the effect of
pregrinding coal by hand in the presence of tetralin, the donor solvent, as
opposed to dry grinding in air, had a pronounced effect on conversion of
Kentucky bituminous coal; total conversion to cresol solubles showed an
increase of 8 to 10 percent. The effect was as great as 20 percent with a Wyoming
subbituminous coal (Amax). On the other hand, recent work by the author,
using a Wilsonville (SRC-I) recycle solvent, showed that the conversion of
Illinois #6 coal was not improved by solvent-aided grinding. Results are shown
in Table 8.5 for Illinois coal preground to -200 mesh in an enclosed ball mill
under dry helium and under solvents. Conversion to oils (measured as pentane
solubles) is actually lower when solvent grinding is used and conversion remains
about the same, contrary to Chang's result. These yield differences have been
attributed to interactions occurring between the coal and the heteroatom
constituents in the recycle solvent; tetralin, with no heteroatoms, was used in
Chang's experiment.

Advanced Comminution Technology. Other strategies for achieving coal
comminution in the presence of a solvent have been discussed. Sellers[53] mixed
coarse coal with a light hydrocarbon oil to form a fluid suspension. When the
mixture was heated rapidly to above the vaporization point of the solvent, a

**TABLE 8.5   Effect of Solvent-aided Grinding on the
Liquefaction[a] of Illinois #6 Burning Star Coal**

Percent, DAF Coal	Pulverization Conditions	
	Dry Helium	Process Solvent
Total conversion	78.3	81.3
Product distribution		
Gases	6.8	6.3
Oils	13.3	8.2
SRC	58.1	66.9
IOM	21.7	18.7

[a]425°C; 850 psig H_2 cold; 1 hour; solvent:coal = 2:1.

significant reduction in particle size was obtained. The comminuted coal was mixed with hydrogen donor solvent and heated at liquefaction conditions.

In a different approach, Winkler[54] suggested that Kentucky bituminous coal can be ground in a solvent containing certain metals such as Fe, Al, or Zn. A slight conversion of the coal at temperatures as low as 300°C can be effected after such treatment. The experiments were performed by charging and grinding predried Ky. coal with either 13 weight percent Fe, 7 percent Al, or 20 percent Zn metal to a ball mill reactor. The temperature was controlled at 300°C by the addition of process solvent. After treatment for 6 hours, the conversion (based on pyridine solubility) was 85, 88, and 95 percent, respectively.

More exotic comminution technology has been considered for coal conversion. One such technique is *jet milling* where coal particles are entrained in a gas or vapor stream at high velocity and directed to impact a solid obstruction. Nelson[51] and others[37] have studied the use of nozzle devices to effect pulverization by achieving a high-energy impact after which the coal is brought into immediate contact with a solvent in a high pressure zone.

A related new technology referred to as *explosive comminution* appears to be applicable as a pretreatment for coal conversion. Strong internal stresses within the micropores of the coal are created by forcing a fluid into the pores at an elevated temperature and at a very high pressure (well beyond supercritical conditions). The system is then rapidly depressurized. The expanding fluid causes the particles to *explode* into ultrafine powders. Massey et al.[55-57] used supercritical methanol as the comminuting agent for explosive shattering of coal. He demonstrated, using Pittsburgh seam and Illinois #6 coals, that methanol comminution altered a number of important characteristics as evidenced by the increased solubility in organic solvents. The comminuted coal exhibited two to six times greater solubility in methanol than similarly sized coal prepared by ball milling. Although these methods appear to be promising from a technical point of view, the high pressure required makes them less feasible commercially.

8.3   CHEMICAL PRETREATMENT

8.3.1   Thermal Pretreatment

Pretreatment at elevated temperatures has been used to effect chemical changes in coal prior to liquefaction. There is ample evidence[58-60] to suggest that thermal reactions of certain functional groups, mainly oxygen containing, are significant at elevated temperatures (but below that normally considered for liquefaction, i.e., < 350°C). Hupfer[61] demonstrated that thermal pretreatment of highly caking coals resulted in enough preswelling to improve operability in the preheater stage of a liquefaction process.

To study the direct effect on liquefaction, Whitehurst et al.[42] thermally pretreated noncaking bituminous and subbituminous coals in donor solvents

TABLE 8.6 Effect of Thermal Pretreatment of Coal on Short Contact Time Liquefaction[a]

	Ky. #9/14 Bituminous		Wyodak Subbituminous	
	No Pretreat	249°C He; 17 Hour	No Pretreat	249°C He; 6 days
Carbon, percent DAF	79.6	78.6	72.9	74.3
Oxygen				
percent dry coal	10.0	9.2	17.1	21.3
Percent of SRC	6.8	8.5	10.2	12.3
Percent of IOM	3.9	12.5	—	—
Conversion, percent DAF	80.6	69.9	60.3	62.2
Products, percent DAF				
Gases	NR	NR	NR	NR
Oils	13.0	9.6	14.2	13.7
SRC	51.7	59.3	38.1	42.7
IOM	30.1	19.4	39.7	37.8

[a]427°C, 1500 psi H_2, 2 minutes. NR = not reported.
Source: D. D. Whitehurst, J. O. Mitchell, and M. Farcasu, *Coal Liquefaction: The Chemistry and Technology of Thermal Processes*, pp. 115,119. Academic Press, New York, 1980.

under helium and showed that the pretreatment did not improve conversion (Table 8.6). The coals were pretreated at 249°C in helium and liquefied at 427°C under short contact-time conditions. Thermal treatment of the bituminous coal reduced the THF-soluble's conversion by 10 percent compared to the untreated, although the selectivity to SRC (i.e., pentane insolubles, THF-soluble fraction) was similar. No major differences were found for the subbituminous coal. However, as indicated in Table 8.6, differences in the oxygen content of the pretreated coals were observed; these reflected rather large changes in the oxygen distribution of the liquefaction products. Therefore, it cannot be concluded from these experiments that thermal pretreatment alone alters the liquefaction yield.

Researchers have shown[58,60,62] that lignites lose an appreciable amount of oxygen when pretreated at 250 to 350°C. Schafer's data,[60] Figure 8.7, shows that lignites lose most of their carboxylic groups at a maximum rate between those temperatures. Later, Hippo et al.[63] showed that decarboxylation at elevated temperatures (110 to 230°C) increased the reactivity of a coal in liquefaction. Research in Japan[64] also showed that thermal treatment (under N_2) at 350°C preferentially removed oxygen and produced a 14 percent increase in the yield of liquefied products compared to untreated coal. It was shown[63a] in another study that decarboxylation was complete at 250°C; but for some coals, there was a concomitant loss of light ends and sufficient thermal polymerization to lower the oil and asphaltene yields. Therefore, it was recommended that thermal pretreatment be conducted at temperatures below 200°C.

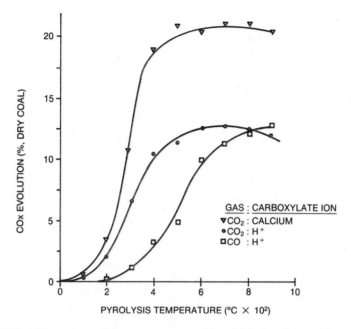

Figure 8.7 Effect of pyrolysis temperature on the yield of carbon oxide gases from Australian brown coal.

It may be argued that overheating during storage, grinding, drying, or other preparative procedures might be detrimental to liquefaction. The thermal pretreatment data thus far indicate that the effects will not be major, at least for high-rank coals not treated above the pyrolysis temperature. An experiment conducted by the author established that grinding Illinois #6 coal in the presence of a coal-derived solvent at a temperature of 170°C and 0°C with the careful exclusion of air produced no differences in the liquefaction product distribution (including both total conversion and pentane-soluble oils).

8.3.2 Pretreatment with Acidic Gases

Since any type of preparation involves exposing the coal to some sort of gaseous atmosphere, the nature of gas interactions and their influence on liquefaction has been of interest. Most of the work has focused either on the reducing gas environments, which incorporate hydrogen, carbon monoxide, or their mixtures, or has been concerned with the influence of oxygen on liquefaction. Relatively less attention has been given to treatment of coals with other gases not intimately involved in the conversion, particularly those that are potential by-products from the liquefaction plant. A number of studies have considered pretreatment of coal with acidic gases such as H_2S, SO_2, SO_3, and CO_2.

Pretreatment with acidic gases from gasifier effluents tends to prevent mineral scale deposits from forming in reactor lines during the liquefaction of low-rank coal. Brunson[65] used SO_2 to pretreat Wyodak subbituminous coal to prevent the build-up of calcium carbonate. Poddar,[66] following the work of Brunson, used both SO_2 and SO_3 with oxygen or ozone in the pretreatment reactor to enhance the preventive effect. Urban[67] pretreated low-rank coal to reduce scale formation using mixtures of SO_2 and acetic acid, SO_2 alone or H_2S. Apparently SO_2 interacts with the calcium ions exchanged onto the carboxyl groups in the coal. Their decomposition upon liquefaction results in the formation of a calcium sulfate salt (instead of the carbonate) and does not contribute to scale formation. Neavel et al.[68] found that pretreatment with CO_2 also reduces scaling.

Pretreatment with other gases has been studied with the intent of increasing conversion. Most emphasis has been on H_2S[69-71] because of its availability in the liquefaction plant, its ability to react with metals to form catalytically active sulfides, and its potential as a hydrogen transfer agent. The studies of Abdel-Baset and Radcliffe[69] and Gatsis[70] showed the direct benefits of an H_2S pretreatment in increasing conversion. Bearden and Aldridge[71] used H_2S pretreatment to presulfide metals in the coal. Mixtures of H_2S and H_2 were used to pretreat Wyodak coal; subsequent liquefaction resulted in some improvement in liquid yields but only when the pretreatment temperatures were high. When Wyodak coal was pretreated in hydrogenated creosote oil between 300 and 385°C in the presence of 18 percent H_2S in hydrogen, the oil yield from a subsequent liquefaction at 438°C was 82.9 percent compared to only 74.3 percent when the coal was liquefied in the presence of the H_2S/H_2 without pretreatment. However, when the same experiments were repeated, with the exception that the pretreatment was conducted below 300°C, the positive effect diminished. Temperature again appears to be an extremely important factor in the effectiveness of various treatments on coal conversion.

8.3.3 Solvent Pretreatment by Presoaking

It is well known[72] that coals tend to swell when immersed in certain organic solvents (or solvent vapors). The extent of swelling varies with the type of coal as well as the type of solvent. In some cases, this phenomenon results in a weight gain of as much as 2 to 3 times the original weight of the coal.[73] Whitehurst et al.[42] demonstrated that Kentucky coal could be swelled 20 percent with a process-recycle solvent and as much as 80 percent with a synthetic solvent containing tetralin, cresol, and methylnapthalene. Using the concept of solubility parameter, Angelovich et al.[74] determined that the effectiveness of a donor solvent for liquid-phase hydrogenation of coal was related to its effectiveness as a swelling solvent. A maximum was reached when the theoretical solubility parameter of the liquefaction solvent was similar to that of the solubility parameters of the best swelling solvents for coal (i.e., around 9.5). It appears that the ideal coal liquefaction solvent functions with the dual purpose of both swelling and hydrogen transfer. If the adsorption of large amounts of solvent takes place prior to liquefaction, the mass transfer of donatable hydrogen at the critical onset of liquefaction can be potentially improved.

Presoaking in solvent should result in a much greater conversion from a mass transfer standpoint. Improvements have been observed in a number of cases where coals were pretreated in this manner.[75,76] Major effects were observed only where the pretreatment was conducted at elevated temperatures, that is, between 290 and 400°C where thermal reactions of the coal become significant.

Long et al.[77,78] studied solvent-vapor swelling at lower temperatures by exposing coals to tetralin, tetrahydroquinoline (THQ), or quinoline vapors maintained between 260 and 270°C. Hydrogen gas was used to sweep away the excess vapor prior to liquefaction at 425°C. Conversion of the tetralin-treated coal (reported as cyclohexane solubles) showed no change compared to liquefaction of untreated coal. On the other hand, THQ-treated coal gave an increase in the oil yield of 12 percent after liquefaction, and about 6 percent of the THQ was irreversibly adsorbed. In another study, Long[79] preswelled Wyodak coal in THQ solvent at 270°C for 1 hour, followed by treatment in hydrogen at 550°C. The overall conversion was found to be nearly 20 percent greater than when no pretreatment was used. From Long's work it is apparent that intimate contacting due to solvent swelling is not as important as the effects of heteroatom interactions between solvent and coal. Employing solvent pretreatment with coal-derived solvents at 200°C, Narain et al.[80] observed small increases in conversion and suggested that the effect was the result of greater solvent-aided liquefaction relative to pyrolytic decomposition.

The importance of heteroatomic effects in solvent extraction is now well established. Marzec et al.[81] provided a fundamental basis for predicting the effectiveness of a solvent for coal. Materials rich in basic nitrogen compounds provide the best solvents, and Long's work has shown that they will indeed impart a strong influence on the liquefaction.

8.3.4 Solvent Pretreatment by Extraction

Extraction of coal under thermally nondestructive conditions to remove soluble organic components has been traditionally practiced to understand coal structure. Comprehensive reviews on extraction methodology were presented in the early 1960s by Dryden[82] and Van Krevelen[83] and were updated by Wender et al.[84] and Pullen.[85] Because of the generally small yield from low-temperature extraction and the high costs of the better solvents, only a few commercial applications have been proposed. Bay[86] offered a process for nonspecific extraction of coal using an admixture of water and an organic solvent (CCl_4) to produce a fraction that could be used as a gasoline additive. There is no reference to the use of the extraction residue, which constitutes over 90 percent of the coal. Zinniel et al.[87] also considered a solvent extraction process with hexane and chloroform to extract *resinous* substances. They have not considered the use of the residue that likewise accounts for the bulk of the material.

Stiller et al.[88] extracted bituminous coal at higher temperatures with certain nitrogen-containing solvents to obtain exceptionally high yields and considered the reactivity of both the extract and the residue. They reported that the residues tended to exhibit an increased combustion reactivity. This was attributed to opening the structure of the solid matrix of the coal by removal of the trapped tarry material. It is possible that this phenomenon will also have an effect on liquefaction reactivity as the opening of the structure would allow better dispersion of either donor solvent or catalysts into the coal.

There is much evidence[89-94] to support the concept first proposed by Bone[95] and Wheeler[96] suggesting that coal consists of low molecular-weight compounds retained as "mobile phase" within the micropores of a "macromolecular network." These compounds are assumed to be extractable by appropriate solvents. Marzec et al.[81] postulated that the extraction depends on the ability of the solvent to disrupt the molecular interactions between the mobile-phase and the "backbone" structure. If the solvent possesses the appropriate electron-donor properties, it causes release of the mobile phase and opens the structure to the surrounding medium without causing major degradation to the three-dimensional network. Harris and Peterson[97] measured CO_2 and N_2 surface areas to show conclusively that as much as twice the original surface areas could be generated by extracting coal. It appears then that pretreating coals with certain organic solvents would have an effect on subsequent liquefaction. Most of the extractable material is composed of aliphatic and alicyclic structures that tend not to be effective as hydrogen donor molecules in liquefaction. Their removal and subsequent replacement by a preferred donor material should presumably produce a pronounced effect on coal conversion.

A related technique for pretreating coal is extraction with a supercritical gas. However, most of the activity in this area[98-104] has focused on extraction at high temperatures. The structural integrity of the coal is destroyed in the process.

8.3.5 Chemical Pretreatment for Improved Catalysis

Catalyst Impregnation. Hawk and Hiteshue[1] described in a review the massive efforts in catalyst screening but noted that the methodology for applying catalysts to coal was limited. In earlier work, catalyst pretreatment was done either by physically mixing finely divided powders with pulverized coal or impregnating coal with aqueous solutions of salts. It has been established that transition metals exhibit the highest hydrogenation activity with the maximum activity occurring when they are impregnated rather than dry-mixed into the coal.[105,106] Many novel impregnation methods for enhancing dispersion of different catalysts have evolved, namely, dispersing heteropolyacid catalysts via aqueous soaking[107]; using process solvent to apply iron or molybdenum catalysts as soluble organic salts (e.g., iron napthenate),[2,3] adding metal-organic catalysts such as $Ni(SCN)$[2] dissolved in liquid ammonia;[108] impregnating catalysts by soaking in metal carbonyl vapor;[12,109] dispersing Ziegler-type catalysts in a process oil;[110,111] and using a water–oil emulsion method for dispersing water-soluble catalysts (e.g., ammonium molybdate) into process solvents.[112]

Several pretreatment methods for *in situ* preparation of catalysts have been considered. Aldridge and Bearden[113] impregnated molybdenum naphthenate into coal using process oil and then presulfided the coal with H_2S prior to liquefaction to convert the metal into its most active form, the sulfide. In related work, Hodgson[114] pretreated coals for liquefaction with aqueous $MoCl_4$ followed by treatment with H_2S and found a 7 percent increase in the distillate oil yield (C_4 to $-400°C$ fraction) over that of impregnated unpresulfided coal. Hodgson[115] also surveyed impregnation with other metals such as V, Ni, and Cu, and it was found that presulfidation resulted in improvements in conversion (to MEK solubles) of as much as 13 to 29 percent.

An efficient way to disperse a catalyst prior to liquefaction is to treat the coal with the catalyst in its vapor state, for example, with volatile metal carbonyls. Dispersion of Ni and Fe into coal using this technique has been successful in improving liquefaction.[109,116] Other investigators[117-119] have considered attaching the catalytic material directly by ion-exchange methods. It is well established that low-rank coals modified by ion-exchange of alkali metals are much more reactive in coal gasification than coals with metals physically admixed.[120] Attempts to add metals to bituminous coals by ion-exchange have been unsuccessful in gasification studies, principally because of the absence of natural exchange sites. Neavel and Long,[117] however, have succeeded in inducing ion-exchange sites for alkali metals by preoxidizing bituminous coals in air; coal gasification reactivity of preoxidized coal containing Na^+ and Ca^{++} added by ion-exchange was found to be 4 to 10 times greater than gasification of the preoxidized coal containing an equivalent amount of the metals added via physical admixture.

Ion-Exchange of Metals as Liquefaction Catalysts

Low-rank Coals. Pretreatment of low-rank coal by ion-exchange techniques has received considerable attention. Durie et al.[121-123] showed that low-rank coals are rich in carboxylic acid groups associated with metal cations, and Miller[124] determined that the carboxyl groups of the low-rank coals in their natural state are associated predominantly with Ca, Mg, Na, and K. These cations are exchangeable with other metals, acids, and ammonium salts. There is an interest in replacing the naturally occurring metal groups with more catalytically active ones to enhance liquefaction.

Given et al.[118] were first to report the benefits of cation exchange in coal liquefaction. A lignite stripped of metals by HCl leaching was exchanged with sodium cations and liquefied in the presence of a standard Co/Mo catalyst at 385°C in hydrogen. The conversion to benzene solubles was 70.1 percent for the sodium-loaded coal compared to only 43.2 percent for the sodium-free, acid-extracted coal. Moreover, the oils from the liquefaction of the sodium-loaded coal were of significantly lower viscosity and richer in simple aromatics. Given et al. suggested that Na^+ was catalyzing the reverse water-gas-shift reaction to produce a mixture of CO and H_2, known to be more effective for hydrogenating lignites. Given[125] conducted follow-up runs using as many as eight different coals to confirm his earlier result. Hydrogen was excluded from the reactor and, as predicted, no clear effect emerged. Given added that the presence of molecular hydrogen in the reactor is necessary to drive the reverse shift reaction. Thus, the major effects of sodium exchange would seem to be highly dependent on the reaction conditions.

Baker et al.[126] tested the effects of metal exchange on liquefaction behavior in the SRL Process. Both raw and acid-extracted lignites were pretreated by soaking in solutions of Na, K, Fe, Ca, Co, Ni, and Mg. Conversion yields were unaffected when the Ca- and Mg-exchanged coals were liquefied, but both the Na- and the K-exchanged coals gave consistent increases in the yield of distillates and a better product quality. The results with the other metals were inconclusive, though significant increases in conversion were never observed.

Hatswell et al.[127] were very successful in using transition metal ion-exchange to study tetralin liquefaction of Victorian brown coal. In their work ion-exchange was effected by pretreating the coal with solutions of iron and tin cations at various acidic pH's. The Fe-exchanged coal (1.16 weight percent Fe added) gave a significant improvement in conversion yield over untreated coal (from 59 to 81 percent MEK solubles) and a slightly higher yield of oils (from 32 to 38 percent soluble in a light petroleum distillate). The hydrogenation character of the Sn-exchanged coal (1.9 weight percent as Sn added) was even superior to that of the Fe-exchanged coal or to a comparable run using a standard CoMo catalyst. Hatswell et al.[128] used a better ion-exchange technique for Fe to further improve conversion in later experiments.

Others have used ion-exchange to improve the operability of lignites in coal conversion reactors. Gorbaty and Tauton[119] ion-exchanged a calcium-loaded coal with sodium to help prevent reactor-scale buildup in EDS liquefaction, for example.

High-Rank Coals. Exchange of metals onto bituminous coals to improve liquefaction has not been studied; but as noted, bituminous coals are without the necessary functional groups to effect aqueous exchange of metal cations. Nevertheless, the work of Neavel and Lang[117] suggests the possibility of using preoxidation to promote the exchange of desirable liquefaction catalysts. Experiments by this author have failed to show positive effects for liquefaction. An Illinois #6 bituminous coal (Burning Star) was preoxidized in air at 170°C long enough to cause a 3 percent incorporation of oxygen, and the coal was then ion-exchanged in solutions of ferrous sulfate at a pH ranging from 2 to 5. The preoxidized coal failed to absorb significant amounts of Fe and no enhanced liquefaction effect was found.

8.3.6 Pretreatment of Coal with Mineral Acids

There has always been widespread interest in the effects of the inorganic constituents in coal on liquefaction, both from an operations and a catalytic viewpoint. Addition of specific minerals to coal has been tried, though none, with the exception of pyrite, has been found to be effective. Removing minerals using physical methods of separation was considered, but the intimate association of minerals within the coal matrix makes study of the effects of any one mineral difficult. This suggests the removal of mineral matter from coal by chemical methods. Mineral acid extraction has been the subject of most attention because different acids can be used to remove individual mineralogical groups selectively. Dilute HCl or dilute H_2SO_4 removes carbonates, other soluble salts, and exchangeable cations; HCl with HF eliminates silicate minerals such as clays and quartz. None of these acids attack pyrite nor have a significant effect on the organic structure. Standard methods for acid extraction are well documented[129,130] and have served as a basis for studies of coal conversion; practice of any of these techniques on a commercial scale would be prohibitively expensive.

A sulfuric acid treatment[106] used simply to neutralize the alkalies in a bituminous coal (from Wyoming) and a lignite (from North Dakota) resulted in a marked increase in hydrogenation activity when the coals were reacted at elevated temperatures in the presence of an ammonium molybdate catalyst. It was reported that distillate oil yields increased from 41.4 percent to 62.6 percent

TABLE 8.7 Summary of Studies Showing the Effect of Acid-Demineralization on Coal Liquefaction

Reference	Liquefaction Conditions	Coal Type/Rank	Pretreatment Conditions	Net Effect on Conversion
Wu and Storch[106]	catalytic hydrogenation	Rock Springs/Bit. Beulah/Lignite	H_2SO_4 H_2SO_4	+21% distillate +11% distillate
Whitehurst et al.[42]	427°C; 3 min. process solv. H_2 427°C; 90 min. tetralin H_2	Wyodak/Sbb. Burning Star/Bit. Wyodak/Sbb.	HCl HF/HCl HCl/HF/HCl HCl HF/HCl	+41% SRC +57% SRC +4% SRC −8.9% SRC −16.5% SRC
Given et al.[131]	400°C; 5 hrs. tetralin	N. Dakota/Lign. Montana/Lignite Texas/Lignite Illinois/Bit. Kentucky/Bit. Wash. Queen/Bit. Wash. Roslyn/Bit. Ill., Wash. and Ky. Bituminous	HCl HCl HCl HCl HCl HCl HCl HCl/HF/HCl	+8–9% BZ sols. +8–9% BZ sols. 0 +8% BZ sols. −3% BZ sols. −2% BZ sols. +8% BZ sols. +/−3 to 5% BZ sols.
Mochida et al.[134]	440°C process solv.	Lignites	HCl HCl/MeOH	+10% BZ sols. +30% BZ sols.
Baker et al.[126]	420°C; 20 min. solvent CO/H_2	Beulah Lignite Indianhead/Lig.	HCl HCl	−22% dists. −12% THF sols. −3% dists.

[a]SRC = pyridine sols., +343°C distillates.

for the Wyoming coal and from 22.6 percent to 44.0 percent for the lignite. The reason for these increases is not understood.

Several groups[42,131-136] have studied the effects of acid-demineralization of coal for liquefaction using HCl or mixtures of HCl and HF, but with highly inconsistent results (see Table 8.7). The data show only minor differences when bituminous coals are acid-treated and indicate that no minerals other than pyrite have an effect on conversion. Lignites are quite different, and major effects are evident.

Calcium is the primary metal cation removed during acid treatment of lignites where it is exchanged by H^+ to convert calcium salts of organic acids to the free-acid form (—COOH). Whitehurst[42] postulated that the removal of calcium from a low-rank coal alters the swelling properties and thus changes the dissolution behavior in the early stage of liquefaction. The large increases in conversion observed in short contact-time experiments are thus explained. It has been shown,[3] moreover, that addition of calcium as the oxide decreases conversion. Calcium addition to highly caking coal destroys the coking and swelling properties.

There are inherent variations in the content and nature of inorganic constituents between coals so that removal of the soluble constituents by mineral acids produces variable effects on conversion. For lignites these effects appear to be significant. Attention should be given to the method of processing lignite coals in coal conversion since a strong possibility exists of altering the content of important exchangeable inorganic species.

8.4 STRUCTURAL MODIFICATION AND OXIDATION

8.4.1 Oxidative Coal Pretreatment

Methodology of Coal Oxidation. Historically, most of the effort in coal oxidation had been directed to structural or combustion studies, although it has long been recognized that oxidation can significantly change the behavior of coal in various conversion processes—gasification, pyrolysis, carbonization, and liquefaction. Substantive activity has been devoted to studying effects of oxidation on liquefaction processes.

Oxidation studies have been of four generic types based on the level of reaction:

Type 1. Studies pertaining to coal aging and storage that focus on oxidation in air at ambient conditions.

Type 2. Studies focused on air-oxidation at slightly elevated temperatures common to routine preparative procedures for liquefaction, such as comminution, drying, washing, and other coal beneficiation unit operations.

Type 3. Studies of oxidative depolymerization of coal using aqueous alkali. Studies of this nature have focused on structural depolymerization of the coal to convert it into a pyridine-soluble material that can then be upgraded via a coal liquefaction process.

Type 4. Oxidative degradation studies aimed at direct liquefaction of coal to produce low-molecular weight aromatic or aliphatic acids.

Because air-oxidation is the most important concern from a coal conversion standpoint, this section will focus on Type 1 and Type 2 studies, though brief mention will be made of Type 3 studies that have considered oxidative depolymerization as a pretreatment for coal conversion.

Mechanism of Air-Oxidation. Jones and Townsend[137] published an excellent paper on reactions of coal in air. They showed that oxidation involved chemisorption of molecular oxygen that in the presence of water was followed by the formation of a stable coal–peroxygen–water complex. They further found that 70°C appeared to be a critical temperature for oxidizing bituminous coals, for it was above that temperature that the complex broke down to form secondary oxygenated functional groups that remained on the coal. The importance of both water and temperature in air-oxidation of coals was established. A typical example of the magnitude of the effect of temperature on the rate of air-oxidation of a bituminous coal is illustrated in Figure 8.8.

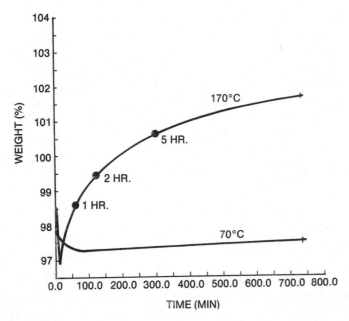

Figure 8.8 Effect of temperature on the rate of air-oxidation of Illinois #6 Burning Star coal as determined by thermogravimetric analysis.

Infrared spectroscopic studies[138] and others[139] have conclusively shown that oxygen incorporation is mainly in the form of carbonyl, carboxylic, or hydroxyl, accompanied by a significant loss of aliphatic CH-groups. That loss is attributed to the attack of oxygen on methylene linkages to aromatic rings at benzylic positions. These sites are susceptible to oxidation in model organic reactions and, based on the spectroscopic evidence, probably represent the initial sites of oxidative attack in coals.[138]

Theoretical Effects of Oxidation on Liquefaction.

Several explanations have been offered to account for the detrimental effects of mild oxidation on coal conversion.

1. From an operational point of view, the release of oxygen as gas will increase the consumption of hydrogen. If the oxygen is not fully removed during the hydrogenation, the resultant product will be of much lower quality.

2. The oxygenated functional groups formed during oxidation are susceptible to condensation reactions that cause a decrease in conversion. Moschopedis and Speight[140] observed that preoxidized bitumens gave very poor yields when hydrogenated. This was caused by repolymerization reactions involving the oxygenated groups. Many of the oxygen groups formed during preoxidation of coal will react with active sites generated during liquefaction and form refractory condensation residues at the expense of the desired liquid products.

3. Changes in the coal structure by preoxidation alters the conversion chemistry and results in different product distribution. For example, the attack on aliphatic CH_2 or hydroaromatic sites during oxidation may alter the hydrogen-transfer reactions and lead to lower conversions. It has been suggested[139,141] that air-oxidation causes the formation of a cross-linked network structure involving other linkages, and this structure inhibits the conversion.

4. The addition of more heteroatomic sites onto the coal through oxidation will increase the propensity for solvent adduction effects and may lead to decreased conversion yields.

Studies of the Effects of Air-Oxidation on Liquefaction.

Liquid yields from both high- and low-rank coals oxidized in air are drastically lower than yields from fresh coals. A detailed summary of the results is presented in Table 8.8.

Change et al.[142] were concerned with the possibility that grinding coal in air at elevated temperatures during routine preparation for liquefaction produced detrimental effects, and so they preoxidized a Kentucky #9–14 bituminous coal in air for various times at 175°C to match the conditions of coal preparation. The pretreated coals were liquefied in tetralin at 425°C with and without the

TABLE 8.8 Summary of Studies Showing the Effect of Preoxidation on Coal Liquefaction

Reference	Coal Type	Oxidation Conditions			Liquefaction Conditions					Conversion Results (Percent DAF Coal)			
		Media	Time	T, °C	T, °C	Time	Solvent	H	Cat.	Method	Base	Oxidized	Net Effect
Krichko et al.[144]	Lignites	H2O2	?	150	425	2 hrs.	+240°C petro	yes	Fe/Mo	?	76.8	65.7	−11.1
Chang et al.[142]	KY #9 Bit.	Air	12 hrs.	175	415	30 min.	tetralin	yes	no	cresol sols.	92.0	55.0	−37.0
								no		cresol sols.	83.0	37.0	−46.0
Whitehurst et al.[42]	KY #9 Bit.	Air	17 hrs.	146	427	2 min.	synthetic	yes	no	pyridine sols.	80.6	79.1	−1.5
			5 days	149						pyridine sols.	80.6	55.7	−24.9
Neavel[143]	HVC Bit.	Air	?	177	400	30 min.	tetralin	no	no	benzene sols.	60.0	30.0	−30.0
Cronauer et al.[25]	Belle Ayre Sbb.	Air	2 hrs.	160	450	5 min.	SRC-II heavy distillate	yes		pyridine sols.	79.0	71.3	−7.7
			2 hrs.	160		12 min.				pyridine sols.	73.7	62.4	−11.3
			2 hrs.	160		30 min.				pyridine sols.	67.9	61.8	−6.1
			30 min.	50		30 min.				pyridine sols.	67.9	68.4	+1.0
			5 min.	100		5 min.				pyridine sols.	67.9	72.0	+4.1
			2 hrs.	160		12 min.				benzene sols.	47.0	27.7	−19.3
			2 hrs.	160		30 min.				benzene sols.	42.5	27.4	−15.1
			2 hrs.	160		30 min.				benzene sols.	47.8	36.3	−11.5
Severson[145]	Lignites	Air	70 wks.	amb.	400	30 min.	process-derived	yes+CO		pyridine sols.	91.2	87.6	−3.6
										light distillates	24.4	20.5	−3.9
										heavy distillates	42.0	37.1	−4.9
EDS Report[32]	TX Lignite	10% O2 90% N2	2 min.	149	450	40 min.	EDS Recycle	yes	no	cyclohexane sols.	41.6	40.1	−1.5
			2 min.	316						cyclohexane sols.	41.6	34.0	−7.6
			10 min.	316						cyclohexane sols.	41.6	25.1	−16.5
	Wyodak Sbb.	10% O2 90% N2	2 min.	316						cyclohexane sols.	41.3	37.6	−3.7
			2 min.	316						cyclohexane sols.	41.3	31.3	−10.0
			10 min.	316						cyclohexane sols.	41.3	19.0	−22.3
Skinner[34]	KY #6 Bit.	Air	?	82	454	30 min.	Hydrogenated creosote oil	yes	no	pyridine sols.	90.1	91.9	+1.8

presence of molecular hydrogen. As seen in Figure 8.9, the most severe preoxidation (at 12 hours) resulted in a marked drop in the yield of cresol solubles from 80 percent to 45 percent in the absence of H_2 and from 90 to 55 percent in the presence of H_2. Under less severe conditions of preoxidation (less than 3 hours) cresol-soluble yields were lower but only when the liquefaction was run without hydrogen.

Neavel[143] showed the effect of oxygen incorporation on liquid yields in a related study giving very similar results. Preoxidation in air at 177°C has pronounced effect on the yield of benzene-soluble products beyond about 2 percent oxygen incorporation (Figure 8.9). Cronauer et al.[24-28] found that partial oxidation occurred upon air-drying subbituminous coal at 175°C, and the conversion to pyridine- and toluene-soluble material was lower by 10 percent and 18 percent, respectively. Krichko et al.[144] used hydrogen peroxide to oxidize coal at 150°C, but studied liquefaction under catalytic conditions. Again, the conversion of oxidized coal was significantly worse; the gas yield increased from 17 to 22 percent, the insoluble organic material (IOM) increased from 7 to 13 percent, and the yield of liquids dropped from 77 to 66 percent.

Liquefaction of oxidized coals results in a significant incorporation of oxygen into the products irrespective of yield changes. The work of Skinner and Givens[34] showed a 1.2 percent oxygen increase in the SRC product from liquefaction of preoxidized Kentucky coal, although the yields of SRC were unaffected. Whitehurst et al.,[42] who also studied the preoxidation of Kentucky coal, noted that the oxygen increase in the lighter ends was largely a result of a higher phenolics content. The increase in the residual material was attributed to repolymerization reactions incorporating oxygen.

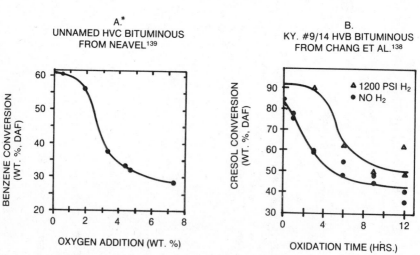

LIQUEFACTION 400-415°C; 30 MINUTES TETRALIN

A.*
UNNAMED HVC BITUMINOUS
FROM NEAVEL[139]

B.
KY. #9/14 HVB BITUMINOUS
FROM CHANG ET AL.[138]

Figure 8.9 Effect of air-oxidation at 175°C on the liquefaction of bituminous coals.

It has been consistently shown that mild oxidation of coals is detrimental to conversion; both yields and product quality are reduced. The net effect on yield is significant beyond several percent oxygen incorporation.

Impact of Aging on Coal Conversion. In the above cases, preoxidation conducted over a period of several hours at an elevated temperature (100–200°C) was shown to cause a serious effect on coal liquefaction. An important consideration is that of the effect of long-term exposure of coal at ambient conditions. Many feedstocks are likely to be stockpiled for extended periods before use. One pertinent study was conducted by Severson[145] to test the effect of long-term storage on the liquefaction of lignites. Pulverized lignite was stored under air and under nitrogen (as a control) for up to 70 weeks. Liquefaction data from the 70-week-old samples indicated no appreciable differences in either the light-oil yields or the gas yields, with only minor reduction in the yield of heavy liquids. No differences were observed in samples aged less than 36 weeks. Aging studies conducted by Cronauer et al.[25–28] revealed only modest changes in toluene-soluble yields when pulverized subbituminous coals were aged (Figure 8.10).

Definitive studies are lacking for bituminous coals, though in view of the minor effects of aging on low-rank coals, it is unlikely that the effects would be significant for high-rank coals. Moreover, it is the practice to stockpile coarse ROM rather than pulverized coal, the former being more resistant to oxidation.

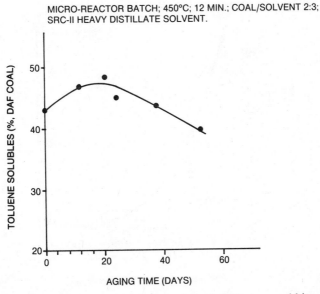

MICRO-REACTOR BATCH; 450°C; 12 MIN.; COAL/SOLVENT 2:3; SRC-II HEAVY DISTILLATE SOLVENT.

Figure 8.10 Effect of storage on liquefaction of Belle-Ayre subbituminous coal (85 × 230 mesh).

Effect of Preoxidation on Other Coal Conversion Processes. The subject of preoxidation and its impact on the operability or behavior of a coal in other coal conversion processes has been of interest. There is now recognition that preoxidation or weathering (especially on highly caking bituminous coals) has a marked effect on carbonization and gasification processes. It is well known[146-148] that even very mild preoxidation destroys the plastic properties of caking bituminous coals, as well illustrated by the work of Maloney[146] showing the loss in swelling character (Figure 8.11). For the effect of oxidation on carbonization processes, see Chapter 6.

The loss of caking properties has been shown to have a positive effect on coal gasification. When a caking coal is heated, it first becomes plastic and swells as the volatiles are lost. It then agglomerates to produce a coherent mass. In a gasification reactor, this process can lead to reactor shutdown caused by the char build-up in the reactors and feed lines. The agglomerated mass has a drastically reduced permeability, forcing the rate of gasification to be unsatisfactorily slow. Oxidative pretreatment of such coals can alleviate these problems.

The rate of production of gas in coal gasification is limited by the relatively slow step of reacting the char; by controlling the agglomerating characteristics

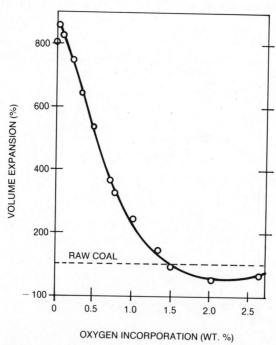

Figure 8.11 Effect of preoxidation on the swelling properties of a caking bituminous coal.

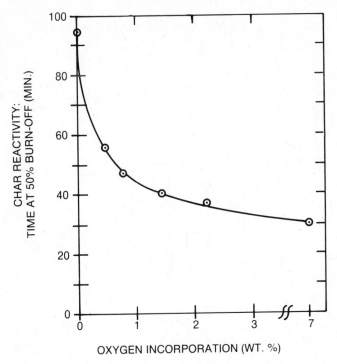

Figure 8.12 Effect of different levels of preoxidation on char gasification reactivity at 470°C.

in the initial devolatilization stage, the rate of gasification can be improved. Mahajan et al.[149] have shown that preoxidation not only improves operability but also markedly enhances the reactivity of subsequent chars (Figure 8.12). Mahajan et al. also measured sharp increases in the surface areas of chars produced from preoxidized coals and concluded that this effect undoubtedly contributed to an increased gas permeability in the char and hence a faster gasification rate.

The effects of preoxidation on the rate and amount of the volatile components of hot pyrolysis and gasification processes produced under high temperature conditions have been observed.[150-152] In a pyrolysis process, condensable volatiles are the desired products. It has been shown that pretreatment of certain coals with air significantly reduces the yield of pyrolysis liquids. Data of Maloney and Jenkins,[150] Figure 8.13, typify the magnitude of this effect for a bituminous coal. Furimsky et al.[152] showed that preoxidation results in a decrease of liquid hydrocarbon yields from Fisher Assay pyrolysis. These findings are consistent with observations on the donor-solvent liquefaction of oxidized coals.

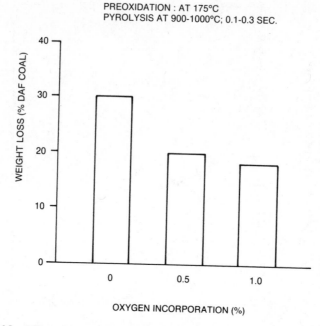

Figure 8.13   Effect of preoxidation on the yield of volatile products from pyrolysis.

Oxidative Depolymerization: Alkali Pretreatment Studies.

Treatment of coals in aqueous alkali or alcoholic alkali has been viewed as a potential coal liquefaction method. It has been established that certain coals can be completely converted to a pyridine-soluble material directly by treatment at temperatures ranging from 100 to 350°C.[153-165] The application of the technology is limited to low-rank coals because they have the functional groups that are susceptible to alkaline hydrolysis. Furthermore, the nature of the reaction is such that alkali-soluble products are the principal components. Conversion to benzene solubles or oils is negligibly small even for the low-rank coals. Attempts to treat bituminous coals with alkali have resulted in only modest increases in the amount of pyridine-soluble material, though it is possible that certain chemical changes have taken place that may be beneficial to liquefaction.

Some investigations have utilized alkali pretreatment prior to liquefaction but with mixed results. Kasehagen[154] found that when a bituminous coal was pretreated in 0.5N NaOH at temperatures between 275 and 375°C and catalytically hydrogenated with MoS_2 and H_2 at 400°C for 24 hours, the conversion to petroleum ether or benzene solubles showed only marginal improvement and actually decreased when the pretreatment temperature exceeded 300°C. Chang et al.[166] investigated the effects of ambient alkali treatment of Kentucky #9–14 bituminous coal and found that the liquefaction to cresol solubles was about 20 percent worse for nonpretreated coal. The worst

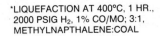

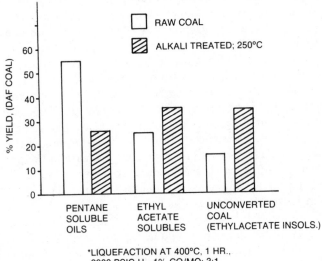

*LIQUEFACTION AT 400°C, 1 HR.,
2000 PSIG H₂, 1% CO/MO; 3:1,
METHYLNAPTHALENE:COAL

Figure 8.14 Effect of alkali pretreatment on the donor-solvent hydroliquefaction of Kentucky #9 bituminous coal.

conditions were for pretreatment times longer than 3 hours or alkali concentrations in excess of 0.5N.

In contradiction to earlier work, Chow's data[167] indicated that the alkali treatment was beneficial in increasing coal conversion. Chow pretreated Kentucky #9–14 bituminous coal with 5N NaOH at temperatures up to 200°C and found a 5–6 percent improvement in the yield of benzene-soluble material from liquefaction. He determined that the alkali treatment promoted the oil yields and reduced the yields of preasphaltenes and residue. The increase in oil yield for three different coals was 17 percent for a Kentucky bituminous, 6 percent for a Forrestburg subbituminous, and 10 percent for a Zap lignite.

Work by this author, Figure 8.14, confirms the results of Chang[166] and Kasehagen[154] but not those of Chow[167] in that the alkali treatment of Kentucky bituminous coal lowers the liquefaction yields considerably. The oil yield, shown in Figure 8.14, is the most seriously affected, being decreased by nearly half. The reasons for the inconsistency of the data among investigators are not known. Different liquefaction conditions were employed in the different studies and this leads to a degree of uncertainty. Nevertheless, it is plausible that the detrimental effects of alkali treatment on coal conversion can be attributed to retrogressive reactions involving oxygen-containing groups (such as OH) formed during the alkali-catalyzed hydrolysis of benzylic ethers in the coal.

8.4.2 Effect of Organic Reactions of Coal as Pretreatments for Liquefaction

Chemical derivatization for modifying functional groups has been exhaustively investigated in fundamental studies of coal structure (review by Wender et al.,[84]) but only a few of the methods have been seriously considered as pretreatments for coal conversion. The high cost of chemical reagents has precluded progress toward commercial development.

Several reactions are known to have a profound influence on conversion simply by altering the nature of heteroatomic functional groups on the coal. Schlosberg et al.,[168-170] in their attempt to find an improved conversion process, demonstrated that the use of acylation and alkylation (using alkyl halides under Friedel–Krafts conditions) to pretreat Illinois #6 and Kentucky #9 bituminous coals resulted in a significant increase in cyclohexane-soluble oils. On an alkyl-free basis, the increase is as much as 10 to 21 percent. Schlosberg postulated improved solvent-coal contacting and a reduction in heteroatom reactions as being principally responsible for the observed yield increases. Liotta et al.[171-174] and others[175,176] have used O-alkylation successfully as a pretreatment to improve liquefaction of coals rich in oxygen. Under ambient conditions, quaternary ammonium hydroxide and alkyl halides will form alkylated derivatives of acidic —OH and —COOH groups in coals. In the liquefaction of alkylated coal, both the viscosity and the boiling range of the liquid distillate products will be considerably lower, and the overall liquid quality improved. The solubility of the pretreated coal will be enhanced significantly even without liquefaction. Liotta concluded that derivatization of the polar hydroxyl groups (to form nonpolar ethers and/or esters) serves two functions: the first is to diminish intermolecular forces sufficiently to open the structure for improved solvent contacting, and the second is to passivate reactive oxygenated sites that would otherwise undergo repolymerization reactions resulting in low liquid yields.

Horowitz and Dickert[177] discussed the potential benefits to a liquefaction process of using pretreatment reactions where coal is activated with free-radical or ionic-catalyst systems to incorporate olefinic structures as polymeric side chains. Solubility studies have revealed marked increases in benzene solubility as a result of this type of pretreatment.

8.4.3 Pretreatment with Irradiation Methods

A few papers have considered irradiative methods such as microwaves, ultrasound, and x-rays as pretreatments. The first application of using ultrasound to enhance extraction of coal was reported by Berkowitz[178] in 1949. Since then others[179,180] have shown the benefits of using ultrasonic energy to increase the dissolution of coal in solvents. Kesseler et al.[179] dissolved 79 percent of Pittsburgh bituminous coal by irradiating a coal–quinoline slurry at ambient temperatures for a day. An obvious extension of this work would be to pretreat the coal-solvent slurry prior to initiating the thermal liquefaction reaction, but no tests of this nature have been published.

Stone[181] has used microwaves for enhancing coal liquefaction, and Flagg et al.[182] have demonstrated that pretreatment with x-rays (from 0.001 to 100 Å) modestly increases conversion.

8.5 CONCLUSIONS

A review of the subjects of coal pretreatment and preparation and their effect on coal conversion has been presented with abstraction of a number of arguments from the literature attesting to both beneficial and detrimental effects. Many chemical and physical methods of coal treatment will produce significant effects in coal conversion but the magnitude depends strongly on the nature of the coal.

For low-rank coals (i.e., lignites, brown, and subbituminous), pretreatments have the most pronounced effect, especially procedures considered for routine preparation of coal (such as comminution or drying methods). The data indicate that these effects are more often detrimental than beneficial to liquefaction. Bituminous coals are not as sensitive to pretreatments as low-rank coals.

The major effects of important pretreatments are summarized as follows:

1. *Physical beneficiation* produces noticeable change in coal conversion processes affecting both operations and yield. Liquefaction yields can be tailored by altering the amount of pyrite, reactive macerals, and inert materials.

2. *Comminution* of coal does not generally produce a significant effect on conversion unless the coal is exposed to air during the grinding operation or is pulverized to ultrafine particle sizes of less than a few micrometers. Even grinding the coal in the presence of a solvent does not seem to have much of an effect on liquefaction, especially when distillates are the desired product. However, comminution is still practiced in coal conversion to facilitate operational procedures in the handling of the coal-solvent slurry.

3. *Moisture removal* from low-rank coals, a necessity in conversion processes for optimizing thermal efficiency, causes irreversible changes in the physical structural constitution. This is detrimental to conversion. The observed effects on conversion yields are particularly bad when the thermal drying operation is accompanied by exposure to oxygen, but the effects of drying alone are negligible at higher rank.

4. *Thermal pretreatment* of coal prior to liquefaction does not have much effect unless the procedures are carried out at high temperatures approaching those that actually initiate liquefaction. Low-rank coals are more susceptible to thermal pretreatment; relatively low temperatures cause the decomposition of their abundant oxygenated groups.

5. *Solvent soaking and solvent extraction* are pretreatments whose net effect on liquefaction depends strongly on controlling the heteroatomic interactions that occur between the coal, the extracted material, and the solvent.

6. *Catalyst impregnation* to obtain a dispersion on a molecular level is extremely effective in optimizing catalytic liquefaction. The coal modification technique, that of ion-exchange of catalytic metals onto functional groups in the coal, is effective for only low-rank coals because of the absence of the appropriate exchange sites in the high-rank coals.

7. *Acid demineralization* with aqueous mineral acids or pretreatment with acidic gases does little to change the liquefaction behavior of bituminous coal. Low-rank coals undergo severe alterations when subjected to these pretreatments and show marked changes in conversion. The direction and magnitude of the effect is dependent on the nature of the mineral matter contained within each coal.

8. *Preoxidation* of coal in air is clearly detrimental to the conversion of both high- and low-rank coals, and the effect is particularly significant when pretreatment temperatures exceed 70°C or where oxygen incorporation exceeds 1–2 percent. Since oxidation is likely to accompany any number of coal preparation methods or pretreatments, the combined effect can have an even greater influence on the outcome of the liquefaction. On the other hand, preoxidation is extremely beneficial in gasification processes that use caking bituminous coals as feedstocks.

9. *Chemical derivatization* is an expensive but effective pretreatment for increasing conversion because it can open the coal structure and passivate heteroatomic sites that would otherwise enter into retrogressive reactions.

The focus of pretreatment for liquefaction has been on increasing overall conversion based on residue yields rather than products. An increased yield means little if the quality of the product has been destroyed.

A relatively clear picture has emerged that the overriding effects of pretreatment, preparation, or oxidation are those related to coal rank and to the characteristics of coals. More research is needed to delineate many of the observed phenomena and to assess their importance among coals of different geological origin and rank.

The probability of achieving significant advances in coal liquefaction performance by preparation and chemical pretreatment technology is undoubtedly dependent on developing a means to effect novel changes in the internal physical and chemical structure of the coal. The following areas are suggested as the foci for future fundamental studies in coal:

1. Opening the structure (especially in high-rank coals) to permit intimate contacting of the H-transfer vehicle and a catalyst prior to initiation of the thermal reactions. Opening of the coal structure by physical or chemical means must involve technology capable of dissociating intermolecular bonds between mobile species and the backbone structure without causing significant alteration that would be detrimental to conversion.

2. Modifying the surface properties of the coal matrix to facilitate distribution of reagents, catalysts, and or solvents. Of particular importance is modification of reactive heteroatomic sites.

3. Obtaining means for maximizing intimate dispersion of catalytic agents to accelerate oil-forming reactions early in the liquefaction.

Certainly for establishing homogeneity in feedstock quality and for eliminating problems associated with handling of inert materials, coal preparation technology will undoubtedly be practiced in a commercial liquefaction plant. However, the kind of technology used must guard against the detrimental effects on liquefaction performance, as discussed herein. Moreover, consideration of pretreatment technology aimed at improving conversion will most likely have to be developed and integrated within the coal preparation stage.

REFERENCES

1. C. O. Hawk and R. W. Hiteshue, Hydrogenation of coal in the batch autoclave. *Bull.—U.S. Bur. Mines* **622**, 12 (1965).

2. D. Garg and E. N. Givens, Effect of catalyst distribution in coal liquefaction. *Fuel Process. Technol.* **7**, 59–70 (1983).

3. D. Garg, E. N. Givens, and F. K. Schweighardt, *Evaluation of Coal Minerals and Metal Residues as Coal Liquefaction Catalysts,"* Final Rep. Vol. 1, DOE/TIC-14806-27-I. U.S. Dept. of Energy, Washington, DC, 1982.

4. P. H. Given, W. Spackman, A. Davis, and R. G. Jenkins, Some proved and unproved effects of coal geochemistry on liquefaction behavior with emphasis on U.S. coals. *ACS Symp. Ser.* **139**, (1980).

5. D. Gray, G. Barrass, J. Jezko, and J. R. Kershaw, Relations between hydroliquefaction behavior and the organic properties of a variety of South African coals. *Fuel* **59**(3), 146–150 (1980).

6. G. J. Perry, D. J. Allardice, and L. T. Kiss, Variation in Victorian brown coal characteristics and hydrogenation potential. *Fuel* **61**, 1058–1064 (1982).

7. R. F. Yarzab, P. H. Given, W. Spackman, A. Davis, and A. Rabinovich, *Dependence of Coal Liquefaction Behavior on Coal Characteristics: Cluster Analyses for Characteristics of 104 Coals*, Tech. Rep. No. 2, FE-2494-TR-2. U.S. Dept. of Energy, Washington, DC, 1979.

8. M. Abdel-Baset, R. F. Yarzab, and P. H. Given, Dependence of coal liquefaction behavior on coal characteristics. 3. Statistical correlations of conversion in coal-tetralin reactions. *Fuel* **57**, 89–95 (1978).

9. D. S. Hoover, Correlation of coal quality to coal liquefaction. *ACS Prep.* **28**(5), 48 (1983).

10. P. H. Given, Dependence of liquefaction behavior of coal characteristics. Part I. A selective and critical review of the literature. *Energy Res. Abstr.* **8**(16), Abstr. No. 38348 (1983).

11. R. A. Durie, Coal properties and their importance in the production of liquid fuels—an overview. *Fuel* **61**, 883–888 (1982).

12. B. J. Mayland, Process using nickel carbonyl in hydrogenation, desulfurization and gasification of carbonaceous materials. U.S. Patent 2,756,194 (1956).

13. D. S. Hoover, A. E. Bland, M. C. Laneet, J. A. Terrible, and S. V. Joshi, *Effect of Feed Coal Variation on Demonstration Plant Performance*, Final Rep., DOE/OR/03054- 74. U.S. Dept. of Energy, Washington, DC, 1983.

14. D. Gray, M. Neuworth, and G. Tomlinson, Incentives for coal pretreatment in direct liquefaction processes. *Proc (ITSL) Integrated Two-Stage Liquefaction Meet.* (1982).

15. R. P. Anderson, Disposable catalysts in the solvent refined coal process. *Intersoc. Convers. Eng. Conf., Am. Inst. Aeronaut. Astronaut.* **2** (1980).

16. R. P. Killmeyer, Coal preparation for synfuels—Where do we stand? *Energy Prog.* **2**(1) (1982).

17. L. Lorenzi, Effect of coal cleaning on coal conversion and utilization. PETC Internal Report on Western Kentucky Coal Studies (1979).

18. U.S. Department of Transportation, *Coal Cleaning Effects During H-Coal Catalytic Liquefaction of a Western Kentucky Coal*, DOE Rep. EW-78-C-22-0262. USDOE, Washington, DC, 1976.

19. N. J. Mazzocco, E. B. Klunder, and D. Krastam, *Study of Catalytic Effects of Mineral Matter Level on Coal Reactivity*, DOE/PETC/TR-8011. U.S. Dept. of Energy, Washington, DC, 1981.

20. L. G. Benedict, Maceral and mineral concentrations in chance cone products. M.S. Thesis, Pennsylvania State University, University Park, 1962.

21. J. L. Bayer, Maceral segregation in commercially prepared coal products. M.S. Thesis, Pennsylvania State University, University Park, 1960.

22. D. S. Hoover and A. Davis, *The Development and Evaluation of an Automated Reflectance Microscope System for the Petrographic Characterization of Bituminous Coals*, DOE/TR-23 Contract No. AC01-76ET10615. U.S. Dept. of Energy, Washington, DC, 1980.

23. W. Spackman, A. Davis, P. L. Walker, Jr., H. L. Lovell, R. Stefanko, L. H. Essenhigh, F. J. Vastola, and P. H. Given, *Evaluation and Development of Special Purpose Coals*, ERDA Final Rep., FE-0390-2. Pennsylvania State University, University Park, 1976.

24. D. S. Hoover and E. N. Givens, The impact of coal selection on the SRC-1 process, oral presentation at *AIChE Natl Meet.* (1981).

25. D. C. Cronauer, R. G. Ruberto, R. S. Silver, R. J. Jenkins, and A. Davis, *Investigation of the Liquefaction of Partially-Dried and Oxidized Coals*, EPRI Final Rep., AP-1625. Elec. Power Res. Inst., Palo Alto, CA, 1980.

26. D. C. Cronauer, R. G. Ruberto, R. G. Jenkins, I. M. K. Ismail, and D. Schlyer, Liquefaction of partially dried and oxidized coals. *Fuel* **62**, 1116–1123 (1983).

27. D. C. Cronauer, R. G. Ruberto, R. G. Jenkins, A. Davis, P. C. Painter, D. S. Hoover, M. E. Starsinic, and D. Schlyer, Liquefaction of partially dried and oxidized coals. 2. Coal characteristics. *Fuel* **62**, 1124–1132 (1983).

28. D. C. Cronauer, R. G. Ruberto, R. S. Silver, R. G. Jenkins, A. Davis, and D. S. Hoover, Liquefaction of partially dried and oxidized coals. 3. Liquefaction results. *Fuel* **63**, 71–77 (1984).

29. A. W. Gauger, Condition of water in coals. In *The Chemistry of Coal Utilization* (H. H. Lowry, ed.), 1st edition Chapter 17, pp. 600–627. Wiley, New York, 1945.

30. H. Gan, S. P. Nandi, and P. L. Walker, Jr., Nature of the porosity in American coals. *Fuel* **51**, 272 (1972).

31. M. L. Gorbaty, Effect of drying on the absorptive properties of subbituminous coal. *Fuel* **57**, 796–797 (1978).

32. EDS Coal Liquefaction Process Development, Phase V. Quarterly Technical Progress Report for July 1 to Sept. 30, 1982, FE-2893-99, U.S. Dept. of Energy, Washington D.C. pp. 49–51. 1983.

33. R. A. Konyashima and T. S. Nikiforova, Effect of drying coal in vortex chambers on technological properties of coal for hydrogenation. *Khim. Tverd. Topl. (Moscow)* **5**, 33–34 (1978) *Chem. Abstr.* **90**; 41160d.

34. R. W. Skinner and E. N. Givens, *Effects of Accumulated Solids and Solvent IBP on Dissolver Performance*, SRC-I Q. Tech. Rep., DOE/OR/03054-4, Vol. 2. U.S. Dept. of Energy, Washington, DC, 1981.

35. F. D. Hoffert, Crushed coal drying system. U.S. Patent 3,953,927 (1976).

36. H. Kakunai, K. Montonaga, J. Nada, Y. Nakata, T. Ohzawa, and S. Yokota, Mfg. slurries for the hydroliquefaction of coal. U.S. Patent 4,344,837 (1982).

37. T. Komiyama, K. Oi, N. Ohnishi, S. Mori, and I. Kubo, Process for the dehydration and liquefaction of water-containing coal. U.S. Patent 4,344,837 (1982).

38. G. A. Smith, Sr., Coal cleaning. U.S. Patent 4,263,013 (1981).

39. W. Dolkmeyer, A. Giehr, K. H. Keim, and E. Meisenberg, PCT Int. Appl. WO 82 02,057 (1982); *Chem. Abstr.* **97**, 185284q.

40. Kawasaki Heavy Ind. Ltd., Jpn. Kokai Tokyo Koho, JP 82/08,286 (1982); *Chem. Abstr.* **96**, 165441q.

41. M. I. Greene, Short residence time low pressure hydropyrolysis of carbonaceous materials. *Fuel Process. Technol.* **1**, 169–185 (1977–1987).

42. D. D. Whitehurst, J. O. Mitchell, and M. Farcasiu, *Coal Liquefaction: The Chemistry and Technology of Thermal Processes*, pp. 85–122. Academic Press, New York, 1980.

43. G. P. Curran, R. T. Struck, and E. Gorin, The mechanism of the H-transfer process to coal and coal extract. *Prepr. Pap.—Am. Chem. Soc., Div. Fuel Chem.* **10**(2), 130–148 (1966).

44. V. Rank, *Influence of the Fineness of Coal Grinding on the Results in Liquid Phase Hydrogenation*, Rep. No. TOM-302-2788. Texas A&M Univ., College Station, 1944.

45. R. Yoshida, K. Ishida, T. Yoshida, S. Veda, I. Sekiguchi, Y. Nakata, S. Yokoyama, T. Okutani, Y. Jomoto, and Y. Yoshida, Effect of particle size on the coal hydrogenation reaction—in relation to the plasticity of coal. *Fuel Process. Technol.* **3**, 1–5 (1980).

46. F. J. Riedl, R. S. Corey, and R. E. Svacha, Coal liquefaction process. U.S. Patent 3,536,608 (1970).

47. J. A. Solomon and G. J. Mains, A mild, protective and efficient procedure for grinding coal: Cryocrushing. *Fuel* **56**(3), 302–304 (1977).

48. R. T. Yang, R. Smol, G. Farber, and L. M. Naphtali, Mechanochemical hydrogenation of coal. U.S. Patent 4,250,015 (1981).

49. R. H. Wolk, Coal hydrogenation using pretreatment reactor. U.S. Patent 3,791,957 (1974).

50. D. R. Skidmore and C. J. Konya, Chemical comminution of coal. *W. Va. Univ. Sch. Mines, Coal Res. Bur., Rep.* **134**, 1–14 (1976).

51. E. F. Nelson, Method of treating coal. U.S. Patent 3,477,941 (1969).

52. C. Y. Chang, J. A. Guin, and A. R. Tarrer, Effect of mechanical-chemical pretreatment on coal liquefaction. *J. Chim. Inst. Chem Eng.* **12**(3), 119–25 (1981).

53. R. B. Sellers, U.S. Patent 2,572,061 (1951).

54. J. Winkler, Depolymerization of bituminous coal utilizing friable metal reactants. U.S. Patent 3,282,826 (1963).

55. L. G. Massey, R. I. Brabets, and W. A. Abel, Method for separating undesired components from coal by an explosion type comminution process. U.S. Patent 4,313,737 (1982).

56. L. G. Massey, R. I. Brabets, and W. A. Abel, Selective comminution of hydrocarbonaceous solids. U.S. Patent 4,364,740 (1982).

57. W. A. Abel, R. I. Brabets, and L. G. Massey, U.S. Patent 4,365,740 (1982).

58. H. Juntgen and K. H. Van Heek, *Research in the Field of Pyrolysis During the Last 15 Years at Bergbauforschung*, pp. 1–42. IEA Working Party on Coal Technology, National Coal Board, Statre Orchard, U.K., 1977.

59. H. Juntgen and K. H. VanHeek, An update on German non-isothermal coal pyrolysis. *Fuel Process. Technol.* **2**, 261–293 (1979).

60. H. N. S. Schafer, Pyrolysis of brown coals. 2. Decomposition of acidic groups on heating in the range of 100–900°C. *Fuel* **58**, 673–679 (1979).

61. H. Hupfer, *Pretreatment of Bituminous Coal for Hydrogenation (Older Experiments)*, Rep. No. TOM-256-343-344. Texas A&M Univ., College Station, 1943.

62. M. Z. Sze and G. J. Snell, Liquefaction of subbituminous and lignitic coal. U.S. Patent 4,028,221 (1977).

63. E. Hippo, W. L. Goldman, G. R. DeVaux, and A. G. Comolli, Drying and deoxygenating coal destined for liquefaction. Ger. Offen. DE 3,339,139 (1984).

63a. H. Shimada, T. Sato, M. Kurita, Y. Yoshimora, A. Nishijima, and N. Todo, Effect of preheating on direct coal liquefaction. *Sekiyu Gakkaishi* **26**(3), 195–200 (1983); *Chem. Abstr.* **99**(10), 73559T.

64. Agency of Industrial Sciences and Technology, Coal liquefaction, JP Kokai Tokoyo Koho 81/59,893 (1981); *Chem. Abstr.* **95**, 83575b.

65. R. J. Brunson, Liquefaction of calcium-containing subbituminous coals and coals of lower rank. U.S. Patent 4,161,440 (1979).

66. S. K. Poddar, Liquefaction process. U.S. Patent 4,304,655 (1981).

67. P. Urban, Liquefaction of calcium containing coals. U.S. Patents 4,401,449 and 4,401,550 (1983).

68. R. C. Neavel, R. J. Brunson, and J. J. Chaback, CO_2 pretreatment prevents calcium carbonate deposition. U.S. Patent 4,205,033 (1980).

69. M. B. Abdel-Baset and C. T. Radcliffe, Novel approach to coal liquefaction utilizing H_2S and CO. *Prepr. Pap.—Am. Chem. Soc. Div. Fuel Chem.* **25**(1), 1–7 (1980).

70. J. G. Gatsis, Coal liquefaction process. U.S. Patent 3,503,863 (1970).

71. R. Bearden, Jr. and C. L. Aldridge, Coal liquefaction process. U.S. Patent 4,094,765 (1978); 4,149,959 (1979).

72. J. W. Larsen and J. Kovac, Polymer structure of bituminous coal. *ACS Symp. Ser.* **71**, 36–50 (1978).

73. N. Y. Kirov, J. M. O'Shea, and G. D. Sergeant, Determination of solubility parameters for coal. *Fuel* **47**, 415 (1968).

74. J. M. Angelovich, G. R. Pastor, and H. F. Silver, Solvents used in the conversion of coal. *Ind. Eng. Chem. Process Dev. Des.* **9**(1), 106–109 (1970).

75. E. L. Wilson and R. E. Pennington, Coal liquefaction at staged temperatures. U.S. Patent 3,692,662 (1972).

76. R. Horn, Solvent-vapor treatment of coal prior to high-pressure hydroliquefaction. German Patent 3,045,338 (1982).

77. R. B. Long, M. L. Gorbaty, and R. H. Schlosberg, Coal liquefaction process. U.S. Patent 4,252,633 (1981).

78. R. B. Long, M. L. Gorbaty, and L. W. Vernon, Coal liquefaction process. U.S. Patent 4,250,014 (1981).

79. R. B. Long, Coal liquefaction process. U.S. Patent 4,253,937 (1981).

80. N. K. Narain, H. R. Appell, and B. R. Utz, The effect of solvent pretreatment on coal liquefaction. *Prepr. Pap.—Am. Chem. Soc., Div. Fuel Chem.* **28**(1), 161–162 (1983).

81. A. Marzec, M. Juzwa, K. Betlej, and M. Sobkowiak, Bituminous coal extraction in terms of electron-donor and -acceptor interactions in the solvent/coal system. *Fuel Process. Technol.* **2**, 35–44 (1979).

82. I. G. C. Dryden, Chemical reactions and consitution of coal. In *The Chemistry of Coal Utilization* (H. H. Lowry ed.), Suppl. Vol., Chapter 6, pp. 237–252. Wiley, New York, 1963.

83. D. W. Van Krevelen, *Coal-Typology, Chemistry, Physics and Constitution.* pp. 177–199. Am. Elsevier, New York, 1961.

84. I. Wender, L. A. Heredy, M. B. Neuworth, and I. G. C. Dryden, Chemical reactions and constitution of coal. In *Chemistry of Coal Utilization* (*Second Supplementary Volume*) (M. A. Elliott, ed.), Vol. 2, Chapter 8, pp. 425–523. Wiley, New York, 1981.

85. J. R. Pullen, *Solvent Extraction of Coal*, IEA Coal Res. Rep., pp. 1–124. IEA, London, 1981.

86. E. H. Bay, Coal extraction and fuel addition made therefrom. U.S. Patent 4,089,658 (1978).

87. F. R. Zinniel, E. R. Blome, and P. D. Kimball, Resin extraction process. U.S. Patent 3,637,639 (1972).

88. A. H. Stiller, J. T. Sears, and R. W. Hammack, Coal extraction process. U.S. Patent 4,272,356 (1981).

89. M. Vahrman, Smaller molecules derived from coal and their significance. *Fuel* **49**, 5 (1970).

90. J. A. Spence and M. Vahrman, Aliphatic hydrocarbons in a low-temp. tar. *Fuel* **49**, 395 (1970).

91. R. Hayatsu, R. E. Winans, R. G. Scott, L. P. Moore, and M. H. Studier, Trapped organic compounds and aromatic units in coal. *Fuel* **57**, 541 (1978).

92. A. Jurkiewicz, A. Marzec, and N. Pislewski, Molecular structure of bituminous coal studied with pulse NMR. *Fuel* **61**, 647–650 (1982).

93. N. A. Peppas, M. E. Hill-Lievense, D. T. Hooker, III, and L. M. Lucht, *Macromolecular Structural Changes in Bituminous Coal During Extraction and*

Solubilization, DOE Prog. Rep. under Contract No. DE-FG22-80PC30222. Purdue University Dep. Chem. Eng., West Lafayette, IN, 1981.

94. M. Radke, H. Willsch, D. Leythaeuser, and M. Teichmuller, Aromatic components of coal: Relation of distribution pattern to rank. *Geochim. Cosmochim. Acta* **46**(10), 1831–1848 (1982).

95. W. A. Bone, A. R. Pearson, and R. Quarendon, Research on the chemistry of coal. Part III. The extraction of coals by benzene under pressure. *Proc. R. Soc. London* **105**, 608–625 (1924).

96. R. V. Wheeler, The volatile constituents of coal liquefaction: Role of molecular hydrogen. *Fuel* **59**, 102–106 (1913).

97. E. C. Harris, Jr. and E. E. Peterson, Change in the physical characteristics of Roland Seam coal with progressive solvent extraction. *Fuel* **58**, 599 (1979).

98. K. D. Bartle, T. G. Martin, and D. F. Williams. Chemical nature of a supercritical-gas extract of coal at 350°C. *Fuel* **54**, 226 (1975).

99. R. R. Maddocks, J. Gibson, and D. F. Williams, Supercritical extraction of coal. *Chem. Eng. Prog.* **75**, 49–55 (1979).

100. E. F. Williams, Extraction with supercritical gases. *Chem. Eng. Sci.* **36**(11), 1769–1788 (1981).

101. J. E. Blessing and D. S. Ross, Supercritical solvents and the dissolution of coal and lignite. *ACS Symp. Ser.* **71**, 171–185 (1978).

102. J. Josephson, Supercritical fluids. *Environ. Sci. Technol.* **16**(10), 548A–551A (1982).

103. J. R. Kershaw, Solvent effects on the supercritical gas extraction of coal. The role of mixed solvents. *Fuel Process. Technol.* **5**, 241-246 (1982).

104. J. C. Whitehead, Development of a process for the supercritical gas extraction of coal. *Coal Process. Technol.* **7**, 1–8 (1981).

105. S. Weller and M. G. Pelipetz, Coal hydrogenation catalysts: Studies of catalyst distribution. *Ind. Eng. Chem.* **43**(5), 1243–1246 (1952).

106. W. R. K. Wu and H. H. Storch, Hydrogenation of coal and tar. *Bull.—U.S. Bur. Mines* **633**, 19–20 (1968).

107. J. G. Gatsis, Solvent extraction of coal using a heteropolyacid catalyst. U.S. Patent 3,813,329 (1974).

108. K. Higashiyoma, A. Tomita, S. Takizawa, and Y. Tamai, Gasification of coal impregnated with nickel salt in liquid ammonia. *Nenryo Kyokaishi* **60**(654), 842–847 (1981); *Chem. Abstr.* **96**, 71543x.

109. C. R. Porter and R. L. Bain, Low-valent organometallic catalysis in coal liquefaction. *AIChE Annu. Meet.* Feb.-Mar. (1982).

110. J. L. Cox and W. A. Wilcox, U.S. Patent 4,155,832 (1979).

111. G. A. Cremer, Catalytic hydrogenation of model-coal compounds by soluble transition metal hydrides. Ph.D. Dissertation, University of California, Berkely, 1982.

112. N. G. Moll and G. J. Quarderer, Emulsion catalyst for hydrogenation processes. U.S. Patent 4,136,013 (1979).

113. C. L. Aldridge and R. Bearden, Staged hydroconversion of an oil-coal mixture. U.S. Patent 4,111,787 (1978).

114. R. L. Hodgson, Hydroconversion of coal with combination of catalysts. U.S. Patent 3,532,617 (1970).

115. R. L. Hodgson, Hydrogenation of coal employing impregnation of in-situ prepared catalyst. U.S. Patent 3,502,564 (1970).

116. R. E. Wood, W. H. Wiser, L. L. Anderson, and A. G. Oblat, Process for minimizing vaporizable catalyst requirements for coal hydrogenation-liquefaction. U.S. Patent 4,134,822 (1979).

117. R. C. Neavel and R. J. Lang, Distributing coal liquefaction or gasification catalysts on coal. British Patent 1,599,932 (1978).

118. P. H. Given, D. C. Cronauer, W. Spackman, H. L. Lovell, A. Davis, and B. Biswas, Dependence of coal liquefaction behavior on coal characteristics. 1. Vitrinite-rich samples. *Fuel* **54**(1), 34–39 (1975).

119. M. L. Gorbaty and J. W. Taunton, Liquefaction of calcium-containing subbituminous coals and coals of lower rank. U.S. Patent 4,227,989 (1980).

120. J. L. Johnson, The use of catalysts in coal gasification. *Catal. Rev.*—Sci. Eng. **14**(1), 131–152 (1976).

121. M. S. Burns, R. A. Durie, and D. J. Swaine, Significance of chemical evidence for the presence of carbonate minerals in brown coals and lignites. *Fuel* **41**, 373 (1962).

122. R. A. Durie, Inorganic constituents in Australian coals (III). Morwell and Yallourn brown coals. *Fuel* **40**, 407 (1961).

123. H. N. S. Schafer, Carboxyl groups and ion-exchange in low-rank coals. *Fuel* **49**, 197–213 (1970).

124. R. N. Miller, A geochemical study of the inorganic constituents in some low-rank coals. Ph.D. Thesis, Pennsylvania State University, University Park, 1977.

125. P. H. Given, W. Spackman, A. Davis, P. L. Walker, Jr., H. L. Lovell, M. Coleman, and P. C. Painter, *The Relation of Coal Characteristics to Liquefaction Behavior*, Tech. Prog. Rep., DOE-FE-2494-3/6, p. 40. Pennsylvania State University, University Park, 1979.

126. G. G. Baker, T. C. Owens, and J. R. Rindt, Effects of several disposable catalysts on liquefaction of lignite. *Am. Soc. Civ. Eng. Annu. Meet.* CONF-8210360-1, pp. 1–22 (1982).

127. M. R. Hatswell, W. R. Jackson, F. P. Larkins, M. Marshall, D. Rash, and D. E. Rodgers, Hydrogenation of brown coal. *Fuel* **62**, 336–341 (1983).

128. M. R. Hatswell, W. R. Jackson, F. P. Larkins, M. Marshall, D. Rash, and D. E. Rodgers, Hydrogenation of ion-exchanged Victorian brown coal. *Fuel* **59**, 442–444 (1980).

129. M. Bishop and O. L. Ward, The direct determination of the mineral matter in coal. *Fuel* **37**, 191–200 (1958).

130. W. Radmacher and P. Mohrhauer, The direct determination of the mineral matter content of coal. *Brennst.-Chem.* **36**, 236 (1955).

131. P. H. Given, W. Spackman, A. Davis, P. L. Walker, Jr., and H. L. Lovell, *The Relationship of Coal Characteristics to Liquefaction Behavior*, Semiannu. Rep., RANN Grant No. GI-38974, pp. 1–43. Natl. Sci. Found., Pennsylvania State University, University Park, 1974.

132. P. H. Given, W. Spackman, A. Davis, P. L. Walker, Jr., H. L. Lovell, M. Coleman, and P. C. Painter, *The Relation of Coal Characteristics to Liquefaction Behavior*, Q. Tech. Prog. Rep. DOE-FE-2494-7/8, p. 36. Pennsylvania State University, University Park, 1979.

133. J. J. Dickert, Jr., T. O. Mitchell, and D. D. Whitehurst, Liquefaction of acid-treated coals. U.S. Patent 4,257,869 (1981).

134. I. Mochida, Y. Moriguchi, K. Iwamoto, H. Matsuoka, Y. Korai, H. Fujitsu, and K. Takeshita, Thermal degradation of coal in their liquefaction and coking. *Proc. Int. Conf. Coal Sci., 1981* pp. 580–585 (1981).

135. I. Mochida, T. Tahara, K. Iwamoto, Y. Korai, H. Fujitsu, and K. Takeshita, Liquefaction, carbonization and modification of low-rank coals. *Nippon Kagaku Kaishi* 6, 899–907 (1980); *Chem. Abstr.* 93; 152840g.

136. I. Mochida, Y. Moriguchi, T. Shimohara, K. Yazo, H. Fujitsu, and K. Takeshita, Enhanced reactivity of some lignites by deashing pretreatment in H-transfer liquefaction under atmospheric pressure. *Fuel* 62, 471–473 (1983).

137. R. E. Jones and D. T. A. Townsend, The oxidation of coal. *J. Soc. Chem. Ind., London* 68; 197–201 (1949).

138. P. C. Painter, M. Coleman, R. Snyder, O. Mahajan, M. Komatsu, and P. L. Walker, Jr., Low-temperature air oxidation of coking coals. Fourier transform infrared studies. *Appl. Spectrosc.* 35, 106 (1981).

139. R. Liotta, R. G. Beons, and J. Isaacs, Oxidative weathering of Illinois #6 coal. *Fuel* 62; 782–791 (1983).

140. S. E. Moschopedis and J. G. Speight, Oxidation of a bitumen. *Fuel* 54, 210–212 (1975).

141. K. Ouchi, Y. Maeda, H. Itoh, and M. Makabe, Effect of air-oxidation on coal hydrogenation. *Fuel* 63(1); 35–38 (1984).

142. C. Y. Chang, J. A. Guin, and A. R. Tarrer, An investigation of the effect of air-oxidation on coal liquefaction. *J. Chin. Chem. Soc. (Taipei)* 28(3); 155–160 (1981).

143. R. C. Neavel, Liquefaction of coal in hydrogen-donor and non-hydrogen donor vehicle. *Fuel* 55; 237–242 (1976).

144. A. A. Krichko, R. A. Konyashima, and T. S. Nikiforova, Effect of oxidation on technological properties of coal for hydrogenation. *Khim. Tverd. Topl. (Moscow)* 1; 65–67 (1982); *Chem. Abstr.* 96; 165304x.

145. D. E. Severson, Process development for solvent refined lignite: Laboratory autoclave studies. Part II. Project lignite. *N. Dak. Univ., Grand Forks Eng. Exp. Stn. R&D Rep.* 106, 1–65 (1978).

146. D. J. Maloney, R. G. Jenkins, and P. L. Walker, Jr., Low-temperature air-oxidation of caking coals. 2. Effects on swelling and softening properties. *Fuel* 61(2); 175–181 (1982).

147. A. Y. Y. Kam, Effects of oxidation of bituminous coal on its caking properties: A study of kinetics, modelling and correlations. Ph.D. Thesis, pp. 1–222. University of Pennsylvania, Philadelphia, 1975.

148. M. Kaiho, "Research into the pressurized gasification of coal: Effect of air-oxidation on coal caking properties. *Proc. Cont. Coal Sci.* 17, 212–218 (1980).

149. O. P. Mahajan, M. Komatsu, and P. L. Walker, Jr., Low-temperature air-oxidation of caking coals. 1. Effect on subsequent reactivity of chars produced. *Fuel* 59; 3–10 (1980).

150. D. J. Maloney and R. G. Jenkins, Effects of preoxidation on pyrolysis behavior and resultant char structure of caking coal. *Prepr. Pap.—Am. Chem. Soc., Div. Fuel Chem.* 27(1); 25–30 (1982).

151. D. J. McCarthy, Flash pyrolysis of Australian coals: Effects of preoxidation and of oxygen in the inlet gases on formation of agglomerated material. *Fuel* **60**(3); 205–209 (1981).

152. E. Furimsky, J. A. MacPhee, L. Vancea, L. A. Ciavaglia, and B. N. Nandi, Effect of oxidation on the chemical nature and distribution of low-temperature pyrolysis products from bituminous coal. *Fuel* **62**(4); 395–400 (1983).

153. F. G. Parker, J. P. Fugasi, and H. C. Howard, Conversion of coal to simple compounds: Action of aqueous alkali on a subbituminous coal at elevated temperature. *Ind. Eng. Chem.* **47**(8), 1586 (1955).

154. L. Kasehagen, Action of aqueous alkylation of a bituminous coal. *Ind. Eng. Chem.* **29**; 600–604 (1937).

155. M. Makabe, Y. Hirano, and K. Ouchi, Extraction increase of coals with alcohol-sodium hydroxide at elevated temperatures. *Fuel* **57**, 289–292 (1978).

156. J. G. Huntington, F. R. Mayo, and N. A. Kirshen, Mild oxidation of coal fractions. *Fuel* **58**; 24–36 (1979).

157. P. E. Araya, R. Badilla-Ohlbaum, and S. E. Droquett, Study of the treatment of subbituminous coals by NaOH solutions. *Fuel* **60**; 1127–1130 (1981).

158. R. J. Camier and S. R. Siemon, Colloidal structure of Victorian brown coals. 1. Alkaline digestion of brown coal. *Fuel* **57**; 85–88 (1978).

159. J. D. Brooks and S. Sternhell, Brown coals (I): O-containing functional groups in Victorian brown coals. *Fuel* **37**; 124 (1958).

160. D. S. Ross and J. E. Blessing, Alcohols as H-donor media in coal conversion. *Fuel* **58**(6); 433–444 (1979).

161. P. Urban, Coal liquefaction process. U.S. Patent 3,796,650 (1974).

162. K. Shimizu, Process for liquefaction of coal. U.S. Patent 4,222,849 (1980).

163. J. H. Hickey, U.S. Patent 2,556,496 (1951).

164. E. P. Staumbaugh, Deashing and desulfurizing carbonaceous solids. U.S. Patent 4,121,910 (1978).

165. R. Swanson, Procedure for conversion of coal into gaseous hydrocarbons. Belgian Patent 884,644, pp. 1–30 (1980).

166. C. Y. Chang, J. A. Guin, A. R. Tarrer, and M. H. Lee, A study of the effect of aqueous alkali pretreatment on coal liquefaction. *J. Chin. Inst. Chem. Eng.* **13**(3), 113–121 (1982).

167. C. K. Chow, Reactivity of hydrolyzed coals in liquefaction. *Fuel* **62**; 317–322 (1983).

168. R. H. Schlosberg, P. S. Maa, and R. C. Neavel, Treatment of coal by alkylation t increase liquid products from coal liquefaction. U.S. Patent 4,092,235 (1978).

169. R. H. Schlosberg, R. C. Neavel, P. S. Maa, and M. L. Gorbaty, Alkylation, a beneficial pretreatment for coal liquefaction. *Fuel* **59**; 45–47 (1980).

170. R. H. Schlosberg, M. L. Gorbaty, and R. J. Lang, Method for the preparation of non-caking coals from caking coals by means of electrophilic aromatic substitution. U.S. Patent 4,059,410 (1977).

171. R. Liotta, Pretreatment of solid, naturally-occurring carbonaceous material. U.S. Patent 4,259,084 (1981).

172. R. Liotta, Selective alkylation of acidic hydroxyl groups in coal. *Fuel* **58**; 724–728 (1979).

173. R. Liotta, K. Rose, and E. Hippo, O-alkylation chemistry of coal and its implications for the chemical and physical structure of coal. *J. Org. Chem.* **46**; 277–283 (1981).

174. R. Liotta, O-alkylated/O-acylated coal and coal bottoms. U.S. Patent 4,372,750 (1983).

175. Y. Sanada, S. Yokoyama, M. Shimohara, S. Ono, and H. Maritomi, Solubilization of coals with olefins and cracked light oils from petroleum. *Proc. Int. Conf. Coal Sci., 1981*, pp. 92–97 (1981).

176. F. Mondragon, H. Itoh, and K. Ouchi, Solubility increase of coal by alkylation with various alcohols. *Fuel* **61**(11), 1131–1134 (1982).

177. C. Horowitz and M. Dickert, Process for treating coal and products produced thereby. U.S. Patent 4,033,852 (1977).

178. N. Berkowitz, Physical behavior of coal in ultrasonic fields. *Nature (London)* **163**; 804 (1949).

179. I. Kessler, R. A. Friedel, and A. G. Sharkey, Jr., Ultrasonic solvation of coal in quinoline and other solvents. *Fuel* **49**, 222–223 (1970).

180. K. Littlewood, The application of ultrasonic irradiation to solid-liquid systems with particular reference to the extraction of coals with pyridine. *Fuel Soc. J.* **11**; 27–39 (1960).

181. R. D. Stone, Coal liquefaction process. U.S. Patent 3,505,865 (1970).

182. J. F. Flagg, R. W. Johnson, and B. Gatsis, Coal liquifaction process. U.S. Patent 4,326,945 (1982).

Commentary

A study of aging and weathering would not be complete without consideration of microbial action on coal. Not only can microbes be a mediating factor in a number of oxidative and decomposition processes, they also can be useful for desulfurization, demineralization, and gasification. The general subject will continue to be of interest as genetically altered microbes are developed for enhanced biotreatment programs. Chapter 9 presents an overview of the biomechanisms involved in microbial action on coal. Considering its importance, we, the editors, believed it would be useful to include the topic of microbial action on coal in this book.

9

MICROBIAL ACTION ON COAL

R. S. Hsu-Chou, H. C. G. do Nascimento, and
T. F. Yen
University of Southern California

9.1 INTRODUCTION

In studying the origin of coal formation, it is a matter of practical importance to understand the role that microorganisms play in the coal-forming process. As microorganisms cause plant remains to degrade, anaerobic conditions are created by aerobic microbial processes and burial. Anaerobic microbial metabolism continues until toxic metabolic products accumulate or water availability decreases. These processes are carried out by microorganisms in every stage of coal formation, so a contemporary microbiological investigation should assist in the proper understanding of biogeochemical roles of microorganisms during each period.

Even though coal is an end product of extensive biodegradation, it is feasible that it can be further utilized by microbes as a substrate. Since very few studies have been conducted to explore the interactions between coal and microbiota, the history of microbial conversion of coal is worthy of discussion. In this context, the application of mechanisms toward possible degradation of coals, such as those of the lignin nature, coal's aromatic ring containing components, the heterocyclic linkages or bridges, and the montan wax portion are evaluated. The possible use of a crude genetic engineering approach is also included. It has been confirmed that by selecting certain species of microorganisms, biodegradation of coal can have specific pathways which yield desirable end products.

Microbial beneficiation is the process of selectively removing sulfur and nitrogen heteroatoms from coal and generally minimizing the concentration of minerals. This process is capable of eliminating many of the engineering problems associated with high-pressure and high-temperature processing because reactions can proceed at ambient pressure and temperature, requiring little or no additional thermal energy.

In nature, coal comes in contact with soil and air that contain varied and large quantities of microbes. Handling and storing of coal may even increase microbial action, which could induce harmful results. The three most common problems caused by microbial contamination of coal are (1) lowering of the heating value of coal, (2) acid drainage of stockpiled coal, and (3) spontaneous combustion of coal.

Currently, several processes are being used to convert medium- to high-rank coals into liquefied and gasified fuels. Chemical modification in conjunction with biodegradation seems to be promising because both processes are capable of converting coal into useful energy products with both low cost and low environmental impact.

9.2 ASSOCIATION OF MICROORGANISMS WITH COAL DEPOSITS

Coal represents an end product that has already undergone extensive biological processing. Early in the coal-forming process, plant remains are degraded by microorganisms. Aerobic microbial processes and burial create anaerobic conditions, and anaerobic microbial metabolism of the material continues until it is stopped by the accumulation of toxic metabolic products or reduction in water availability.[1,2] Some investigators suggest that microbial cell debris may make up a large part of fossil fuels, which include coal.[3]

On the basis of the information available at this time, it seems that the same microorganisms that are now capable of carrying out activities in soil and peat are those that were present in every period of coal formation in the past. Therefore, the physiology and ecology of modern microorganisms should be a reliable guide to their biogeochemical role in the past.[4]

There are numerous indigenous microorganisms in the soil. These microorganisms are inherent in the ecological niche and are often found in fossilized systems. A good example is that of *Micrococcus nishinomiyaenisis* and *M. sedentarius* isolated from the core of a piece of oil shale as old as 50 million years.[5] Other examples indicate that carbon and graphite can be biodeteriorated through processes similar to those commonly occurring in soil.[6] Even gold can be solubilized by bacteria.[7]

This section discusses the role of microorganisms throughout the process of coalification and in the origin of coals. Analysis of peat, the precursor of all coals, has shown the presence of varied microflora. In general, microorganisms found in peat are comprised of several groups such as aerobic and anaerobic microorganisms active in primary decay, facultative and obligate anaerobes accomplishing the decomposition of organic complexes, and nitrogen-fixing (N_2-NH_3) bacteria. Fuchs emphasized the importance of microbiological processes in coalification evolution.[8] The frequently occurring microbial species associated with coal or related to coal studies are listed in Table 9.1.[9-18] The microorganisms shown there are capable of physiological oxidative activities by direct oxidation or oxidative dehydrogenation processes. The functions of some microorganisms are summarized as follows:

1. *Actinomycetes*, and especially *Thermoactinomycetes*, are ubiquitous soil forms found abundantly in peat.[18] They are closely associated with the process of humification and are known to decompose humic acid isolated from coal. They can also decompose mellitic acid and graphitic acid.[19]

2. The genus *Pseudomonas* is also ubiquitous in soil. One of the genera thoroughly studied for its role in the degradation of polynuclear aromatic hydrocarbons,[20,21] it is highly adaptive and attacks petroleum, oil fractions, and even asphalts.

3. The yeast *Candida* and its subspecies are known to attack coal-derived products such as coal tar fractions,[22] oxidized coal,[16] and the low-temperature lignitic tar.[23] Most of these yeasts are producers of single-cell protein from hydrocarbon and wax fractions.

4. Molds of the genera *Penicillium*, *Aspergillus* and *Trichoderma* reportedly oxidize paraffins and humic acid. They can also decompose cellulose.

As indicated by Ehrlich,[2] reports on the isolation of microorganisms from different kinds of coal have claimed the detection of fungi such as *Aspergillus*, *Penicillium*, *Verticillium*, *Trichoderma*, *Fusarium*, and *Mucor*. Similar claims exist for the detection of bacteria such as *Sarcina*, *Clostridium welchii*, *Bacillus megaterium*, *B. subtilis*, *Pseudomonas fluorescens*, *Galionella*, and *Thiothrix*. However, because most of these organisms are common in soil and water, it is not clear if they represent an indigenous flora of coal. The most frequent claims for microbial association with coal are in connection with lower-rank coals, for example, lignites, whose flora includes *Psuedomonas*, molds, and *Actinomycetes*.[9] Despite a claim to the contrary,[24] anthracite coal probably does not harbor an indigenous flora.[10,25]

Iron-oxidizing *thiobacilli* have been found in bituminous coal. These microorganisms grow at the expense of pyrite or marcasite associated with the coal seams.[26,27] FeS_2 is undoubtedly formed during the anaerobic phase of coal formation, and its development is at least in part, if not wholly, the result of microbial activity. The ecology of iron-oxidizing bacteria associated with coal has been studied.[28]

Ward[29,30] isolated a number of coal-degrading fungi from a Mississippi lignite outcrop (see Table 9.2). Mississippi lignite serves as sole source of organic carbon. The biodegradation for some fungal isolates could be enhanced by 30 percent when supplemented with dilute basal mineral solution. Deletion of sulfur from the mineral medium does not affect growth, and this indicates a potential of using fungi for desulfurization. The preliminary data show that one fungal isolate, LW-20, can utilize organic sulfur compounds.

TABLE 9.1 Microorganisms Found to Be Associated with Coal or Related to Coal Studies

Genus and Species	Habitat	Faculty	Reference
	Bacteria		
Sarcina	Soil, mud, swamp, water	Methane producers	9
Bacillus magatherium	Soil, water, decomposing organic matter	Saprophyte	10
Bacillus subtilis	Soil, water, decomposing organic matter	Aerobic spore former	11
Gallionella	Fe-containing water	Stalks of ferric oxides	9
Thiothrix	Decaying vegetation	Sulfur-bacteria	12
Clostridium perfringers	Soil, sewage	Gas producer	9
Pseudomonas aeruginosa	Soil, sewage	Hydrocarbons aromatics	9
Pseudomonas rathonis	Soil near coal mines	Aromatics	13
Escherichia freundii	Soil near coal mines; intestinal tract	Citrate utilizers	13
Thermoactinomycetes	Soil–peat	Decomposing organic compounds	14
	Fungi, Yeast		
Aspergillus minor	Dry vegetable matter	Tolerates high osmotic pressure	15
Candida tropicallis	Decomposing vegetation	Protein producer	16
Candida lipolytica	Soil	Protein producer	17
Trichoderma	Soil	Cellulose decomposition ammonification	9
Fusarim	Soil	Air contaminant	12
Mucor	Manure and fruits	Flex wetting	18
Humicola	Soil	Humus	18

TABLE 9.2   Lignite Degrading Fungal Isolates

Organism[a]	Relative Growth Rate
LW-2 (yeast)	Slow
LW-5 (mycelial fungus)	Slow
LW-13 *Candida* sp.	Fast
LW-18 *Penicillium* sp.	Fast
LW-20 *Penicillium* sp.	Fast
LE-24 *Mucor* sp.	Moderate
LE-24 *Mucor* sp.	Slow
LE-27 *Mucor* sp.	Moderate
LE-30 (mycelial fungus)	Fast
LE-31 *Penicillium* sp.	Fast
LE-33 *Paecilomyces* sp.	Fast
LE-34 *Penicillium* sp.	Fast

[a]LW = Isolated by H. B. Ward; LE = Isolated by T. E. Enslin.
Source: B. Ward, Coal degrading fungi: Isolated from a Mississippi lignite outcrop. In *Proceedings of Symposium on Biological and Chemical Removal of Sulfur and Trace Element in Coal and Lignite*, p. 149. Louisiana Tech. University, Ruston, 1982.

9.3   MICROORGANISMS AND COAL FORMATION

Coal originated through the accumulation of plant debris that was later covered, compacted, and changed into the form of organic rock that we know today.[31] The sequential mechanism of coal formation based on end products and intermediates can be illustrated in the following stages:

plant <u>humification</u> ⟶ humic acid <u>peatification</u> ⟶ peat <u>coalification</u> ⟶ coal

Humification and peatification are accomplished mainly by the biodegradation processes. Biochemical processes ceased gradually and were replaced in the initial stage of coalification by endothermic processes subjected to mild or severe pressure influence. The activity of bacteria, particularly anaerobic ones, continued and the plant material was increasingly compressed and reduced in volume as a result of water loss.

Four groups of biochemical processes are involved in the decomposition of plant matter:[32]

1. *Rotting* occurs in free air. It is a complete decomposition;
2. *Mouldering* occurs under insufficient air access, where the plant material is saturated with water and compressed;

3. *Putrefaction* is a predominantly anaerobic process that decomposes remains of plant and faunal organisms in the presence of water, without air;

4. *Peatification* occurs below the water level in the region of aquatic and subaquatic growth, that is, in places with partial access to air.

Each one of these biochemical processes yields specific end products. Table 9.3 summarizes possible end products after the action of the above mentioned biodegradation processes. The destruction processes—putrefaction, mouldering, and peatification—are combined in different ratios to contribute to the generation of organic rock.

The principal components of plant matter are cellulose and lignin. The average content of cellulose in peat is about 15–25 percent. Cellulose is decomposed by aerobic and anaerobic bacteria (*Bacillus cellulosae methanicus* and *Bacillus cellulosae hydrogenicus* are the well-known ones), mycobacteria, and fungi.[32] The main portion of humic acids in peat and coal is derived from lignin.[33] Investigations have indicated that white-rot fungi (*Basidiomycetes phanerochacte chrysosporium*[34,35] and closely related litter-decomposing fungi are dominant lignin degraders; brown-rot fungi (*Basidiomycetes*) causes limited degradation of lignin.[36,37] It is known that soft-rot fungi (*Fungi imperfecti*), various soil *Ascomycetes*, *Fungi imperfecti* (*Fusarium, Aspergillus*), and also some bacteria (*Norcardia, Streptomycetes*[38]) partially degrade lignin.[39] Under anaerobic conditions, lignin can be degraded by *Flavobacterium* and *Pseudomonas*.[32]

Since cellulose and lignin are deficient in nitrogen, microorganisms must take it from proteins and amino acids, especially during humification. Figure 9.1[40] shows a proposed mechanism of humification. Among other factors, humification is accelerated by high nitrogen content, high temperature, alkalinity, and aeration.

TABLE 9.3   End Products of Biodegradation of Plant Matter

	Biological Process			
	Rotting	Mouldering	Putrefaction	Peatification
Gaseous products	CO_2 H_2O	CO_2 H_2O	CH_4, NH_3 N_2O, N_2 H_2S, CO_2	CO_2, H_2O CH_4 less NH_3
Remaining residue	resins waxes	humic matter resins waxes suberin cutin mineral	bitumen kerogen	peat resins waxes mineral

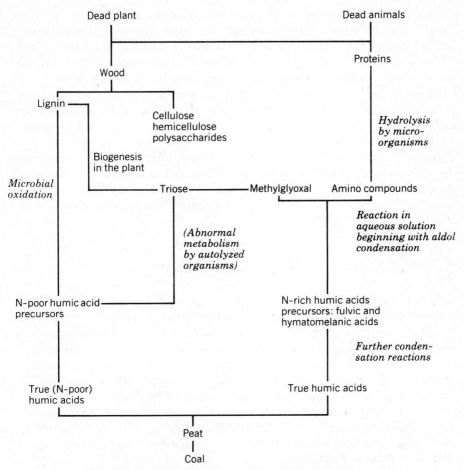

FIGURE 9.1   Mechanism of humification (after van Krevelen[40]).

With regard to the peatification process, information on specific organisms involved in the formation of peat is sparse at this time.[41] Studies were made on peat deposits in Florida, south of Lake Okeechobee. Peat samples were taken, and populations of microorganisms in surface layers (0 to 8 cm depth) were studied from plate counts. Investigation of peat fragments showed that those cultivated for sugar cane crops were colonized by hyphae of fungi and/or *streptomycetes* and few of the fragments were sterile. Conversely, 40 to 90 percent of the undisturbed peats were sterile, showing profuse growth of *streptomycetes* rather than fungi. This finding indicated that *streptomycetes* were probably more important agents for decomposing the plant material than other types of microorganisms.[4]

Microorganisms were essential to the early stages of the coalification process because they performed the function of converting dead plant material to the partially decayed organic matter called peat. Reasons for partial rather than total decay of the plant material are still somewhat open to debate. However, studies of present peat deposits in Florida indicate two important factors: the inherent resistance to decay of certain plant materials such as lignin and the chemical condition of the swamp water. For example, pH, oxygen content, and minerals would have determined whether the dominant bacterial population of the water was aerobic or anaerobic and thus would have controlled the rate of decomposition.[42]

In the coalification process, physical and chemical effects become pronounced. Fuchs[43] postulated that redox potential depends almost completely on the depth to which the material is submerged. Redox potential, a factor that changes with depth, is stabilized by the action of microorganisms. With the aid of a thermodynamical concept, Fuchs proved that coalification reactions may proceed spontaneously.

9.4 MICROBIAL CONVERSION OF COAL

In contrast to petroleum, very few studies have been done on the interactions between coal and microbiota. Coal is usually referred to as refractory due to the presence of graphite layers.[31,44] In the past, many factors discouraged microbiochemical studies related to coal, for example, insolubility, steric hindrance of enzymatic action due to physical structure, and presence of antibiotic materials.

All the historical developments before 1960 under the general topic of microbiology of coal have been critically reviewed by Rogoff et al.[9] Prior to 1960, a number of research projects were centered on the occurrence of microorganisms in coal.[10,12,15] The microflora found in coal were thought to be related to the coalification process. The study of that biota may provide insights into coal origin and structure.

In the early 1950s, small fractions of biologically active components such as *vitricin* were fractionated from the vitrain of coal,[45] and the nature of this fraction was studied by a number of investigators.[11,46-49] Phenolic compounds were found to be largely responsible for the observed antimicrobial activity.

Study of microbial oxidation of coal in the laboratory started six decades ago. Potter,[50] employing soil enrichment techniques, used a *Diplococcus* to interact with coal. The interaction resulted in the evolution of CO_2 and a slight temperature change, both of which were indications of microbial oxidation. Fisher and Fuchs were able to use brown coal as a substrate for the growth of *Penicillium* mold.[51] Microbial growth was found on both raw and treated coal. Subsequently, the U.S. Bureau of Mines reported that North Dakota lignite in a mineral salt and agar medium would support the growth of *Penicillium* sp. and *Trichoderma* sp.[9] Marginal growth of *Escherichia freundii* and *Pseudomonas rathonis* was observed in coal slurries.[13]

From 1957 to 1966, almost all the effort in microbiological research on coal was directed toward the development of single-cell protein sources. For example, different coal-derived materials such as coal tar fractions,[22] Fischer-Tropsch synthetic liquid fuel fractions,[17] low-temperature lignite tar,[23] coal acids,[23] Kogasin fuel,[52] and mode polycyclic hydrocarbons[53,54] were all investigated for that purpose. Most of these studies were done with yeast cultures, *Candida lipolytica* and *C. tropicalis*.

From 1965 to 1967, the Metallurgy Research Center at College Park, Maryland, undertook the task of growing microorganisms on oxidized coal[16,55] in an attempt to exploit the possibility of growing protein-rich microorganisms for food. They found that yeast strains of *Candida tropicalis* could oxidize leonardite to a slight degree in a Warburg respirometer, though growth was actually observed only in the water extracts.

Excellent work on the microbiological oxidation of polynuclear aromatic hydrocarbons, used as model compounds to represent coal, was done at the U.S. Bureau of Mines. The catabolic pathways of methylnaphthalenes,[53,56] anthracene, and phenanthrene[54] from *Pseudomonas* sp. were also studied and documented.

Recently, microbial coal liquefaction and gasification and upgrading of solid coal by removal of O, S, or N functional groups have gained increasing attention in the field of energy and fossil fuels research. Cohen and Gebriele[57] have found that crushed lignite added to agar medium can be partially degraded by fungal species of *Polyporus versicolor* and *Poria monticola*. Black liquid drops were found to appear on agar. Infrared spectra of material showed an increase in the absorption intensity of the carboxylic group. Ward[58] reported that a dark water-soluble liquid digestion product was produced from solid lignite by the soil fungus *Penicillium waksmani* and the yeast *Candida sp.* Infrared and ultraviolet analysis of the liquid product revealed a complex aqueous solution of apparent organic components. Wilson et al.[59-61] used subbituminous coal and lignitic coal as substrate for the growth of *Polyporus cryophilus ver vulpinus* (ATCC 22318) and *Poria placenta* (ATCC 11538), respectively. Yen et al.[62,63] stated that *P. versicolor* was capable of growing solely on the surface of the raw lignite and also on lignitic aqueous alkaline fraction. The latter suggests that alkaline soluble organic compounds trapped in the porous structure of lignite are responsible for sustaining *P. versicolor* growth. Alkaline-soluble organic compounds are believed to amount to only 5 percent of moisture-free lignite. Mild chemical modification processes of coal, by means of CuO oxidation,[64] nitration followed by ammoniation,[65,66] or sodium metal-methanol modification,[67-69] were applied to break down or modify lignite into smaller molecules that could be easily utilized by microbes, so that the yield of biodegraded products derived from lignite was increased.

Many studies were done on the growth of microbes on lignite. Scott et al.,[70,71] for example, used *Trameters versicolor, Poria placenta (monticola), Penicillium waksmani, Candida* sp., *Aspergillus* sp., *Sporothrix* sp., and *Paecilomyces* sp. in their investigation. The water-soluble product obtained was found to be composed primarily of a mixture of polar organic compounds with moderate to high molecular weights and a high degree of aromaticity.

9.5 BIODEGRADABILITY OF COAL

The general approach to biodegradation of coal is that if an appropriate organism is grown under proper conditions on a suitable substrate, it will produce a desired product in a manner satisfying both yield and rate requirements.

In any biological process, one of the most critical factors is the acquisition of an appropriate organism, since it provides the enzymatic system or catalyst series for the desired reaction. For example, laccase, believed to be the enzyme of the group of phenol oxidases primarily responsible for the extracellular breakdown of lignin and lignite, was found to be produced by *P. versicolor* grown on lignite.[72,73] Even when an appropriate strain of microorganisms is obtained to perform a given task, considerable study must be devoted toward the development of conditions to optimize the growth of microbial culture. Factors such as pH, salt concentration, aeration, substrate, and temperature must be adapted to promote maximum yield within the shortest processing time and the expenditure of the least amount of enzymes.

Some general classes of organisms and their respective substrates, of importance when considering biodegradation of coals, are:

1. *Species capable of degrading lignin:* Lignite contains large amounts of humic acid (brown, colloidal substances formed from lignin and cellulose components of various plants) and uncoalified lignite. Fungal species known to be capable of degrading lignin and wood are shown in Table 9.4.

2. *Species capable of degrading phenolic compounds:* Phenolic structure is an important attribute of coal. Species such as *Azotobacter* sp., *Pseudomonas* sp., *Gloeoporous dichrous*, *Rhodotorula glutinis*, and *Vibrio cyclosites* are responsible for ring cleavage of mono- and multihydric phenols.

3. *Species capable of degrading aromatic ring compounds: Pseudomonas* sp. such as *Ps. Fluorescens* and *Ps. aeruginosa* are known to cleave the ring system of aromatics, hydroaromatics, naphthenics, and heterocyclics in nature. The degradative metabolic pathway of mononuclear systems (such as benzene) and polynuclear systems (such as naphthalene, phenanthrene, and anthracene), and the degrading characteristics of *Pseudomonas* sp. are well documented.[9,74-76]

4. *Species capable of attacking specific linkages or heterocyclic ring components:* Desulfurization of coal, especially the removal of pyritic sulfur by sulfur and iron bacteria, has been successfully demonstrated.[77-87] Heterocyclic rings containing sulfur such as thiophene can be cleaved and the sulfur linkages connected to aromatic rings can be attached. Other heterocyclic systems such as pyridine (e.g., nicotinic acid),[88] proline (e.g., hydroxyl-L-proline), and furan (e.g., hydroxyl-2-furoic acid), can also be cleaved (Figure 9.2).

5. *Population selected by plasmid-assisted molecular breeding (PAMB):* PAMB is a technique where well-documented bacteria harboring known degradative plasmids are combined with a mixture of bacteria from locations

TABLE 9.4 Fungal Species Capable of Degrading Lignin and Wood[a]

Lignin	Wood
Bispora betulina 36124	*Acrospeira levis* 16214
Bjerkandera adustus 28315	*Brevibacterium* sp. 29895
Chaetomium piluliferum 32781, 32782	*Dendryphion nanum* 16226
Chaetomium reflexum 16213	*Doratomyces stemonitis* 16313
Doratomyces nanus 16219	*Mammaria echinobotyoides* 16301
Doratomyces purpureo-fuscus 16224	*Preussia funiculata* 16294
Gilomastix murorum 16277	*Preussia vulgaris* 16285
Humicola grisea 16298	
Hyalodendron sp. 16292	
Phanaerochaete chrysosporium 32629	
Pleurotus ostreatus 32783	
Preussia fleischhaki 32784	
Scopulariopsis brumptii 16278	
Scopulariopsis chartarum 16279	
Sopulariopsis koningii 16280	
Stachybotrys chararum 16275	
Stachybotrys cylindrospora 16276	
Trichocladium opacum 16273	

[a]ATCC (American Type Culture Collection) numbers are designated.

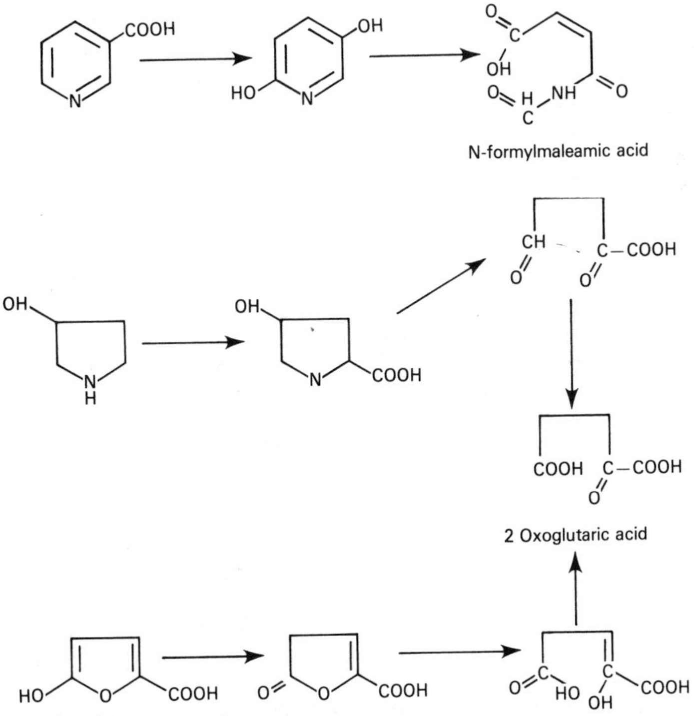

FIGURE 9.2 Biodegradable heterocyclic systems.

417

such as waste dumping sites. By subsequently exposing the microorganisms to increasing concentrations of the substances to be degraded while decreasing the readily metabolized substrates, a population of bacteria capable of utilizing the unfamiliar substances will eventually appear from that process. Thus, by using this technique, it is possible to accelerate the biodegradation of a number of polyaromatic hydrocarbons found in either coal or products derived from coal oxidation.

The general mode of microbial ring splitting is via orthohydroxyl intermediates with subsequent cleavage. The intermediates of mononuclear derivatives are catechol and protocatechuic acid, shown in Figure 9.3. Ring opening yields β-ketoadipic acid via the mechanism. The polynuclear derivatives can be again degraded stepwise to the catechol type intermediates (Figure 9.4). Since compounds such as phenanthrene, biphenyl ether, tetralin,

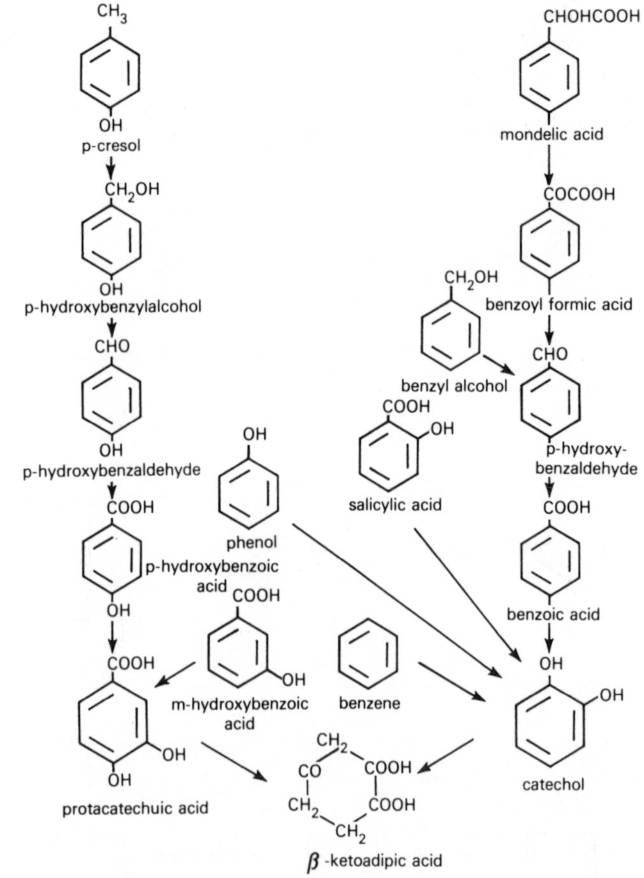

FIGURE 9.3 Catechol and protocatechuic acid as key intermediates of biological pathways (after Chapman[76]).

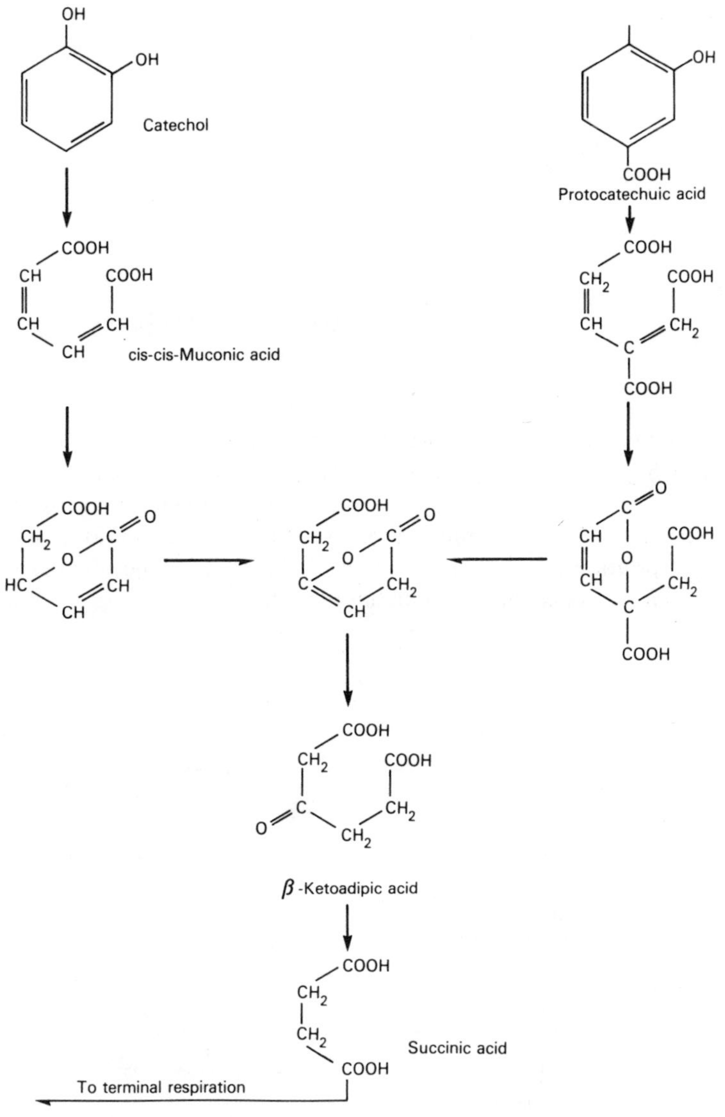

FIGURE 9.4 Fate of the benzene ring after cleavage (after Dagley[74]).

acenaphthene, and anthracene have nuclei that bear structural relationships to the coal molecules, the results of the biochemical degradation of aromatic molecules are of importance for investigating the biochemical degradation of coal.

Ring cleavage is also expected in reactions involving kata-condensed naphthenics, for example, cyclohexane would yield valeric acid, and decalin would yield adipic acid, as shown in Figure 9.5.[20]

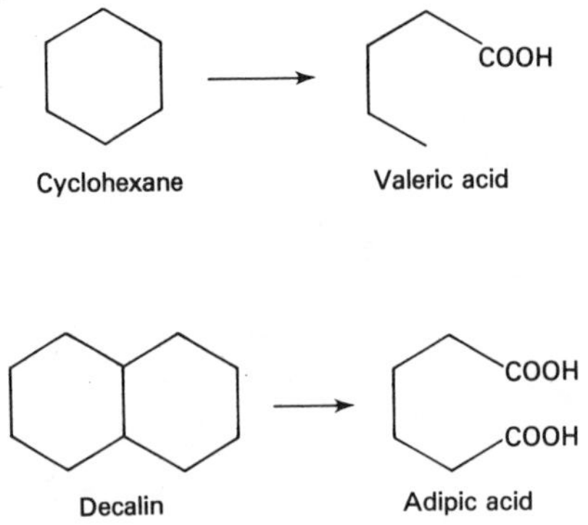

Cyclohexane → Valeric acid

Decalin → Adipic acid

FIGURE 9.5 Ring cleavage of cyclohexane and decalin.

Pseudomonas sp. is well known for the transformation of the catabolic products such as fatty acids, glycolipids, esters, ketones, epoxides, and alkenes. Alkanes from methane to n-tetratriacontane are known to be oxidized by *Pseudomonas* either through monoterminal oxidation (β-oxidation) or through diterminal oxidation (β- and ω-oxidation).[89,90] There are no alkanes in coals except in lignitic coals that contain montan wax fractions as a mixture of

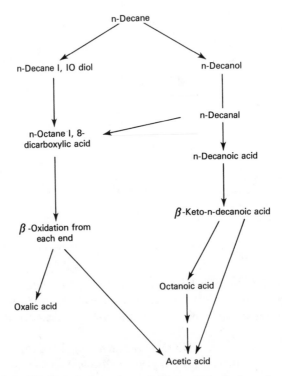

FIGURE 9.6 Biodegradation scheme for long chain hydrocarbons and acids.

aliphatic and naphthenic hydrocarbons. Usually the end products from paraffins are acetic acid and dicarboxylic acid (Figure 9.6).

In addition to *Pseudomonas*, other hydrocarbon oxidizers and utilizers may be developed so that strains suitable for symbiotic relations can be established. These microorganisms will include *Flavobacteria, Achromobacter* sp., *Coryne bacteria* sp., *Mycobacterium* sp., *Arthrobacter* sp., *Micrococcus* sp., *Torulopsis* sp., *Nocardia* sp., and so on. Table 9.5[90-93] illustrates the results of paraffins that were attached by these organisms.[90]

The conversion of heptane to propionic acid by *Pseudomonas* sp. KSLA 473, was investigated in a resting cell suspension of 2.7 mg/ml of hexane-grown *Pseudomonas*. After 5 hours of incubation, it yielded 0.54 mg/ml of propionic acid (60 percent conversion). This result was consistent only if chloramphenicol, an inhibitor of protein synthesis, was added to the medium.[94] Unfortunately, special considerations regarding the use of a promoter or inhibitor, and the optimization of the process were not provided. According to this work, one can envision that the two processes, ring cleavages of aromatic systems and fatty acid synthesis (oxidation), proceeded simultaneously. The latter process may involve further degradative breakage of the chainlike molecules into simple organic acids.

Basically, the aerobic biodegradation process involving oxidative degradation by *Pseudomonas* is an important step in degrading carboxylic acids. *Clostridia* and other fermentative bacteria can also degrade carboxylic acids. Laboratory acclimatization and plasmid-assisted molecular breeding have shown that in most of the cases, ethanol has been the highest yield product. Conventionally, it is difficult to convert simple acids biochemically to ethanol via acetyl-CoA by way of the EMP pathway. Currently, studies of solvent fermentation from a nonglucose source have achieved some success.[95]

9.6   MICROBIAL BENEFICIATION OF COAL

In general, microbial beneficiation is the process for the selective removal of sulfur and nitrogen heteroatoms from coal and the minimization of minerals. Microbial processes offer several potentially attractive features compared to conventional physical and chemical beneficiation processes, which include thermal treatment, oxidation, reduction, dielectric heating, magnetic seeding, and iron carbonyl treatment to increase magnetic susceptibility and to facilitate the separation of minerals from coal.[96] One of the advantages of microbial processes over conventional ones is that reactions in microbial processes proceed at ambient pressure and temperature, thus requiring little or no additional thermal energy. Many of the engineering problems associated with high-pressure and high-temperature processing in conventional methods could be eliminated.

Much effort has been spent on trying to desulfurize coal. The removal of pyritic sulfur by iron bacteria has been successfully demonstrated.[77-87] Some of

TABLE 9.5 Degradation of Paraffins by Various Microorganisms

Products	Substrate	Organism	Yield	Remarks
Ethanol Acetaldehyde Acetate	Ethane	*Pseudomonas methanica*	4.6 mg/l ethanol 49 mg/l acetaldehyde, 300 mg/l acetate	Cells co-oxidized substrate while growing on methane. Acid, alcohol, and ketones also produced from propane or butane.
Propionic acid	0.17% wt/v Heptane	*Pseudomonas aeruginosa* KSLA 473	0.54 mg/ml	Cells were grown on hexane and resuspended to 2.7 mg/ml in buffer. Chloramphenicol also added.
Hexanoic acid	Hexane	*Pseudomonas* KSLA 473	204.5 mg/liter	Oxygen supply reduced.[91] Only trace quantities of other fatty acids.
Propionic acid Pentanoic acid Heptanoic acid	Heptane Heptane Heptane	*Pseudomonas* KSLA 473	35 mg/l 145 mg/l 360 mg/l	Oxygen supply reduced 25 percent toward the end of the fermentation. Fatty acids accumulated simultaneously
2-Pentanone	0.33% v/v Pentane	*Mycobacterium smegmatis*	83 mg/l	Cells grown on propane and resuspended in nitrogen-free medium with butane.
Fatty acids	Liquid paraffin	*Pseudomonas aeruginosa*	5.3 percent conversion	Continuously grown cells.
Isovaleric acid	0.13% wt/v Methylhexane	*Pseudomonas* KSLA 473	0.12 mg/ml	Used heptane-grown resting cells plus chloramphenicol.

Product	Substrate	Organism	Amount	Comments
Fatty acids	13% v/v Paraffin	*Candida*	~21 mg/ml	Employed nitrogen limited medium. Cells and supernatant extracted for acid determination. See Ref. 92 for listing of other organisms producing fatty acids from alkenes.
Hexanoic acid 2-Pentanone	1% v/v Hexane	*Pseudomonas aeruginosa* KSLA 473	0.68 mg/ml	Heptane-grown resting cells (1–2 mg/ml) incubated with substrate and 0.1 percent acrylic acid.
Hydroxy fatty acids (glycolipids)	1% wt/v An alkene	*Torulopsis gropengresseri*	Trace to 7 mg/ml	See Ref. 93 for the identity of the various lipids found.
Octadecylstearate Octadecylpalmitate	1% v/v Octadecane	*Micrococcus cerificans*	3.3 mg/ml	Equal quantities of both esters produced simultaneously.

Source: B. J. Abbott and W. E. Gledhill, The extracellular accumulation of metabolic products by hydrocarbon-degrading microorganisms. *Adv. Appl. Microbiol.* 14, p. 249 (1971).

these microorganisms have shown their ability to solubilize pyrite and marcasite, and it has been suggested that these organisms may be useful in biological treatment processes for removal of sulfur prior to combustion. Biodesulfurization of coal is especially promising for coals containing very finely disseminated pyrite generally not removable by mechanical techniques.[97,98] The anticipated disadvantages regarding desulfurization of coal by microbial means relate to the relatively long bioprocessing time required (days to weeks) and to the production of acidic corrosive leaching solutions. However, bioprocessing of stored coal or slurry pipeline coal would make the longer times acceptable if problems due to corrosion could be solved. Additionally, thermophilic microbial activity, occurring at $>50°C$, is capable of accelerating the process. Other recent evidence shows that microorganisms are capable of rendering coal pyrite hydrophilic in a matter of minutes, thereby, making it amenable to separation from coal by techniques such as oil agglomeration.[97,99,100]

A thermophilic *Thiobacillus*-like strain (TH1) can remove both pyritic and organic sulfur from Turkish lignite.[101] Unfortunately, the products obtained through organic sulfur solubilization were not identified, mass balances were not reported, nor was any study done regarding pure organic compounds derived from the degradation of sulfur. Thermophilic *Sulfolobus* also removed pyrite and organic sulfur from coal.[79] Later, Kargi reported that it was possible to oxidize dibenzothiophene by way of *Sulfolobus*;[102] and an organism grown in a medium enriched with dibenzothiophene was found to remove 20 percent of organic sulfur from an Indian coal.[103]

Rai found that *Pseudomonas putida* was more effective than *P. aeruginosa* in reducing the pyritic sulfur content of coal. *P. putida* reduced the pyritic sulfur content of Illinois #6 coal and lignites by 69 to 76 percent in a period of 5 to 7 days, whereas *P. aeruginosa* reduced 26 to 32.5 percent of pyritic sulfur content of Illinois #6 coal.[104] Rai also reported that about 80 to 85 percent pyritic sulfur removal was achieved by *Thiobacillus ferrooxidans* in a slurry pipeline reactor at room temperature in a period of 7 to 14 days.[105]

In Section 9.7.2, we discuss the role of bacteria that oxidize iron and iron pyrite to cause the formation of acidic coal mine drainage. In addition to pyrite, *Thiobacillus ferrooxidans* solubilizes a variety of metal sulfide minerals.[106,107] Consequently, metals as well as sulfur can be removed from coal by this organism. As yet, few reports have provided detailed investigation of the scope of these processes. However, it was notably reported that up to 90 percent reduction in the arsenic content of a high sulfur and arsenic coal from Czechoslovakia could be achieved. This was accomplished by treating the leaching material with *ferrooxidans*.[108]

9.7   CONTAMINATION DUE TO COAL STORAGE

During handling and storing, large quantities of coal are in contact with soil and air, hence microbes may contaminate the coal. As a result, several problems may arise, such as lowering of the heating value of coal, acid drainage of stockpiled coal, and spontaneous combustion of coal.

Contamination of coal can occur either by contact with the soil of the mine itself or by deposition of dust from the atmosphere. Both soil and dust contain varied and large amounts of microbes. The rate of deterioration of coal due to microbial activity depends not only on the rank of a given coal and ambient conditions but also on the specific chemical composition and reactivity of the coal. The lower the rank, the easier it is for that coal to absorb oxygen. Generally, the lower the rank, the higher its moisture and oxygen content, amount of volatile matter, and other factors contributing to rapid deterioration and high reactivity. Exposure of freshly mined coal to air results in oxidation and loss of moisture. Microbes greatly enhance the effect of oxidation.

The three most common problems caused by microbial contamination of coal are discussed below.

9.7.1 Lowering of the Heating Value of Coal

Many microorganisms, such as fungi and soil bacteria,[58,109] are capable of utilizing hydrocarbons (especially n-alkanes) present in coal as a source of energy and organic carbon. The growth of a microorganism on stored coal reduces its heating value and gradually destroys its coking properties.

The heating value and the coking quality are reduced during storage. Stored high-rank coals are known to lose about 1 percent of their thermal value per year. Investigators have found that winter-stored coals lose 1.4 percent of their heating value, whereas summer-stored coals lose 2.1 percent. Both figures are taken for the first winter and summer of storage.[110]

9.7.2 Acid Drainage of Stockpiled Coal

Generally, coal contains a high percentage of sulfur in the form of pyritic sulfur. Coal can easily trap moisture and oxygen from air and undergo oxidation of pyrites, induced by microorganisms (*Thiobacillus ferrooxidans* and *Thiobacillus thiooxidans*). Pyritic sulfur is oxidized to sulfuric acid. The overall biological pyrite oxidation sequence is

$$2FeS_2 + 2H_2O + 7O_2 \rightarrow 2FeSO_4 + 2H_2SO_4 \quad \Delta H = -217 \text{ kcal/mol}$$

or

$$FeS_2 + 3O_2 \rightarrow FeSO_4 + SO_2(\text{dry})$$

$$2SO_2 + O_2 + 2H_2O \rightarrow 2H_2SO_4(\text{wet})$$

Direct microbial oxidation of ferrous sulfate via interaction with oxygen and water produces additional acidity. The formation of sulfuric acid results in acid drainage, a severe environmental problem.

9.7.3   Spontaneous Combustion of Coal

Spontaneous combustion results from the oxidation of coal under conditions in which the rate of heat generated by complex chemical and biological reactions exceeds the rate of heat dissipated into the surroundings. This phenomenon typically involves the development of *hot spots* in a coal stockpile that eventually reach temperatures at which the coal begins to burn. It has long been thought that spontaneous combustion is triggered by microbial oxidation of pyrites or by surface peroxide complexes present in coal. Mapstone[111] has postulated the oxidation of FeS to Fe^{+2} and Fe^{+3} sulfates. These reactions would ultimately release 840 KJ/mole and create regions near coal-to-pyrite interfaces at which coal oxidation could very easily become self-accelerating.

Bacteria can often contaminate coal products, coal derived wastes, and effluents of coal processing, especially when these substrates are stored for an extensive period.

Jones and Carrington[112] studied microorganisms isolated from a sludge that were used successfully in treating the waste liquors of a coal carbonization plant. The liquors consisted mainly of solutions of phenols and thiocyanate. Eight strains of bacteria were isolated from the sludge, three of which were found to be responsible for the destruction of major constituents of the waste liquor. These three strains of bacteria were studied further for identification purposes. One organism, which was grown on thiosulphate or thiocyanate, was autotrophic, did not produce polythionate, and lowered the pH of the medium to about 5.0. This organism was found to be similar to *Thiobacillus thioparus*. The other two bacterial strains found to grow on phenols were *Comamonas* sp. and a member of the *Moraxella actinetobacter* group. The former was known to grow on a wide range of phenols, including cresols and xylenols. The latter was found to be taxonomically close to *Achromobacter* sp. NCIB8250.

Stafford and Callely[113] reported the presence of bacteria in a tip-lagoon used in purifying coke-oven effluents that contained phenols and thiocyanate. These bacteria were aerobic, gram negative *bacilli* that could metabolize thiocyanate and phenols. The organisms isolated in this experiment were not used to a high level of efficiency due to the lack of aeration in the lagoon.

9.8   MICROBIAL PROCESSING OF COAL—AN OPTION

Presently a variety of processes are used to convert medium- to high-rank coals into liquefied and gasified fuels.[114-116] These methods require the use of solvents, high pressures, and high temperatures because of the characteristics of the chemical structure of medium- to high-rank coals. Since the chemical structure of low-rank coals is not as recalcitrant as that of medium- and high-rank coals (e.g., lignites and subbituminous coals), it seems extremely attractive to apply biological processes to degrade low-rank coals. Following are some advantages that low-rank coals exhibit over the other types of coal regarding microbiological application:

1. Lignite contains large amounts of *humic acids* (up to 65 percent on dry basis), which are brown, colloidal substances formed from lignin and cellulose. These humic acids can be extracted by alkaline or ammonia solution. In addition, there are bonded humic salts that can be further released from lignite by acid treatment (giving an additional yield of 20 percent humic material). Humic acids and lignins are known to be attacked by microorganisms.[117-122]

2. Lignitic coals usually contain 5 to 20 percent wax (montan wax components), which can be extracted by solvents such as benzene, benzene-alcohol, 1-propanol or 1-butanol. Petroleum wax fractions are widely biodegraded by hydrocarbon utilizers.[90]

3. The moisture content of lignite is in the range of 50 percent. This alone is an important factor for the growth and interaction of microorganisms.

4. Phenolic structures are more important attributes of lignitic coal than of other types of coals. Methods for the bacterial degradation of phenolic compounds have been developed.[76,123] *Azotobacter* and *Pseudomonas* species are principally responsible for the ring cleavages of mono- and multihydric phenols. Industrial waste disposal of phenols can be achieved by organisms specially developed to withstand a phenol concentration of 0.16 percent. It has been claimed that a barrel culture of bacteria could consume more than 100 g of phenol per day. In a processing plant, it is equivalent to 1000 lbs/day capacity.

5. The extent of aromatic systems in lignitic coals seldom exceeds a number of three aromatic rings. The frequently occurring aromatic systems are 1-mono ring, 2-membered ring, and 3-membered ring. Extensive microbial work has been done utilizing *Pseudomonas* sp. for the biodegradation of benzene, naphthalene, anthracene, and phenanthrene.

6. In recent years, miscellaneous bituminous materials related to coals and coal derivatives have been found to be attacked by microorganisms. These findings are summarized in Table 9.6.[124-129]

Regarding the above-mentioned advantages of low-rank coals over other types of coals, lignite and subbituminous coals are the best candidates for microbiological application. There is still another alternative, that is, to modify raw materials by chemical pretreatment followed by biological processes. This idea, suggested by Yen et al.,[91] is illustrated in Figure 9.7, showing a diagram of the production of fuel and useful chemicals.

Raw coal can either be subjected to direct biological processes or be pretreated by chemical processes followed by biodegradation. The purpose of chemical processes is not only to break down weak linkages of the aromatic cluster to yield predominantly some simple acids and humic acids but also to modify the characteristics of the surface of coal, producing suitable sites for microbial attack.

Modification of coals has been accomplished in a variety of ways such as (1)

TABLE 9.6   Microbiological Attack of Miscellaneous Bituminous Materials

Bitumens	Microorganism	Reference
Coal tar	*Candida*	22
Crude oil	*Cladosporia*	124
Humic acid	*Streptomycetes*	117
Mellitic acid	*Mycobacteria*	19
Graphitic acid	*Cephalosporium ascrimonium*	19
Graphitic acid	*Aspergillus penicilloides*	19
Asphalt	*Mycobacteria*	125
Asphalt	*Norcardia*	126
Kerogen	*Alternaria*	127
Lignin	*Pseudomonas*	118
Petroleum waxes	*Corynebacteria*	128
Tar sands	*Desulfovibrio*	129

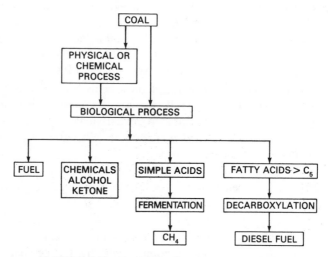

FIGURE 9.7   Procedure for bioconversion of coal.

potassium permanganate oxidation,[130] (2) hydrogen peroxide–acetic acid oxidation,[131] (3) oxygen oxidation, (4) ozone oxidation, (5) electrochemical oxidation,[132] (6) cupric oxidation,[62,64] (7) nitration followed by ammoniation,[65,66] and (8) sodium metal–methanol modification.[67-69]

Young and Yen[130] successfully oxidized a sample of North Dakota lignite to a mixture of normal dicarboxylic acids by stepwise potassium permanganate oxidation. The production of water-soluble acids was found to be 50 percent. Virtually no monocarboxylic acids were detected. Schnitzer and Skinner[133] have shown that a mixture of acetic acid–hydrogen peroxide oxidizes humic acids under mild conditions while preserving phenolic groups. Using this procedure, Studier et al.[131] oxidized lignite and bituminous coals to achieve over 80 percent conversion to methanol-soluble acids. Further, the acids produced were methylated.

Belcher[132] used a Cu gauze anode and a sheet Pb cathode in 3N NaOH to oxidize vitrain to yield 24–32 percent H_2O-soluble-Et_2O-soluble acids. The H_2O-soluble acids obtained appeared to be similar to those produced by $KMnO_4$ oxidation of coal. Yen et al.[64] employed modified cupric oxidation to obtain 79.3 percent conversion to alkaline soluble acids, which were further fractionated into 48.9 percent methanol-soluble and 30.4 percent methanol-insoluble acids. The conversion percentage was based on moisture-free and ash-free conditions. Methanol-insoluble acids were further oxidized by sodium dichromate to increase the conversion percentage of coal to methanol-soluble acids.

Since the nitrogen content of coal is a limiting factor for biodegradation, Yen et al.[65] impregnated nitrogen into the structure of lignite by nitration followed by ammoniation. Elemental analysis showed that the nitrogen content of nitrogen-enriched lignite was five times higher than that of the original untreated raw lignite. Yen et al.[67] also reported that lignite could be significantly converted into water-soluble products by reacting lignite with sodium metal and methanol. The modified lignite yielded 80 percent water solubles on a moisture-free and ash-free basis.

Most of the biological processes discussed here are aerobic processes involving the oxidative degradation of coal and its derivatives by the action of various microbes. Products such as fuel, useful chemicals (e.g., alcohol and ketone), simple acids, and fatty acids can be produced. Alcohol and ketone can be further recovered as fuels, and simple acids can be fermented to methane. Most of these oxidized products are simple, small molecules such as acetic, propionic, butyric, lactic, succinic, and oxalic acids. These products can be further subjected to anaerobic decomposition. The types of microorganisms suitable for these conversions are, for example, a mixture of closely dependent symbiotes, such as *Clostridia, Propionibacteria, Butyribacteria, Bacillus marcerans*, and specifically, methane bacteria. After fermentation the common end products are methane and carbon dioxide.

Treated coals can also yield oily liquids after chemical oxidation and biological degradation. These are anticipated to be a mixture of fatty acids originated from the waxy fraction, which amounts to 5–20 percent of original coals. These liquids can be easily decarboxylated into diesel fuels.

The possibility of obtaining liquid products (fuel and useful chemicals) and gas products (methane) by chemical modification and biodegradation of coal seems to be promising. For the microbiological point of view, liquefaction of coal can be achieved under aerobic conditions and gasification of coal can be carried out under anaerobic conditions. Both processes are capable of converting coal into useful energy products with low cost and low environmental impact.

9.9 SUMMARY

The relationships between coal and microorganisms are far from being clearly understood, although microbial species have been extensively associated with coal. The starting material in coal formation was plant remains that were slowly degraded by biological activity. The first type of microorganisms appearing in this long process were aerobes. Anaerobic microbes ensued because of the anaerobic conditions promoted by aerobic processes, burial, and submergence. Some basic factors present throughout the unfolding of these processes are no longer available today. For example, environmental conditions have changed and many plant species have become extinct. Surprisingly, species of microorganisms recently isolated from fossilized materials are thought to be the same as those that once actively transformed plant remains into precursors of coal. According to this perspective, these species of microbes are presently active in our environment. Besides being important in the role of processing plant remains, microbes also constitute significant components of coal because cell debris were gradually incorporated into the structure as the genesis of coal progressed.

Biochemically speaking, the activity of most of the microorganisms in the process of generating precursors of coal was through direct oxidation or oxidative dehydrogenation. The results were humification of plant remains, degradation of polynuclear aromatic hydrocarbons, decomposition of cellulose, oxidation of paraffins, oxidation of humic acids, oxidation of iron in pyrites, and degradation of coal tar fractions.

Many species of fungi and bacteria have been isolated from low-rank coals. A significant fraction of these is not considered to be part of the indigenous flora of coal, but contaminants introduced from current soil and water. More studies are needed to determine the original species of microbes that contributed to the process of coalification. During the long process of coalification, from plant remains to coal, microbes were responsible for most of the early transformations undergone by the starting plant matter. These transformations took place in two stages known as humification and peatification. The conditions of the surroundings played an important role as to which biological process would be dominant. Biodegradation of plant remains through processes not favoring the activity of strict aerobes, such as putrefaction, mouldering, and peatification, favored coalification. These processes, unlike rotting, a typical aerobic process, do not completely decompose plant debris.

Biochemical transformation of plant debris into coal precursors was primarily done on cellulose and lignin, since these substrates are the two main components of plant matter. Cellulose can be rapidly consumed by aerobic and anaerobic microbes, whereas lignin remains almost unchanged during the early stages of decomposition of plant matter. Lignin may be later transformed into humic acids. The principal degraders of lignin are white-rot fungi and soil bacteria. After the stage known as peatification, the microbial role was gradually replaced by physical and chemical factors in the coalification process.

Due to the recalcitrant character of coal, studies related to coal microbiology have only recently gained significant attention. The objectives of these studies have been toward simple biodegradation of coal, isolation of microbial species, coal liquefaction, coal gasification, production of single cell protein, and microbial removal of functional groups.

The research on determining the chemical structure of coal and its biodegradability has profited from experimental work utilizing model compounds. However, much work has to be done to clearly determine the structure of low-rank coals and establish the factors responsible for sustaining the growth of microorganisms on coals and their derivatives. The growth of microbes on low-rank coals has been associated with the presence of special enzymatic systems capable of attacking aromatic rings. The group of enzymes most known for its activity on low-rank coals is the group of the phenol oxidases. One of the most studied enzymes of this group has been laccase.

To estimate whether a coal is biodegradable, many factors have to be considered, among which are the characteristics of the substrate, the conditions to be applied during the process of biodegradation, the types of organisms to be used, and identification of the substrates commonly degraded by these organisms. The selection of the right microbial species is critical in attaining successful biodegradation. Microbes capable of degrading lignin, wood, aromatics, specific linkages of aromatic compounds and heterocyclics are the preferred types of organisms to achieve biodegradation of coal. Another way of selecting the appropriate microbial species is through acclimation of potential candidates. It is also possible to select cultures of desirable microbes by plasmid-assisted molecular breeding techniques.

The study of the mechanisms involved in the biodegradation of coals usually concerns microbial ring cleavage. Studies of this type are commonly done indirectly by means of mononuclear and polynuclear model compounds. In the case of lignites, besides ring cleavage, mechanisms related to the degradation of aliphatic hydrocarbons are also of extreme value, since lignites contain montan wax, a mixture of aliphatic and naphthenic hydrocarbons. Many bacteria have been associated with the biodegradation of paraffins to yield acetic acid, ethanol, acetaldehyde, monocarboxylic acids, dicarboxylic acids, fatty acids, and ketones.

Conventional beneficiation processing of coals requires expensive processing conditions. On the other hand, microbiological processes have a potential for considerable economics since they can be performed under mild processing conditions. One of the applications of microbial beneficiation of coal is the removal of specific heteroatoms from the coal structure. Desulfurization and removal of arsenic are examples. Both sulfur and arsenic are of environmental concern.

An undesirable aspect of the interaction of microbes with coal is the contamination of stored coal. This phenomenon is due to handling during transportation of coal from mining sites to the storage place and also due to storage itself. Some factors to be considered during storage of coals are: the rank of the coal, ambient conditions, composition and reactivity of the coal, and microbial species present. Generally, the lower the rank, the easier it is for a coal to be damaged by microbial contamination. Commonly, contamination of coals results in the loss of their original heating value. Contaminated coals are also known to produce acid drainage and to exhibit spontaneous combustion. Fungi and bacteria are responsible for causing these conditions, which are further enhanced by environmental conditions such as moisture and temperature. The heating and coking value of stored coals decreases more severely during the summer. Acid drainage is generated by the action of microorganisms on stored coals with pyrites in the presence of moisture and oxygen. Spontaneous combustion of stored coal is the result of interactions between pyrites, peroxides, and microbes. When the excess heat produced by complex chemical and biological reactions fails to be dissipated into the surroundings, it triggers the development of hot spots in coal stockpiles that may promote spontaneous combustion.

Microbes cause contamination of coal, but on the other hand, these same microbes can be directed to achieving desirable end products. Waste liquors derived from processing of coal can be treated microbiologically. Microbial processes are capable of eliminating or reducing the concentration of many types of phenols and thiocyanate present in waste liquors.

Microbial processing of coal is attractive for substrates such as low-rank coals and their derivatives. Low-rank coals are more compatible with microbiological processes than other types of coals because they exhibit porous chemical structure, relatively high moisture content, presence of biodegradable components (montan wax, phenols, humic acids, and lignin), and relatively low condensation of aromatic rings. Various processes of modifying and fractionating coal have been developed to enhance biodegradation. Some of the products yielded through these processes have proven to be capable of supporting the growth of microorganisms. Unfortunately, very little information is available regarding the chemical composition of the majority of these products, especially in the case of high molecular weight compounds. The most commonly reported products of degradation of coal are humic acids, a variety of carboxylic acids, fatty acids, carbon dioxide, and methane. There is great potential for using coals to produce fuels, useful chemicals, simple acids, and fatty acids, to mention a few. Coal is an abundant resource that is theoretically capable of yielding useful energy products in an economically and environmentally sound fashion. Nevertheless, new techniques need to be developed in order to transform this theoretical framework into reality. So far, no studies have been able to provide solutions to overcome typical limitations pertinent to processing coals through biological means. Examples of these limitations are (1) slow rate of biological reactions, (2) lack of information on the structure of coal, its derivatives, and obtained products, (3) lack of information on biochemical pathways, (4) insufficient amounts of products of biodegradation, and (5) difficulty of inducing the biological process to yield a desirable product. There is hope that positive results will soon be reported regarding the pretreatment of coals prior to subjecting them to biological action. It is hoped that these pretreatments will convert coals into substrates presenting chemical characteristics compatible with the existing enzymatical system of ordinary microorganisms. Depending on the way these substrates are handled, precursors of valuable products could be generated.

REFERENCES

1. W. Francis, *Coals, Its Formation and Composition.* Edward Arnold, London, 1956.

2. H. L. Ehrlich, *Geomicrobiology*, p. 311 Dekker, 1981. New York.

3. G. Ourisson, P. Albrecht, and M. Rohmer, The microbial origin of fossil fuels. *Sci. Am.* **251**, 44 (1984).

4. P. H. Given, Biological aspects of the geochemistry of coal. *Adv. Org. Geochem., Proc. Int. Meet., 5th, 1971* pp. 69–92 (1972).

5. T. G. Tornabene, S. Wu-Hunter, and P. S. Eastman, Production of aliphatic hydrocarbons by microorganisms isolated from shale. In *Chemical and Geochemical Aspects of Fossil Energy Extraction* (T. F. Yen, F. K. Kawakara, and R. Hertzberg, eds.), p. 169. Ann Arbor Sci. Publ., Ann Arbor, MI, 1983.

6. E. A. Shneour, Oxidation of graphitic carbon in certain soil. *Science* **151**, 991–992 (1966).

7. Y. Pares, Solvent extraction of gold by bacteria. *Rev. Ind. Miner.* **50**, 408–415 (1968).

8. W. Fuchs, Lignin and its importance in coal formation. *Brenst.-Chem.* **36**, 354 (1955).

9. M. H. Rogoff, I. Wender, and R. B. Anderson, Microbiology of coal. *Inf. Circ.— U.S., Bur. Mines* **8075** (1962).

10. M. A. Farrell and H. G. Turner, Bacteria in anthracite coal. *J. Bacteriol.* **23**, 155–162 (1932).

11. H. M. Rogoff and I. Wender, Biologically active materials in coal. *Nature (London)* **192**, 378–379 (1961).

12. R. Lieske and E. Hoffman, Investigation of the microbiology of coals as they occur in the coal beds. *Brenst.-Chem.* **9**, 174–178 (1928).

13. J. A. Koburger, Microbiology of coal: Growth of bacteria in plain and oxidized coal slurries. *Proc. W. Va. Acad. Sci.* **36**, 26–30 (1964).

14. E. Kuster and R. Locci, Studies on peat and peat microorganisms. I. Taxonomic studies on thermophilic actinomycetes isolated from peat. *Arch. Mikrobiol.* **45**, 188–197 (1963).

15. T. Iwasaki, Fungi which grow on coal. *Tech. Rep. Tohoku Imp. Univ.* **6**, 85–94 (1926).

16. J. M. Rose, J. M. Carosella, J. P. Corrick, and J. A. Sutton, Growth of *Candida Tropicallis* on a water extract of leonardite. *Technol. Use Lignite* (1967).

17. M. P. Silverman, J. N. Gordon, and I. Wender, Food from coal-derived materials by microbial synthesis. *Nature (London)* **211**, 735–736 (1966).

18. E. Kuster and R. Locci, Studies on peat and peat microorganisms II. Occurrence of thermophilic fungi in peat. *Arch. Mikrobiol.* **48**, 319–324 (1964).

19. H. Thiele and G. Anderson, The microbiological utilization of graphitic, humic, and mellitic acids. *Zentralb. Bakteriol., Parasitenk. Infektionskp. Hyg., Abt. 1: Orig.* **107**(2), 247–250 (1953).

20. S. Sichi, Microbial desulfurization of petroleum. *Petroleum Fermentation*, pp. 548–571. Science Publishers, Peking, 1973.

21. T. F. Yen, Biodeterioration and biodisintegration. In *Recycling and Disposal of Solid Waste* (T. F. Yen, ed.). Ann Arbor Sci. Publ., Ann Arbor, MI, 1974.

22. J. D. Brooks and J. W. Smith, Microbiological oxidation of a coal tar fraction. *Aust. J. Chem.* **19**, 1987–1989 (1966).

23. M. P. Silverman, J. N. Gordon, and I. Wender, Microbial synthesis of food from coal-derived material. *Adv. Chem. Ser.* **57A**, 269–279 (1966).

24. C. B. Lipman, Living microorganisms in ancient rocks. *J. Bacteriol.* **22**(3), 183–198 (1931).

25. V. Burke and A. J. Wiley, Bacteria in coal. *J. Bacteriol.* **34**; 475–481 (1937).

26. Q. P. Granger, Bacterial leaching of minerals. *Colliery Guardian* **232**(6), 212–214 (1984).

27. W. J. Ingledew, Ferrous ion oxidation by *Thiobacillus Ferrooxidans. Biotechnol. Bioeng. Symp.* **16**, 23–24 (1986).

28. R. T. Belly and T. D. Brock, Ecology of iron-oxidizing bacteria in pyritic materials associated with coal. *J. Bacteriol.* **117**, 726–732 (1974).

29. B. Ward, Coal degrading fungi: Isolated from a Mississippi lignite outcrop. In *Proceedings of Symposium on Biological and Chemical Removal of Sulfur and Trace Element in Coal and Lignite*, pp. 149–163. Louisiana Tech. University, Ruston, 1982.

30. H. B. Ward, Lignite degrading fungi isolated from a weathered outcrop. *Syst. Appl. Microbiol.* **6**(2), 236–238 (1985).

31. M. A. Elliott, ed., *Chemistry of Coal Utilization (Secondary Supplementary Volume)*, Vol. 2. Wiley (Interscience), New York, 1981.

32. V. Bouska, *Geochemistry of Coal*, pp. 72–90. Am. Elsevier, New York, 1981.

33. F. Fischer and H. Schrader, The origin and chemical structure of coal. *Brennst.-Chem.* **2**, 213–219 (1921).

34. B. D. Faison and T. K. Kirk, Relationship between lignin degradation and production of reduced oxygen species by *Phanerochaete chrysosporium. Appl. Environ. Microbiol.* **46**, 1140–1145 (1983).

35. T. K. Kirk, E. Schulz, W. J. Connors, L. F. Lorenz, and J. G. Zeikus, Influence of culture parameters on lignin metabolism by *Phanerochaete chrysosporium. Arch. Microbiol.* **117**, 277–285 (1978).

36. T. K. Kirk, W. J. Connors, and J. G. Zeikus, Advances in understanding the microbial degradation of lignin. *Recent Adv. Phytochem.* **11**, 369–394 (1977).

37. T. K. Kirk, T. Higuchi, and H.-M. Chang, *Lignin Biodegradation: Microbiology, Chemistry, and Potential Applications*, Vol. 1, p. 236. CRC Press, Boca Raton, FL, 1980.

38. J. R. Borgmeyer and D. L. Crawford, Production and characterization of polymeric lignin degradation intermediates from two different *Streptomycetes* sp. *Appl. Environ. Microbiol.* **49**(2), 273–278 (1985).

39. T. Higuch, Biodegradation of lignin: Biochemistry and potential applications. *Experientia* **38**, 159–166 (1982).

40. D. W. van Krevelen, *Coal*, p. 108. Elsevier, Amsterdam, 1961.

41. H. M. Braunstein, E. D. Copenhaver, and H. A. Pfuderer, *Environmental, Health, and Control Aspects of Coal Conversion: An Information Overview*, Vol. 1, pp. 2–5. Oak Ridge Natl. Lab., Oak Ridge, TN, 1977.

42. B. C. Parks, Origin, petrography, and classification of coal. In *Chemistry of Coal Utilization* (H. H. Lowery, ed.), Suppl. Vol. 1, pp. 1–34. Wiley, New York, 1963.

43. W. Fuchs, The origin of coal and the change of rank in coal fields. *Fuel* **25**, 132–134 (1946); *Trans. Am. Inst. Min. Metall. Eng.* **149**, 218 (1942).

44. D. W. Van Krevelen, *Coal Science*, 2nd ed. Elsevier, Amsterdam, 1980.

45. W. D. Evans, Mineralogical aspects of mine dusts. *Trans. Inst. Min. Eng.* **119**, Pt. 11, 658–670 (1959).

46. A. A. Mills, Biological properties of coal and coal extracts. *Nature (London)* **184**, 1885 (1959).

47. R. M. Kosanke, A bacteriostatic substance extracted from the vitrain ingredient of coal. *Science* **119**, 214–216 (1954).

48. N. C. Schenck and J. C. Carter, A fungistatic substance extracted from vitrain. *Science* **119**, 213–214 (1954).

49. M. H. Rogoff and I. Wender, Materials in coal inhibitory to the growth of microorganisms. *Rep. Invest.—U.S., Bur. Mines* **RT-6279** (1962).

50. M. C. Potter, Bacteria as agents in the oxidation of amorphous carbon. *Proc. R. Soc. London, Ser. B* **80**, 239–259 (1908).

51. F. Fischer and W. Fuchs, Growth of mold fungi on coals. *Brennst.-Chem.* **8**, 231–233 (1927).

52. W. Hoerburger, Biosynthesis of protein from hydrocarbon. *Forschungsber. Wirtsch.-u Verkehrsminist. Nordrhein-Westfalen* **131**, 22 (1955).

53. M. H. Rogoff and I. Wender, 3-hydroxy-2-naphthoic acid as an intermediate in bacterial dissimilation of anthracene. *J. Bacteriol.* **74**, 108–109 (1957).

54. M. H. Rogoff and I. Wender, The microbiology of coal. I. Bacterial oxidation of phenanthrene. *J. Bacteriol.* **73**, 264–268 (1957).

55. R. A. Heindl and J. A. Sutton, *Microbial Growth on Oxidized Coal*, Coal Res. Rep., 3rd Quarter, p. 165. College Park Metallurgy Research Center, 1967.

56. M. H. Rogoff and I. Wender, Methylnaphthalene oxidations by *Pseudomonas*. *J. Bacteriol.* **77**, 783–788 (1959).

57. M. S. Cohen and P. D. Gabriele, Degradation of coal by fungi *Polyporus versicolor* and *Poria monticola*. *Appl. Environ. Microbiol.* **44**, 23–27 (1982).

58. B. Ward, Apparent bioliquefaction of lignite by fungi and their growth on lignite components. *Proc. Bioenergy '84 World Conf. Exhib.* (1984).

59. B. W. Wilson, E. Lewis, D. Stewart, and S. Li, *Microbial Beneficiation of Coal and Coal Derived Liquids*, Commun. Rep. Batelle Pacific Northwest Lab., Richland, Washington, 1984.

60. B. W. Wilson, R. M. Bean, J. A. Franz, B. L. Thomas, M. S. Cohen, H. Aronxon, and E. T. Gray, Jr., Microbial conversion of low-rank coal: Characterization of biodegraded product. *Energy Fuels* **1**, 80–84 (1987).

61. B. W. Wilson, R. M. Bean, J. Pyne, D. L. Stewart, and J. Fredrickson, Microbial beneficiation of low rank coals. *Proc. Biol. Treat. Coals Workshop* pp. 114–127 (1986).

62. T. F. Yen, H. C. G. do Nascimento, J. R. Chen, K. I. Lee, R. S. Hsu-Chou, and W. C. Wang, *Microbial Screening Test for Lignite Degradation*, Q. Prog. Rep. No. 2, DOE Contract No. DEFG22-84PC 70809. U.S. Dept. of Energy, Oak Ridge, TN, 1985.

63. H. C. G. do Nascimento, J. R. Chen, K. I. Lee, R. S. Hsu-Chou, W. C. Wang, and T. F. Yen, *Polyporus versicolor* growth on lignite fractions. *Process Biochem.* **20**(1), 24 (1987).

64. T. F. Yen, H. C. G. do Nascimento, J. R. Chen, K. I. Lee, R. S. Hsu-Chou, and W. C. Wang, *Microbial Screening Test for Lignite Degradation*, Q. Prog. Rep. No. 3, DOE Contract No. DEFG22-84PC 70809. U.S. Dept. of Energy, Oak Ridge, TN, 1985.

65. T. F. Yen, H. C. G. do Nascimento, K. I. Lee, R. S. Hsu-Chou, and W. C. Wang, *Microbial Screening Test for Lignite Degradation*, Q. Prog. Rep. No. 5, Contract No. DEFG22-84PC 70809. U.S. Dept. of Energy, Oak Ridge, TN, 1986.

66. J. W. Wang, H. C. G. do Nascimento, and T. F. Yen, Biodegradation of nitrogen enriched lignite. *Am. Chem. Soc., 23rd West. Reg. Meet.* (1987).

67. T. F. Yen, H. C. G. do Nascimento, K. I. Lee, R. S. Hsu-Chou, and W. C. Wang, *Microbial Screening Test for Lignite Degradation*, Q. Prog. Rep. No. 6, Contract No. DEFG22-84PC 70809. U.S. Dept. of Energy, Oak Ridge, TN, 1986.

68. R. S. Hsu-Chou, A. S. Lee, and T. F. Yen, A novel chemical solubilization process for beneficiation of lignite. *Am. Chem. Soc., 23rd West. Reg. Meet.* (1987).

69. J. K. Chen, R. S. Hsu-Chou, and T. F. Yen, The nature of fermentation studies to solubilized lignite. *Am. Chem. Soc., 23rd West. Reg. Meet.* (1987).

70. C. D. Scott, G. W. Strandberg, and S. N. Lewis, Microbial solubilization of coal. *Biotechnol. Prog.* **2**(3), 131–139 (1986).

71. C. D. Scott, Microbial coal liquefaction. *Proc. Biol. Treat. Coals Workshop*, pp. 95–113 (1986).

72. T. F. Yen, H. C. G. do Nascimento, J. R. Chen, K. I. Lee, R. S. Hsu-Chou, and W. C. Wang, *Microbial Screening Test for Lignite Degradation*, No. 1, DOE Contract No. DEFG22-84PC 70809. U.S. Dept. of Energy, Oak Ridge, TN, 1985.

73. J. Mayaudon and J. M. Sarkar, Laccases of *Polyporus Versicolor* in soil and litter. *Soil Biol. Biochem.* **1**, 31–34 (1975).

74. S. Dagley, Catabolism of aromatic compounds by microorganisms. *Adv. Microb. Physiol.* **6**, 1–42 (1971).

75. H. W. Doelle, *Bacterial Metabolism*, p. 52. Academic Press, New York, 1969.

76. P. J. Chapman, An outline of reaction sequences used for the bacterial degradation of phenolic compounds. In *Degradation of Synthetic Organic Molecules in the Biosphere*, p. 17. Natl. Acad. Sci., Washington DC 1971.

77. M. P. Silverman, Methane-oxidizing bacteria. *U.S. Bur. Mines, Inf. Circ.* **8246** (1964).

78. M. P. Silverman, M. H. Rogoff, and I. Wender, Removal of pyritic sulfur from coal by bacterial. *Fuel* **42**, 113–124 (1963).

79. F. Kargi and J. M. Robinson, Microbial desulfurization of coal by thermophilic microorganism *Sulfolobus acidocaldrius*. *Biotechnol. Bioeng.* **24**, 2115 (1982).

80. C. M. Detz and G. Barvinchak, Microbial desulfurization of coal. *Min. Congr. J.* **65**, 75 (1979).

81. F. Kargi, Enhancement of microbial removal of pyritic sulfur from coal using concentrated cell suspension of *T. Ferrooxidans* and an external carbon dioxide supply. *Biotechnol. Bioeng.* **24**, 749 (1982).

82. J. Murphy, E. Riestenberg, R. Mohler, D. Marek, B. Beck, and D. Skidmore, Coal desulfurization by microbial processing. In *Processing and Utilization of High Sulfur Coals*, pp. 643–652. Elsevier, Amsterdam, 1985.

83. M. R. Hoffmann, B. C. Faust, F. A. Panda, H. H. Koo, and H. M. Tsuchiya, Kinetics of the removal of iron pyrite from coal by microbial catalysts. *Appl. Environ. Microbiol.* **42**(2), 259–271 (1981).

84. F. Kargi and T. D. Cervoni, An airlift-recycle fermenter for microbial desulfurization of coal. *Biotechnol. Lett.* **5**(1), 33–38 (1983).

85. F. Kargi and J. M. Robinson, Removal of sulfur compounds from coal by the termophilic organism *Sulfolobus Acidocaldarius*. *Appl. Environ. Microbiol.* **44**(4), 878–883 (1982).

86. F. Kargi, Microbiological coal desulfurization. *Enzyme Microb. Technol.* **4**, 13–19 (1982).

87. F. Kargi, Microbial methods for desulfurization of coal. *Trends Biotechnol.* **4**, 293–297 (1986).

88. E. J. Behrman and R. Y. Stanier, The bacterial oxidation of nicotinic acid. *J. Biol. Chem.* **228**, 923–945 (1957).

89. M. J. Klug and A. J. Markovetz, Utilization of aliphatic hydrocarbons by microorganisms. *Adv. Microb. Physiol.* **5**, 1–43 (1971).

90. B. J. Abbott and W. E. Gledhill, The extracellular accumulation of metabolic products by hydrocarbon-degrading microorganisms. *Adv. Appl. Microbiol.* **14**, 249–388 (1971).

91. T. F. Yen, Microbial actions on lignite. *Proc. Biol. Treat. Coals Workshop* pp. 141–171 (1986).

92. C. Ratledge, Microbial conversions of n-alkanes to fatty acids—New attempt to obtain economical microbial fats and fatty acids. *Chem. Ind. (London)* pp. 843–854 (1970).

93. D. F. Jones and R. Howe, Microbiological oxidation of long-chain aliphatic compound. I. Alkanes and 1-alkenes. *J. Chem. Soc. C* **22**, 2801–2808 (1968).

94. G. J. E. Thijsse and A. C. van der Linden, Pathways of hydrocarbon dissimilation by a *Pseudomonas* as revealed by chloramphenicol. *Antonie van Leeunwenhoek* **29**, 89–100 (1963).

95. L. J. Jang, private communication, California State University, Long Beach, 1984.

96. S. C. Tsai, Coal characteristics related to beneficiation. In *Fundamentals of Coal Beneficiation and Utilization*, p. 219. Elsevier, Amsterdam, 1982.

97. C. E. Capes, A. E. McIlhinney, A. F. Sirianni, and I. E. Puddington, Bacterial oxidation in upgrading pyritic coals. *Can. Min. Metall. Bull.* **66**, p. 88 (1973).

98. Z. Volsicky, J. Puncmanova, V. Hosek, and F. Spacek, Bacterial leaching out of finely intergrown sulphur in coal: Methods and features. In *Proceedings of the Seventh International Coal Preparation Congress* (A. C. Partridge, ed), Pap. K. 3. Bankstown, 1976.

99. A. G. Kempton, N. Moneib, R. G. L. McCready, and C. E. Capes, Removal of pyrite from coal by conditioning with *Thiobacillus ferrooxidans* following oil agglomeration. *Hydrometallurgy* **5**, 117 (1980).

100. G. J. Olson, F. E. Brinckman, and W. P. Iverson, *Processing Coal with Microorganism*, Prog. Rep. Elec. Power Res. Inst., Palo Alto, CA, 1984.

101. C. F. Gockay and R. N. Yurteri, Microbial desulfurization of lignites by a thermophilic bacterium. *Fuel* **62**, 1223 (1983).

102. F. Kargi, Microbial oxidation of dibenzothiophene by the thermophilic organism *Sulfolobus acidocaldarius*. *Biotechnol. Bioeng.* **26**, 687 (1984).

103. D. Chandra, P. Roy, A. K. Mishra, J. N. Chakrabarti, and B. Sengupta, Microbial removal of organic sulphur from coal. *Fuel* **58**, 549 (1979).

104. C. Rai, Microbial desulfurization of coals and texas lignites. Presented at the AIChE Spring National Meeting, Houston, Texas (1985).

105. C. Rai, Microbial desulfurization of coals in a slurry pipeline reactor using *Thiobacillus ferrooxidans*. *Biotechnol. Prog.* **1**(3), 200–204 (1985).

106. C. L. Brierley, Bacterial leaching. *CRC Crit. Rev. Microbiol.* **6**, 207 (1978).

107. G. J. Olson and R. M. Kelly, Microbiological metal transformations: Biotechnological applications and potential. *Biotechnol. Prog.* **2**(1), 1–15 (1986).

108. R. Jilek and E. Beranova, Some experiences with bacterial leaching of brown coal. *Proc. Int. Conf. Use Microorg. Hydrometall., 1980* pp. 167–174 (1982).

109. E. Beerstecher, *Petroleum Microbiology*, pp. 185–190. Elsevier, Amsterdam, 1954.

110. H. Colign and E. A. Vitunac, Properties of coal and its effect on storage and handling. *9th Annu. Semin. Bulk Mater.* (1976).

111. G. E. Mapstone, Spontaneous combustion theory. *Chem. Ind. (London)* 658 (1954).

112. G. L. Jones and E. G. Carrington, Growth of pure and mixed cultures of microorganisms concerned in the treatment of carbonization waste liquors. *J. Appl. Bacteriol.* **35**(2), 95–104 (1977).

113. D. A. Stafford and A. G. Callely, The role of microorganisms in waste tip-lagoon systems. Purifying coke oven effluents. *J. Appl. Bacteriol.* **36**(1), 77–87 (1973).

114. P. Nowacki, *Coal Liquefaction Process.* Noyes Data Corp., Park Ridge, NJ, 1979.

115. D. D. Whitehurst, T. O. Mitchell, and M. Farcasiu, *Coal Liquefaction.* Academic Press, New York, 1980.

116. N. Berkowitz, *An Introduction to Coal Technology.* Academic Press, New York, 1979.

117. W. Fuchs, F. Fuchs, and J. J. Reid, Microbiology of coal. Biological decomposition of hydroxycarboxylic acids obtained from bituminous coals. *Fuel* **21**, 96–102 (1942).

118. D. W. Ribbons and J. E. Harrison, Metabolism of methoxy and methylenedioxy-phenol compounds by bacteria. In *Degradation of synthetic Organic Molecules in the Biosphere*, p. 98. Natl. Acad. Sci., Washington, DC, 1972.

119. R. Bunner, A. E. Maccubbin, and R. E. Hodson, Anaerobic biodegradation of the lignin and polysaccharide components of lignocellulose and synthetic lignin by sediment microflora. *Appl. Environ. Microbiol.* pp. 998–1004 (1984).

120. P. J. Colberg and L. Y. Young, Anaerobic degradation of soluble fractions of [C-Lignin] lignocellulose. *Appl. Environ. Microbiol.* **49**(2), 345–349 (1985).

121. J. P. Kaiser and K. W. Hanselmann, Aromatic chemicals through anaerobic microbial conversion of lignin monomers. *Experientia* **38**, 167–176 (1982).

122. P. J. Colberg and L. Y. Young, Biodegradation of lignin-derived molecules under anaerobic conditions. *Can. J. Microbiol.* **28**, 886–889 (1982).

123. N. T. Putilina, Microorganisms employed in industrial waste conduits for de-phenolizing effluents. *Mikrobiologiya* **28**, 757–762 (1959).

124. J. D. Walker, L. Cofone, Jr., and J. J. Cooney, Microbial petroleum degradation: The role of *Gladosporium resinae. API/EPA/USCG Conf. Prev. Control Oil Spills* p. 821 (1973).

125. R. W. Traxler, J. A. Robinson, D. E. Wetmore, and R. N. Traxler, Action of microorganisms on bituminous materials. II. Composition of low molecular weight asphaltic fractions determined by microbial action and infra-red analyses. *J. Appl. Chem.* **16**, 266–271 (1966).

126. R. W. Traxler, P. R. Proteau, and R. N. Traxler, Action of microorganisms on bituminous materials. I. Effect of bacteria on asphalt viscosity. *Appl. Microbiol.* **13**, 838–841 (1965).

127. J. E. Findly, M. D. Appleman, and T. F. Yen, Microbial degradation of oil shale. In *Science and Technology of Oil Shale* (T. F. Yen, ed.), p. 175. Ann Arbor Sci. Publ., Ann Arbor, MI, 1976.

128. H. K. Kyowa (Kyowa Fermentation Industry Co., Ltd.), Fermentative manufacture of amino acids. French Patent 1,577,264, p. 5 (1969).

129. C. E. Zobell, Recovery of hydrocarbons. U.S. Patent 2,641,566 (1953).

130. D. K. Young and T. F. Yen, Oxidation of lignite into water-soluble organic acids. *Energy Resour.* **3**(1), 49 (1976).

131. M. H. Studier, R. Hayatsu, and R. E. Winans, Analysis of organic compounds trapped in coal, and coal oxidation products. in *Analytical Methods for Coal and Coal Products* (C. Karr, ed.), Jr., Vol. 2, p. 68. Academic Press, New York, 1978.

132. R. Belcher, Anodic oxidation of coal. I. Introduction and preliminary experiments. II. Effect of oxidizing vitrain and ulmic acids at various metal electrodes. III. Examination of the water-soluble acids. *J. Soc. Chem. Ind., London* **67**, 213–216, 216–218, 218–221 (1948).

133. M. Schnitzer and S. I. M. Skinner, The low temperature oxidation of humic substance. *Can. J. Chem.* **52**, 1072–1080 (1974).

Commentary

Just as in other technical fields, one of the most critical elements of scientific and engineering studies of coal is that of the experimental measurement of its physical and chemical properties. This theme was addressed both directly and indirectly throughout this book. In the final chapter, D. S. Hoover and F. K. Schweighardt summarize key analytical techniques for monitoring coal sample properties during both handling and storage.

10

ANALYTICAL TECHNIQUES FOR MONITORING COAL SAMPLE STORAGE

D. S. Hoover
and
F. K. Schweighardt
Air Products and Chemicals, Inc.

10.1 INTRODUCTION

Coal is a very reactive fossil fuel that can undergo significant chemical and physical changes during mining, storage, handling, and transport. Despite this problem, very few standard practices are used commercially or in the laboratory for coal preservation. This situation is due in part to the wide variety of end uses for the fuel and also because of its heterogeneous nature.

Coal is a complex mixture of organic species (macerals), inorganic species (minerals and elements), liquids (water and oils), and gases (methane, carbon oxides, and air). The structure of the organic species makes the coal very porous; hence, in the proper environment, both liquids and gases can diffuse in and out. When air diffuses into the organic solids, reactions between oxygen and carbon will occur, commonly referred to as oxidation, weathering, or aging. These reactions are significant because they can reduce the economic value of most of the coal currently being mined, which is used for fuel or metallurgical coke.

The environment to which coal is subjected before mining will affect its level of deterioration due to reactions with air. For example, deep-mined coal is generally below the water table and therefore protected from deterioration due to air infiltration prior to mining. Strip-mined coal or deep-mined coal located above the water table or near the outcrop can weather before mining. In addition to premining effects, processing, transportation, and storage after mining can result in coal deterioration.

The degree of coal aging can be analytically detected and quantified by measuring derived coal properties such as heating value or thermoplasticity or by measuring fundamental properties such as the nature of the oxygen functionality. Because the compositions of unoxidized coals vary considerably, many analytical tests require a baseline against which to measure changes in derived or fundamental properties. To resolve this problem, coal sample banks are now being established in industrial, academic, and government laboratories to supply standard samples from which baselines may be generated.

This chapter concentrates on the effects of coal storage and methods for analyzing coal deterioration during storage, including both commercial and laboratory techniques.

10.2   EFFECTS OF STORAGE ON COAL AGING

The increasing interest in coal, both as a commercial resource and in fundamental studies, has generated many questions regarding the best way to acquire, handle, and store this material. Commercial operations must handle thousands to millions of tons of coal, whereas research studies typically handle much smaller quantities in the kilogram to milligram range. The reactivity of the coal to oxidative effects and the effect of resulting changes on the particular study or operation will influence the approach used for handling and storage.

This section summarizes the current practices used to store coal and the aging effects that are of concern, focusing first on commercial operations and then on laboratory procedures.

10.2.1   Commercial Storage Practices and Their Effects

Commercial operations are managed with a concern for changes in the heating value and coking or washability properties of stored coal and the potential for spontaneous combustion during storage. Large thermoelectric power plants (1000 MW) store substantial amounts of coal, allowing for 30 to 90 days of operation without coal receipt. Because of the large volumes involved, coal is stored primarily in open piles, at costs ranging between $5 and $10 of capital expenditure per ton. Closed piles may cost two to three times more, and silo storage is even more expensive.[1]

One problem that can result from open-pile storage is the loss of fuel heating value due to oxidation. Stockpiled coals will lose heating value as oxygen is adsorbed and carbon/oxygen reactions proceed in localized hot spots. Studies have shown that these losses generally range from 1 to 2 percent (up to 20 percent) of the total heating value available in the pile.[2-4] Rates of change are affected by seasonal climatic changes, with maximum effects experienced in the summer. Severe weathering may also affect the volatile content of the stored coal, adversely influencing its combustion properties.

Several coal properties will affect the rates of change during open-pile storage. Because the reactions of weathering in a coal pile are diffusion-limited (both gas and heat), a pile stacked tightly shows reduction in the rate of reactivity. The initial size of the coal particles and size segregation during stacking also influences gas diffusion into the pile. Coal piles created from some run-of-mine coals that have widely diverse particle sizes will not be amenable to tight stacking. Piles created from coal that is too fine present an increased surface area for reaction. Coal that is approximately 2–3 in. × 0 or 100 mesh, a typical size for washed high-volatile coals, seems acceptable for modern pile construction and handling.

Generally, the piles are constructed so that air diffusion into the pile is limited or avenues are provided so that heat generated can be rapidly released from within the pile. Coals particularly sensitive to weathering reactions can be compacted during pile construction.[5] Kromrey et al.[6] have shown that covering piles by a coating formed from pulverized coal in a wax or plastic binder or by the use of latex emulsions protects them from extreme conditions.

Several additional factors strongly affect the degree to which piled coal oxidizes. Lower-rank coals that are highly porous and contain more oxygen and inherent moisture are more reactive when stored in piles. Bhattacharyya[7] has shown that a change in moisture content, particularly wetting, will contribute to increased particle temperature. Moisture evaporation will decrease particle temperature, but with low-rank coals the drying will also cause particle degradation, resulting in increased fines and increased surface area. Sulfur particularly pyritic sulfur, is believed to participate in air/water reactions leading to heat generation and the potential for spontaneous pile combustion.[2]

One of the extreme results of weathering is the spontaneous ignition of the stored coal. Many studies have attempted to identify the prime factors leading to spontaneous ignition. Current controls are usually based on past experience with the particular coal including, handling and storage practices designed to minimize risks (such as pile construction), and detection of incipient combustion.[8] The most reliable methods of detection involve temperature probes in the pile or monitoring of the oxygen/carbon monoxide ratio in closed storage systems.[8] Stott et al.[9] have been able to induce a subbituminous coal, containing its inherent bed moisture, to heat spontaneously by passing air through a column of coal in a laboratory reactor. Their data have been used to generate an improved mathematical model for predicting spontaneous heating.

Another area of concern for commercial operations involves changes in beneficiation properties due to weathering during coal storage. Weathering reactions can affect the physical properties of particle integrity by the introduction of microcracks. These cracks, whether caused by desiccation or oxidation reactions such as pyrite swelling, increase the friability of the coal and produce more fines.

Weathering also affects the coal's chemical properties by increasing specific gravity, changing pH, and changing the hydrophobicity of the fine coal particles. Kona et al.[10] have shown that under severe conditions of oxidation the specific gravity of a given coal can increase from 1.3 to as much as 1.5. This can affect float/sink separation devices. Changes in particle surface properties due to oxidation are reflected in wettability characteristics. Oxidized coal becomes less hydrophobic, and the performance of flotation systems will degrade unless additional reagents are added.[11,12] The pH of coal–water slurries decreases with oxidation or coal weathering and after flotation cell performance.[13] Beafore et al.,[12] in a study of a hypothetical coal preparation plant, have shown that the introduction of severely weathered coal could decrease plant yields of clean coal by 24 percent.

Perhaps the commercial use of coal that is most sensitive to weathering during storage is the production of metallurgical coke. The thermoplastic properties that make certain coals ideal for the production of high-grade metallurgical coke are rapidly destroyed by weathering. A complete discussion of this topic is included in Chapter 6 and will not be presented here. Techniques to minimize these changes involve minimum pile storage time and many of the same procedures used for general pile construction.

10.2.2   Laboratory Storage Practices and Their Effects

Results of Survey of Coal Research Laboratories. As laboratory studies on coal have proliferated over the past 15 years, so too has interest in how to preserve the characteristics of laboratory samples during storage. A survey of major laboratories involved in coal research was done to determine the current methods used for sample storage and integrity. Analysis of the 30 responses shows that no standard techniques are in practice throughout the research community, apparently because of the diverse interests of the various research groups. Laboratories involved in work that is known to be sensitive to weathering, such as carbonization research, took significant precautions to ensure coal sample integrity, whereas those working in less sensitive areas such as float/sink washability testing were less concerned. In addition, a relationship existed between the size of the sample involved and the storage technique. Groups working with samples weighing 1 kg or less were more likely to use advanced storage techniques than those dealing with samples weighing a ton. However, even though objectives and techniques varied widely, several basic procedures were used to store laboratory samples.

Particle Size and Container. Weathering and coal-particle oxidation are topochemical gas diffusion reactions. Consequently, to limit oxidation to the smallest surface area, the coal to be stored should be of maximum particle size. However, increasing the particle size means that a larger sample has to be stored. For example, 1 g of 60 mesh (0.25 mm) coal may be sufficient for a representative analytical sample, but if the particle size is increased to 50 mm the sample size must be increased to 105 kg. Individual laboratories that were surveyed used the following approaches to this dilemma: 3 kg at −4 mesh,[13] 1.5 kg at −20 mesh,[14] 0.75 kg at −40 mesh,[15] and 5 g at −100 mesh.[16]

The smaller samples were stored in sealed glass jars, sealed cans with plastic liners, sealed glass ampules, or sealed plastic bags. Larger sized samples (particle size and hence total weight) are generally stored under less rigorous conditions of temperature and atmosphere than the smaller laboratory samples (see below). These are generally kept in heavy plastic bags in tightly closed steel drums. Samples in excess of several tons may be stored in large hoppers with limited air circulation.

Preservation Atmosphere. The survey also revealed that the atmosphere introduced into the storage vessel for sample preservation varies from laboratory to laboratory. For example, many laboratories seal their storage containers without attempting to provide a specific atmosphere, especially when the research objectives are not affected by minor amounts of oxidation. On the other hand, Neavel et al.[14] at Exxon used a repeated evacuation and nitrogen flood method to ensure removal of air prior to sample sealing. Larsen and Schmidt[18] have suggested that flame sealed glass ampules are preferable over screw-cap vials as storage vessels. Several other laboratories[15,16,19,20] use an argon purge before sealing the sample containers for storage. Senftle and Davis[21] have shown that using either argon or nitrogen will preserve the thermoplastic properties of a high-volatile Pittsburgh seam coal about equally. Furthermore, this same study showed that a sample stored in boiled distilled water degraded less than those stored in nitrogen or argon.

Huggins et al.[22] have also explored the use of argon as a preservation medium. They report that samples can be preserved for up to two years. However, if longer preservation is required, they suggest cryogenic storage. These authors also postulated that methane might provide a useful atmosphere.

Temperature. Although sample storage temperatures varied from laboratory to laboratory, cold storage is generally accepted as the most desirable for maximum preservation. Typically, only small samples are subjected to cold storage, although in certain cases drum-size quantities have been held under these conditions.[20,21] Cold storage temperatures ranged from just above freezing (35°F)[14,20,23] to approximately 0°F.[14,19] Whether or not such subambient temperatures are necessary when an inert gas atmosphere is present is undetermined.

Laboratory Sample Storage Studies.   Although most laboratories have conducted random tests to determine sample integrity following storage, only a few comprehensive studies have been published. Senftle and Davis[21] reported on the effects of particle size and storage atmosphere on the thermoplastic properties of high-volatile A coals. In the first portion of this study, samples collected from the Sunnyside Seam in Utah, ranging in particle size from 3/8 in. to −40 mesh, were stored in both air and nitrogen for a period of approximately 30 months. The results (see Table 10.1) revealed that the thermoplastic properties of samples stored in air were destroyed irrespective of particle size.

TABLE 10.1 Effect of Storage on the Thermoplastic Properties of Different Size Fractions of Sunnyside Coal (Storage in air and nitrogen)

Sample	Fluidity (DDPM)	Maximum Fluidity Temperature (°C)	Softening Temperature (°C)	Resolidification Temperature (°C)	Plastic Range Temperature (°C)
Original coal	226	438	395	465	70
(−3/8 in.) (air)	3	440	415	458	43
(9.50 mm) (N$_2$)	250	438	396	467	71
(−1/4 in.) (air)	3	439	414	458	44
(6.30 mm) (N$_2$)	222	441	396	469	73
(−4 mesh) (air)	3	440	414	459	45
(4.74 mm) (N$_2$)	251	440	396	466	70
(−10 mesh) (air)	3	438	413	454	41
(2.00 mm) (N$_2$)	232	437	391	466	75
(−20 mesh) (air)	2	437	415	452	37
(850 μm) (N$_2$)	270	439	394	465	71
(−40 mesh) (air)	3	433	414	465	41
(425 μm) (N$_2$)	72	440	396	464	68

TABLE 10.2 Effect of Coal Storage in Different Atmospheres on the Gieseler Fluidity of Pittsburgh Coal

Sample	Fluidity (DDPM)	Maximum fluidity Temperature (°C)	Softening Temperature (°C)	Resolidification Temperature (°C)	Plastic Range Temperature (°C)
Original coal	14,681	438	389	478	89
Sealed in nitrogen	12,053	442	392	479	87
Sealed in argon	11,496	442	392	477	84
Sealed in air	7,660	442	297	478	81
Sealed in boiled distilled N_2O	29,011	436	389	480	91
Unsealed	784	443	397	475	78

However, samples stored in nitrogen retained their plastic properties at all sizes other than −40 mesh. The −40 mesh size fraction displayed reduced fluidity and plastic temperature range. It should be noted that this size fraction did not require additional grinding prior to the plastometer test.

In a second experiment, samples of a high-volatile A Pittsburgh coal were stored in different atmospheres for the same 30-month period. Coal stored under nitrogen and argon suffered only minor reductions in plastic properties (Table 10.2). Coal sealed in an air atmosphere showed a decrease in Gieseler fluidity of approximately 50 percent and some decrease in plastic range. However, the plastic properties of a sample left unsealed for this period were nearly destroyed. A sample stored in boiled distilled water showed an unexplained improvement in plastic properties. These results indicate that storage in inert gases (nitrogen or argon) will retard the degradation of thermoplastic properties. Note that these samples were stored at ambient conditions; cold storage may have further improved these results.

Another study, by Colombo and Scholz[24] at the Bituminous Carbonization Research Association (BCRA), monitored the chemical and thermoplastic properties of a high-volatile coking coal over one year of storage. In this experiment, -2 mm samples were stored in an inert argon atmosphere. Statistical analyses of the coal properties following storage, conducted by nine independent laboratores, showed no degradation of any properties other than maximum dilatation, under the conditions of the experiment. A subsequent study by BCRA suggests that the changes in dilatation were also unaffected by these storage conditions.[25]

10.3 ANALYTICAL METHODS TO QUANTIFY COAL AGING

In order to understand the processes that contribute to the oxidation of coal, the coal science community has established analytical procedures that attempt to quantify changes in the physical and chemical properties of coal. Volbroth[26] in his excellent overview of work on the oxidation of coals refers to technical papers discussing the following analytical methods: ultrasonics,[27] radio frequency heating,[28] low-temperature ashing,[29] ultraviolet spectroscopy,[30] solvent extraction pH and Eh measurements,[30] infrared spectroscopy,[31] x-ray diffraction and scattering studies, electron spin resonance,[32] electrophoresis behavior,[30,33] gas and paper chromatographic techniques,[30] mass spectroscopy and nuclear magnetic resonance,[34] and x-ray microprobe techniques.[35] These studies have centered on oxidation during the coalification process, that is, the metamorphism of organic matter from peat to lignite to bituminous to anthracite. Apparently, in this process such functional groups as carboxyl (—COOH), hydroxyl (—OH), methoxyl (—OCH$_3$), and carbonyl ($>C=O$) are progressively depleted in the coal[36,37] but are never entirely lost, even in the high-rank coals.

In contrast to the reactions occurring during coalification, weathering of fresh-mined coal may involve subsequent reactions that form more crosslinked organic complexes involving primarily ether linkages. The following sections describe analytical methods that have been used to quantify the oxidation of coal, which fall into three major types: (1) chemical techniques, (2) petrographic techniques, and (3) thermoplastic measurements.

10.3.1 Chemical Techniques

The most unambiguous analytical measurement of coal oxidation would be an accurate direct measure of oxygen content before and after weathering. One of the more successful techniques for this measurement has been the use of neutron activation. This method depends on the property of oxygen atoms to become radioactive after exposure to neutron bombardment. Measurement of emitted gamma rays and comparison to standards subjected to the same treatment enables a measurement of total elemental oxygen. Corrections to the determined value are then applied to subtract contributions to the total oxygen by moisture and mineral matter. This method was initially applied to coal by Veal and Cook[38] and has been used in several coal weathering studies, notably Gray et al.[13] A further examination and verification of the experimental technique has been reported by Mahajan.[39]

The chemical aspects of the spontaneous combustion of coal (a potential consequence of severe oxidation) are not well understood because of the complex compositions of coals and the varied heterogeneous surface reactions that can be involved.[40] Lidin[41] believes that the low-temperature stage of oxidation produces coal/oxygen complexes such as carbonyl ($>C=O$) and carboxyl (COOH), which upon further heating liberate carbon monoxide and carbon dioxide. Hence, methods that measure the carbon monoxide or carbon dioxide emission rates or the corresponding oxygen depletion rates have been used as a criterion of coal spontaneous combustion. In fact, the Graham Index,[42] the ratio of carbon monoxide formation to oxygen deficiency (CO/O_2), was one of the first methods suggested around 1900, as a way to predict spontaneous combustion potential.

Carbon monoxide monitoring techniques are applied today in both Europe and the United States for advanced detection of spontaneous combustion in coal mines.[43,44] Kuchta et al.[40] have pointed out two complications with such measurements: (1) the observed carbon monoxide may be due to the desorption of gases formed during earlier coalification stages, and therefore not indicative or self-heating; and (2) the CO/O_2 index can be difficult to analyze when the sampled mine atmosphere has been diluted by ventilation streams of vitiated air, rather than by fresh air.

These in-mine results can be applied to laboratory storage of fresh-mined coal. Kuchta et al.[40] published data on gas emissions from over 25 coals stored in closed containers in air at ambient temperature for 14 days. The gases evolved were carbon monoxide, carbon dioxide, and methane, the amount of which varied with exposure time or oxygen consumption and the composition, dryness, and particle size of the coal. For a Bruceton bituminous (USA, Pittsburgh) coal, the carbon monoxide emission rate closely followed the oxygen reduction rate and was greater for a predried sample (nitrogen/70°C) than an undried sample. We may therefore conclude that coal should be stored under an oxygen-starved condition and kept cold to maintain the coal in its fresh-mined state.

Ignasiak et al.[45] studied the chromatographic analysis of pyrolysis gases as a method to quantify coal weathering. They compared their pyrolysis method to the Ruhr dilatation test (used to measure a coal's thermoplastic properties by determining relative packed-bed expansion and contraction) and found their approach to have broader applicability to coals of differing expansion properties. In addition, pyrolysis gas analysis is easier to conduct and requires far less laboratory time than the Ruhr test.

The method developed by Ignasiak and co-workers is based on the assumption that pyrolysis of oxidized coals should liberate carbon monoxide and carbon dioxide gases. They reasoned that low-temperature ($<200°C$) oxidation of coal should increase —OH, —COOH, and $>CO$ groups and contribute to the CO and CO_2. An important observation made by Ignasiak was that the "difference in volume of *carbon monoxide* evolved in equal weights of coal before and after oxidation" was the most sensitive indicator of low-temperature oxidation/weathering.

Later, Young and Nordon[46] studied the relative susceptibilities of Australian low-rank coals and chars to self-heat. They found it necessary to develop a method to measure the rate of oxygen sorption from flowing air and related the changes to time. Their approach was to measure the volume of oxygen taken up by the sample during periodic interruptions of air flow in their apparatus. Restoring the flow caused the gaseous contents of the reaction vessel to pass through a paramagnetic analyzer. The resulting signal was a wide peak that was integrated. The decrease in oxygen content caused by the interruption of air flow was controlled to less than 15 percent of the initial concentration in order to prevent the rate of oxygen sorption changing in large increments. As Young and Nordon pointed out, the rate of oxidation was nearly dependent on the first power of the oxygen concentration. This method has yet to be fully exploited for a host of coals and compared to methods that measure the resultant oxidation.[42] Nordon has subsequently used his technique in companion studies.[47-49]

Lowenhaupt and Gray[50] developed the alkali-extraction test to detect oxidized metallurgical coal. (See also Chap. 6). The procedure involves boiling 1 g of coal in 100 mL of 1 N NaOH. The assumption is that oxidized coal will dissolve and darken the solution. After 3 minutes of rapid boiling, the resulting filtrate is optically observed at 520 nm. Results are presented as percent transmittance relative to 1 N NaOH. The procedure has been standardized against a petrographic counting technique to identify oxidized coal.

As an analytical procedure, the alkali-extraction test has a detection limit of 3 percent oxidized coal. Compared to the free swelling index (a standardized test for measuring thermoplastic and swelling properties of coal) for quality control purposes, the extraction test can detect oxidized coal that would not affect swelling properties but would affect the behavior of coal in processing. U.S. Steel uses the alkali-extraction test today for quality control purposes.

Ginnard and Corriveau[51] studied lignites and correlated weathering (oxidation) with both pH and depth of cover. They quantified their findings into four relationships: (1) heating value decreases with increased oxygen content; (2) increasing overburden corresponds to decreasing oxygen content; (3) megajoule (Btu) content increases with depth of cover; and (4) megajoule (Btu) content increases with a corresponding increase in the pH of the lignite slurry. The degree of lignite weathering as a measure of a coal's acceptability can be estimated by multiplying its pH times the depth of the coal (DTC): DTC × pH = K. K is the assigned cutoff value considered acceptable for a specific lignite. If the value of K exceeds the cutoff value, then the sample is acceptable for the process under consideration. Once a cutoff value is defined, the technique is easy to use by the field geologist. Twenty-five grams of lignite from a 50 g cutting that has been roughly crushed and sieved to remove large (1/4 in. × 0) pieces is mixed in a 500 mL beaker with 3 drops of isopropanol (to improve wetting) and 200 mL of distilled water. The mixture is stirred, allowed to stand, and stirred again, and the pH of the quite supernatant is measured.

If we apply the relationship derived by Ginnard and Corriveau[48] to quantify the heating valve of a lignite yet to be mined, the value of the method to the coal scientist becomes self-evident. For example, the moist mineral-matter-free Btu/lb can be calculated from the relationship:

$$4991.37 + 4.8 \, (\text{pH}) \, (\text{DTC}) = \text{Btu/lb}$$

If 165 was considered as the cutoff value, that is, the product of pH × DTC, then a slurry pH of 6.6 at 25 ft (7.62 m) and pH of 4.7 at 35 ft would both yield a lignite of 5,782 ± 2 Btu/lb. The value of the constant, 4991.37, was determined for lignites from a large data base.

Volatiles Displacement.   Chandra and Gupta[52] were able to show that the volatile displacement, that is, the difference between the experimentally determined and calculated volatile matter, is dependent on how much a coal has weathered as well as its petrographic composition. Their evidence is restricted to Umaria coals of India. They found that a rectilinear relationship appears to exist when the volatile displacements are plotted against the carbon content (positive slope), hydrogen content (negative slope), and moisture (positive slope).

Hydrogen as an Indicator of Spontaneous Combustion.   Street et al.[53] examined the suitability of measuring the amount of hydrogen gas evolved as an indicator of spontaneous combustion within milling plants. A small quantity of hydrogen is physically adsorbed on coal and may be released by grinding in air or nitrogen. The amount of such hydrogen release is governed by chemical reaction at temperatures that vary with coal rank. However, the temperatures required to release significant volumes of hydrogen are much higher than those normally achieved during grinding. They concluded that "hydrogen is not a viable alternative to carbon monoxide" as an indicator of spontaneous combustion. The original point made by Street et al.[53], which requires further consideration, is to identify the source of hydrogen in mill gas. Such information would provide useful confirmatory evidence for heating.

Spectroscopy.   Infrared spectroscopy has been used by various researchers to quantify the effects of coal oxidation.

The British Carbonization Research Association (BCRA) published "An Infrared Spectroscopic Study of the Influence of Oxygen in Coal on the Plastic Stage of the Coking Process."[54] Their objective was to elucidate the mechanism of the inhibition of coking by oxygen combining with coal. They pointed out that the chemical methods for estimating oxygen functional groups in coal suffer from difficulties associated with the inaccessibility of the functional group in a relatively insoluble material. To overcome these problems, they initiated a program to examine coals and their oxidized coal carbonization products by infrared spectroscopy. The BCRA present a well-documented report on their experimental procedures and state the reasons why they selected specific sample preparation methods.

The results of the BCRA study are consistent with the hypothesis that "oxygen acts via ether-type cross linkages between the constituent structural units of coal." As they point out, the onset of oxidation involves the loss of aliphatic C—H linkages and the subsequent formation of hydroxyl (OH) and carboxyl (COOH) groups. If one proceeds with carbonization of the oxidized coal, the phenolic hydroxyl groups are reduced, particularly in lower rank coals, by the possible "condensation of adjacent hydroxyl groups which would lead to the formation of ether bridges." However, the BCRA study could not find specific infrared evidence "of oxygen in the form of ether groups in any of the oxidized coals or their carbonization products."

It should be pointed out that Ignasiak[55] observed that as little as 0.01 to 0.02 percent hydroxyl oxygen may be the cause of a change in Gieseler fluidity. Such small changes in oxygen functional group distribution would have been difficult for the BCRA to detect with their instrumentation. In addition, mineral matter could mask the ether oxygen infrared absorptions. More work in this area by Fourier transform infrared (FTIR) analysis methods may reveal the hypothesized functional group changes.

The conclusions reached by the BCRA were that the thin film sample preparation techniques for infrared analysis provided much improvement over mulls, KBr disks, and multiple internal reflection techniques. Their spectral studies verified the major changes in the coal structure to be a loss of aliphatic C—H linkages upon oxidiation. In addition, they observed an increase in hydroxyl and carboxyl groups. Subsequent carbonization causes further "loss of aliphatic C—H bonds" and the removal of hydroxyl and carboxyl groups. These major changes occur during the fluid range of the coal. The BCRA content that

this behavior is consistent with "hydroxyl group introduction into coal with oxidation undergoing condensation reactions to produce ether linkages which can bridge coal lamellae, thus reducing their mobility with respect to each other."

The use of FTIR to study coal oxidation is also exemplified in reports by Painter et al.[56] on the weathering of Canadian coals. Fuller et al.[57] and Huffman et al.[58] have used diffuse reflectance to quantify the formation of carbonyl and carboxyl groups during the initial period of oxidation.

In general, the FTIR method has been shown to correlate the relative changes in specific peak assignments to the duration of oxidation. The data base, on a few coals, has been related to variations in the physical properties of the coal. In a recent article, Rhoades et al.[59] used the FTIR method to investigate low-temperature ($60°$ and $140°C$) bituminous coal oxidation. They were able to relate the loss in Gieseler fluidity to loss of aliphatic C—H groups as a direct function of time of oxidation. From their observations, they concluded that the mechanism of oxidation at 60 and $140°C$ is essentially the same. The important factor in the change in fluidity seemed to be "loss of the aliphatic C—H groups that serve as a source of transferable hydrogen and thermally cleavable bridges." Fredericks and Moxon[60] have also shown the ability to predict the relative degree of oxidation of an Australian coal using the area of the C—H stretch band.

In addition to studying the changes in organic structures resulting from oxidation, one can also examine changes to the coal mineral matter. Huggins et al.[61] and Huffman et al.[58] used Mössbauer spectroscopy to follow changes in iron-bearing minerals in a set of well-characterized, weathered coals. Since most coals have pyrite (FeS_2) as a significant component of included mineral matter, studies of pyrite altering to geothite (FeOOH) should give an indication of the state of coal weathering. Huggins et al.[61] found the presence of goethite to be a very sensitive early indicator of coal weathering, however, quantitative ratio of pyrite to goethite did not directly correlate with other measures of weathering.

The presence of bassanite ($CaSO_4 \cdot 1/2H_2O$) has also been postulated as an indicator of early stages of coal weathering. Studies using FTIR[56] have confirmed the presence of bassanite in weathered coals.

10.3.2   Petrographic Techniques

Weathered or oxidized coals may be detected petrographically, by looking for a variety of characteristics.[13,62-68] Crelling et al.[69] have used recognition of these properties in quantitative evaluation of the amounts of weathered coal in coal blends. Characteristics that are used to recognize weathered coal during petrographic analysis include:

- Disclosure around particle edges and cracks,
- Presence of many microcracks, often with irregular and crosshatched appearance,

- High relief around particle edges and maceral boundaries,
- Discoloration of micrinite,
- Weathering and discoloration of minerals,
- Differential coloration of particles by staining techniques,
- Presence of oxidation rims (bright rims indicative of severe oxidation and/or pyrolysis),
- Decrease in vitrinite reflectance (natural weathering), and
- Increase in vitrinite reflectance (thermal oxidation).

The first seven characteristics can be recognized by visual observation and may indicate that a particular coal particle has been weathered. By using these criteria during point-count analysis, one can quantify the amounts of weathered coal particles in a given sample blend. Unfortunately, these characteristics do not quantify the relative amounts of oxidation or weathering. Gray et al.[13] report that differential staining techniques using a saffronin O stain make recognition of weathered particles easier with some coals.

Quantitative measurement of the vitrinite reflectance during petrographic analysis is another measure of coal weathering or oxidation. Low-temperature laboratory oxidation has generally shown an overall increase in vitrinite reflectance,[63,65] whereas natural weathering has generally shown a decrease in particle reflectance.[66,67,70] A high-volatile A Pittsburgh seam coal selectively oxidized in air at 175°C for 2, 5, and 15 hours showed an overall increase in reflectance of 0.1 percent.[71] The increase in reflectance was rapid for the first 2 hours and then progressively slower over the next 15 hours.

Marchioni[66] studied the weathered characteristics of six coals ranging in rank from subbituminous B through low volatile, collected from the Alberta/British Columbia coal fields of Western Canada. In this study, samples of coal were collected, first from the highly weathered outcrop and then working back into the seam into presumably unweathered coal. Four of the six sample sites studied showed decreases in vitrinite reflectance approaching that of the weathered outcrop. These samples were all relatively high-rank bituminous coals (HVA-LV). The two other cases (both relatively low-rank coals, HVC and Subbit B) showed an increase in vitrinite reflectance. The author suggests that this reversal in vitrinite reflectance may be due to the different chemistry of the low-rank coals or possibly weathering conditions. In this study, the author clearly showed that an increase in weathered particles can be detected petrographically by using differential staining techniques. He also demonstrated the decrease in heating value of weathered coal.

Several additional petrographic methods have been used for detecting weathered or oxidized coal. A few workers have examined the quantitative reflectance of cokes or chars produced from oxidized coals.[65,72] Pearson and Creaney[72] have refined this technique to develop an "oxidation index" based on the reflectance properties of carbonized vitrinites from oxidized and unoxidized coals.

Another petrographic method for measuring weathered coal is quantitative fluorescence of liptinite macerals. Babcock and Crelling[73] and Goodarzi[74] have shown that as a subbituminous coal weathers, the wavelength of maximum fluorescence intensity shifts toward the red end of the visible spectrum. In addition, coal microhardness has been shown to relate to the degree of coal weathering or oxidation.[67,75]

10.3.3 Thermoplastic Measurement Techniques

A unique property displayed by certain coals of bituminous rank is their ability to undergo a thermoplastic state during pyrolysis. This property enables coking coals to realign their macromolecular structure into carbon frameworks during the coal-to-coke thermal transformation. Many of the derived properties of metallurgical cokes are dependent upon these macromolecular rearrangements. The two initial coal characteristics that most influence thermoplastic properties are the level of geothermal maturity of the coal (rank) and the amount and character of the liptinite, vitrinite, and inertinite macerals present. The inherent coal property that most affects thermoplastic behavior is the degree of weathering or oxidation to which the coal is exposed either prior to or after mining. Hence, measurement of thermoplastic properties is sensitive to both a coal's coking potential and its level of oxidative degradation.

Over a number of years, the coal community has developed a series of standardized empirical tests for measuring coal thermoplastic properties. The most common tests currently in use include the free swelling index (FSI), Gieseler plasticity, and dilatometry. The FSI[76] is a simple test of thermoplastic and swelling properties of coal, which is often used in exploration and quick screening studies. Many authors have shown that the FSI test is sensitive to coal weathering and oxidation.[13,63,65,66] However, Marchioni[77] has demonstrated that minor lateral changes in coal petrographic composition as well as minor shifts in coal rank can affect FSI determination. Hence, FSI results must be interpreted relative to standards for the particular coal and in light of potential shifts in composition. Unfortunately, these same constraints, to varying degrees, also apply to plasticity and dilatation.

Geiseler plasticity[78] is a test of rheologic properties, and is considered to be one of the most sensitive indicators of loss of thermoplasticity due to oxidation or weathering. Generally, as the degree of weathering or oxidation increases, the maximum coal fluidity decreases. A study completed with a high-volatile Pittsburgh coal stored in air (particle size $-425\,\mu m$) showed that the maximum fluidity was cut by 50 percent after only 10 days.[21] Further aging decreased fluidity more slowly, down to approximately 15 percent of the initial fluidity. The fluid temperature range also generally decreased with increased oxidation or weathering.[13,79,80] Gieseler plastometer measurements are generally considered very sensitive to weathering, but, like the FSI, must be compared to an unweathered reference sample.

Dilatometry (Audibert-Arnu, Rhur, and others) is a technique for measuring a coal's thermoplastic properties by determining relative packed-bed expansion and contraction under a standard set of heating conditions. Minor levels of coal oxidation will result in decreased overall expansion and contraction, and a decreased plastic range.[44,81] Dilatometry, like plasticity, is useful only for coals displaying expansion/contraction properties and requires a reference sample for comparison.

10.4 SUMMARY AND RECOMMENDATIONS

The reactive nature of nearly all ranks, types, and sizes of coal is well documented. Coal undergoes significant chemical and physical changes during mining, storage, handling, and transport. Despite the known effects of weathering or oxidation, standard practices to quantify, monitor, and control these changes have met with limited acceptance, primarily because of the wide variety of end-use objectives, the variable nature of coal composition, and the response of coal to empirical testing. Common methods used in the laboratory to preserve the fresh-mined integrity of the coal would not be cost-effective to the commercial user. Similarly, the product storage, crushing, and grading practices of the commercial coal handler would be inappropriate for the researcher.

A recent aid to the researcher has been the establishment of coal banks, which store selected coals under constant conditions and distribute laboratory size samples along with documented histories. This practice results in a standard set of materials being available for the entire coal science community.

The best way for the coal researchers to store small coal samples (less than 200 kg) is to displace oxygen by an inert gas in a well-sealed, nonreactive container. Refrigeration of the sample appears desirable, but its utility has not been proven. Commercially, coal storage and subsequent oxidation can best be controlled by pile construction and compaction.

Analytical methods that can quantify the changes in coal properties have been developed from semiempirical tests such as the Gieseler plasticity measurement. The Gieseler method is specific for thermoplastic (coking) coals and, although it requires a standard unweathered sample, is very sensitive to oxidative changes. What is needed is a test that can be applied to a broad range of coals to identify the fundamental chemical changes occurring during oxidation and to relate them as a function of degradation to end-use properties. The infrared techniques now being refined appear to contribute most to the development of standard analytical procedures. An organized effort to classify a broad spectrum of coals systematically and to apply a statistical analysis to the data base to arrive at a series of dependent correlations between oxidation and coal properties will significantly advance coal science.

REFERENCES

1. H. Colign, *9th Annu. Semin. Bulk Mater.* (1976).
2. H. Colign and E. A. Vitunac, *9th Annu. Semin. Bulk Mater.* (1976).
3. L. Hegedus, *Mezogazd. Ip.* **3**(5) (1949).
4. Tetra Tech, Inc., Rep. No. TR-76-67. Prepared under Contract No. E(49-18)-2225. Energy Res. Dev. Admin., Oak Ridge, TN, 1976.
5. W. S. Landers and D. J. Donaven, *Chemistry of Coal Utilization* (H. H. Lowry, ed.), Suppl. Vol. 1, Wiley, New York, 1963, p. 299.
6. R. V. Kromrey, R. S. Scheffee, J. A. DePasquale, and R. S. Valentine, Report No. COO-4632-2. Prepared under Contract No. EP-78-C-02-4632. Energy Res. Dev. Admin., Oak Ridge, TN, 1978.
7. K. K. Bhattacharyya, *Fuel* **50**(4), 367–380 (1971).
8. A. G. Kim, *Rep. Invest.—U.S., Bur. Mines* **RI-8756** (1977).
9. J. B. Stott, B. J. Harris, and P. J. Hansen, *Fuel* **66**, 1012 (1987).
10. N. R. Kona, H. N. Fairbanks, and D. W. Leonard, Rep. No. 25. Coal Res. Bur., Morgantown, WV, 1968.
11. J. G. Bayle, W. H. Eddy, and R. W. Shotts, *Rep. Invest.—U.S., Bur. Mines* **RI6620** (1965).
12. F. J. Beafore, K. E. Cawiezel, and C. T. Montgomery, *Proc. Coal Conf. Expo V* (1979).
13. R. J. Gray, A. H. Rhodes, and D. T. King, *Trans.—Soc. Min. Eng. AIME* **260**, 334–341 (1976).
14. R. C. Neavel, E. J. Hippe, S. E. Smith, and R. N. Miller, *Prepr. Pap.—Am. Chem. Soc., Div. Fuel Chem.* **25**(3) (1980).
15. A. Davis to D. S. Hoover, personal correspondence, 1983.
16. D. D. Kaegi to D. S. Hoover, personal correspondence, 1983.
17. J. Haggin, *Chem. Eng. News* **24**, 24 (1984).
18. J. W. Larsen and T. E. Schmidt, *Fuel* **65**, 1310 (1986).
19. J. C. Hower and B. Davis, to D. S. Hoover, personal correspondence, 1983.
20. W. H. Griest to F. K. Schweighardt, personal correspondence, 1983.
21. J. T. Senftle and A. Davis, Final Rep. No. DOE/PC/300133-T2. Prepared under Contract No. AC22-80 PC30013. Energy Res. Dev. Admin., Oak Ridge, TN, 1982.
22. F. E. Huggins, G. R. Dumyre, M. C. Lin, and G. P. Huffman, *Fuel* **64**, 348 (1985).
23. D. S. Hoover, *APCI Lab. Proced.* (1981).
24. A. Colombo and A. Scholz, *BCR Inf. Circ.* (1978).
25. H. C. Wilkinson to F. K. Schweighardt, personal correspondence, 1983.
26. A. Volbroth, Rep. No. COO-2893-3. Prepared under Contract No. E(11-1)-2898. Energy Res. Dev. Admin., Oak Ridge, TN, 1976.
27. B. I. Losev and N. G. Lidina, *Dokl. Akad. Nauk SSSR* **122**(1), 186 (1960).
28. K. Kinson and C. B. Belcher, *Fuel* **54**, 205–209 (1975).
29. H. J. Gluskoter, *Fuel* **44** (1965).

30. S. Manskaya and T. V. Drozdova, *Geochemistry of Organic Substances*, p. 347. Pergamon, Oxford, 1968.

31. S. C. Sun and A. L. Campbell, *Adv. Chem. Ser.* **55**, 363 (1966).

32. D. E. G. Austen, D. J. E. Ingram, P. H. Given, C. R. Binder, and L. W. Hill, *Adv. Chem. Ser.* **55**, 344 (1966).

33. F. M. Swain, *Organic Geochemistry*, p. 87. Macmillan, New York, 1963.

34. B. I. Losev and E. A. Bilina, *Zh. Prikl. Khim.* **32**(10), 2359 (1959).

35. D. G. Murchison, *Adv. Chem. Ser.* **55**, 307 (1966).

36. D. W. Van Krevelen and J. Schuyer, *Coal Science*. Elsevier, Amsterdam, 1957.

37. L. Blom, L. Edelhausen, and D. W. Van Krevelen, *Fuel* **18**(2), 135 (1957).

38. D. J. Veal and C. F. Cook, *Anal. Chem.* **34**, 2 (1962).

39. O. P. Mahajan, *Fuel* **64**, 973 (1985).

40. J. M. Kuchta, V. R. Rowe, and D. S. Burgess, *U.S. Bur. Mines Rep.* **PB81-150039** (1980).

41. G. D. Lidin, ed., *Dokl. Akad. Nauk SSSR, Trans.* **3218**, 260–274 (1962).

42. J. I. Graham, *Trans. Inst. Min. Eng.* **48**, 521 (1914).

43. E. A. C. Chamberlain and D. A. Hall, *Colliery Guardian* p. 100 (1973).

44. H. Eicker and H. J. Kartenberg, *Glueckauf* **3**, 59–63 (1975).

45. B. S. Ignasiak, B. N. Nadi, and D. S. Montgomery, *Fuel* **49**(2) 214–221 (1970).

46. B. C. Young and P. Nordon, *Fuel* **57**, 574 (1978).

47. P. Nordon, B. C. Young, and N. W. Bainbridge, *Fuel* **58**, 443 (1979).

48. P. Nordon and N. W. Bainbridge, *Fuel* **58**, 450 (1979).

49. P. Nordon, *Fuel* **58**, 456 (1979).

50. D. E. Lowenhaupt and R. J. Gray, *Int. J. Coal Geol.* **1**, 63–73 (1980).

51. K. J. Ginnard and M. P. Corriveau, *ASTM Spec. Tech. Publ.* **STP 661**, 41–49 (1978).

52. D. Chandra and U. P. Gupta, *Fuel* **55**, 84 (1976).

53. P. J. Street, J. Smalley, and A. T. S. Cunningham, *J. Inst. Fuel.* (1975).

54. British Carbonization Research Association, *An Infrared Spectroscopic Study of the Influence of Oxygen in Coal on the Plastic Stage of the Coking Process*, Carbon Res. Rep. No. 64. BCRA, 1979.

55. B. S. Ignasiak, D. M. Clugston, and D. S. Montgomery, *Prepr. Pap.—Am. Chem. Soc., Div. Fuel Chem.* **14**, 95–105 (1970).

56. P. C. Painter, R. W. Snyder, D. E. Pearson, and J. Kwong, *Fuel* **59**, 282 (1982).

57. M. P. Fuller, I. M. Hamadeh, P. R. Griffiths, and D. E. Lowenhaupt, *Fuel* **61**, 529 (1982).

58. G. P. Huffman, F. E. Huggins, G. R. Dunmyre, A. J. Pignocco, and M. C. Lin, *Fuel* **64**, 849 (1985).

59. C. A. Rhoades, J. T. Senftle, M. M. Coleman, A. Davis, and P. C. Painter, *Fuel* **62**, 1387 (1983).

60. P. M. Fredericks and N. T. Moxon, *Fuel* **65**, 1531 (1986).

61. F. E. Huggins, G. P. Huffman, D. A. Kosmack, and D. E. Lowenhaupt, *Int. J. Coal Geol.* **1**, 75–81 (1980).

62. R. J. Gray, A. H. Rhoades, and D. T. King, *Trans. ASME* **42**, 185 (1965).

63. L. G. Benedict and W. F. Berry, Report. Bituminous Coal Research, Monroeville, PA, 1964.

64. D. Chandra, *Fuel* **42**, 185 (1962).

65. D. Chandra, *Stach's Textbook of Coal Petrology*, pp. 159–164. Berlin, 1975.

66. D. L. Marchioni, *Int. J. Coal Geol.* **2**, 231–259 (1983).

67. B. N. Nadi, L. A. Ciavaglia, and D. S. Montgomery, *J. Microsc. (Oxford)* **109**(1), 93–103 (1977).

68. G. R. Ingram and J. D. Rimstidt, *Fuel* **63**, 292 (1984).

69. J. C. Crelling, R. H. Schrader, and L. G. Benedict, *Fuel* **58**, 542–546 (1979).

70. R. M. Bustin, *Curr. Res., Part B, Geol. Surv. Can.* pp. 249–254 (1980).

71. D. C. Cronauer et al., *Fuel* **62**,f 1124–1132 (1983).

72. D. E. Pearson and S. Creaney, *Fuel* **60**, 273–275 (1981).

73. D. Babcock and J. C. Crelling, *Annu. Prog. Rep.—Res. Dev. Summ.* (1982).

74. F. Goodarzi, *Fuel* **65**, 260 (1986).

75. B. N. Nadi, A. Ghosh, and L. A. Ciavaglia, *Fuel* **57**, 317–319 (1978).

76. ASTM, Standard Method D720-67: Standard method of test for free-swelling index of coal. *Annu. Book ASTM Stand.* pp. 244–250 (1970).

77. D. L. Marchioni, *Coal Miner* **6**(2), 22–24 (1981).

78. ASTM, Standard Method D2639-74: Standard method of test for plastic properties of coal by the constant-torque Giesler plastometer. *Annu. Book ASTM Stand.* pp. 327–333 (1977).

79. R. Loison, A. Peytavy, A. F. Boyer, and R. Grillot, in *Chemistry of Coal Utilization* (H. H. Lowry, ed.), Suppl. Vol. 1, p. 173, Wiley, New York, 1963.

80. C. C. Russell and M. Pend, *Am. Gas Assoc., Proc.* **31**, 372–389 (1949).

81. D. J. Maloney, R. G. Jenkins, and P. L. Walker, Jr., Techn. Rep. No. 20. Prepared under Contract No. EX-76-C-01-2030. Energy Res. Dev. Admin., Oak Ridge, TN, 1980.

INDEX